PRINCIPLES OF
NON-NEWTONIAN
FLUID MECHANICS

London
New York
St Louis
San Francisco
Düsseldorf
Johannesburg
Kuala Lumpur
Mexico
Montreal
New Delhi
Panama
Paris
São Paulo
Singapore
Sydney
Toronto

G. Astarita
Università di Napoli

G. Marrucci
Università di Palermo

Principles of Non-Newtonian Fluid Mechanics

Library of Congress Cataloging in Publication Data

Astarita, Giovanni.
  Principles of non-Newtonian fluid mechanics.
  Includes bibliographical references.
  1. Non-Newtonian fluids. 2. Fluid mechanics.
I. Marrucci, G., joint authors. II. Title.
QA929.5.A87   532   73-17232
ISBN 0-07-084022-9

Published by McGRAW-HILL Book Company (UK) Limited

MAIDENHEAD · BERKSHIRE · ENGLAND

07 084022 9

PRINTED AND BOUND IN GREAT BRITAIN

CONTENTS

Non-Newtonian fluids form an extremely wide class of different materials, whose only common features are fluidity and a failure to obey Newton's law of friction. It is therefore impossible to deal with the subject of non-Newtonian fluid mechanics unless one of two choices is made: the analysis is either restricted to a specific class of fluids sharing a common form of mechanical behavior, or is restricted to only the *principles* of non-Newtonian fluid mechanics, i.e., to that body of knowledge which is supposedly applicable to all fluids. In this book we have adhered to the second choice, and only in the last two chapters have we attempted to hint at methods of approach that could be chosen for the solution of actual problems relative to specific materials.

Fluid mechanics belongs principally to the field of engineering science. It is a unique characteristic of engineering science that it does not take a stand in the current (and possibly everlasting) dichotomy of axiomatic versus naturalistic science, but draws from the results of both and applies these results to the solution of any problems that are considered relevant. To the classical question of whether mathematics has been created or discovered, the engineering scientist would answer that it does not really matter, provided it works; neither would he be implicated in any discussion on what a satisfactory definition of 'working,' as applied to mathematics, should be. In the specific field of non-Newtonian fluid mechanics, the main results on the general principles have been obtained by mathematicians, and more generally within the framework of an axiomatic approach to science. Many of these results are available only in fragmentary form in the specialized literature, which is not generally accessible to the engineering scientist. Even the one outstanding coherent exposition, the *Principles of Non-linear Field Theories* by Truesdell and Noll to which we acknowledge our great indebtedness, is a very difficult study for the engineering scientist interested in fluid mechanics, since it covers a much wider field and requires, therefore, painstaking work to extract the necessary information. We have attempted to present the results of modern non-linear theory of continuous media in a way that is easily grasped by the engineering scientist and is aimed at the solution of pragmatic problems of non-Newtonian fluid mechanics.

The subject matter of this book cannot be discussed without making use of some mathematical methods which are not usually available to engineers. Rather than confine the exposition of the mathematics to an arid appendix (a practice which unsuccessfully tries to hide the hard fact that mathematics needs to be studied before it is used, rather than afterwards), we have chosen to place it in the text whenever the need arises. Also, we have illustrated the mathematics with no presumption of either formal rigor or completeness, attempting to guide the reader as effortlessly as possible towards an understanding of the basic concepts, and helping him to acquire the skill of using the algorithm when necessary.

Although we have included in each chapter a few illustrations and problems, this book is not aimed primarily at classroom use. We think it can be used for graduate courses, and both of the authors have given courses based on parts of it; but we think the usefulness of the book is essentially to be found by workers who wish to do fundamental research in the field.

There is one field of science, undoubtedly relevant to non-Newtonian fluid mechanics, which we have purposely omitted, and that is the entire field of molecular theories of polymer behavior. Good coherent expositions of this field are available which can be studied easily by engineering scientists, and its inclusion in this book would have been a poor duplication.

We have worked on this book over a period of more than three years, mostly at the University of Naples, but part of the work has been carried out by G. Marrucci at the University of Delaware and at the University of Palermo. Although both authors have worked on all chapters in the various stages of their evolution, it should be mentioned that G. Astarita is responsible for chapters 2, 4, and 5, and G. Marrucci for chapters 3 and 6.

The bibliography in this book is very concise. We soon realized that we could not hope to acknowledge all the literature from which we had drawn our ideas, nor to give due credit to the authors who originated the available information, and we have therefore restricted the references to those works that we suggest the reader should study if he wishes to strengthen his understanding of any given subject. Even in doing this we have restricted the references to works which, by themselves, will inevitably generate an additional list of relevant references.

We wish to acknowledge the contribution of several people with whom we have discussed parts of the manuscript. W. R. Schowalter of Princeton University has read and commented on the early version of the first four chapters and on the overall presentation; our co-workers, D. Acierno, R. Greco, L. Nicodemo, G. C. Sarti, and G. Titomanlio, have studied several sections and have suggested alterations; and technical discussions with J. R. A. Pearson and B. D. Coleman have stimulated revision of parts of the manuscript. Towards the end of our work we discovered the truth of the saying that one does not finish a book, one only gives up the endless task of improving it. We can hope that we have not given up too soon.

GIOVANNI ASTARITA
GIUSEPPE MARRUCCI

Naples, August 1972

THE BASIC EQUATIONS
OF FLUID MECHANICS

1-1 GENERALITIES

Fluid mechanics is the study of the behavior of fluid-like materials in flow. The analysis of the phenomena of fluid mechanics is based on the simultaneous solution of a number of equations representing certain physical laws which are assumed to hold for the phenomena being considered.

These equation can be grouped in two different categories. In the first we include those equations which represent physical laws that are assumed to hold for *every* material. These equations are called balance equations, inasmuch as they represent the mathematical statement of a principle of conservation. There are basically four balance equations, representing the principles of conservation of mass, momentum, moment of momentum, and energy.

A second group of equations represents certain physical laws governing the behavior of specific materials. The form of these equations depends on the *class* of materials being considered; the value of the parameters appearing in the equations depends on the particular material. There are basically four equations in this group. The recent and entirely general approach of Coleman

[1, 2, 3]† considers equations giving explicitly the following four dependent variables: internal energy, entropy, stress, and heat flux. This approach will be discussed in chapter 4. At this stage, we prefer a much less rigorous approach, which makes use of concepts deriving from classical thermodynamics. Such a simplified approach still makes use of four equations representing the behavior of the materials considered: a thermodynamic equation of state, which is a relationship of density, pressure, and temperature; a constitutive equation, relating internal stresses to kinematic variables; a heat transfer equation, relating the heat flux to the temperature distribution; and an equation relating internal energy to the relevant independent variables, called the energetic equation of state.

In principle, any fluid mechanical problem requires the simultaneous solution of the entire set of eight equations mentioned above. In practice, this is a hopelessly difficult task, and a simplified form of one or more of the relevant equations is often used in the solution of certain classes of problem. A particularly important simplification arises when constant density fluids are considered, i.e., when the thermodynamic equation of state takes the very simple form:

$$\text{Density} = \text{constant} \qquad (1\text{-}1.1)$$

This book is concerned only with constant density fluids.

When writing the balance equation, a system must be chosen on which the relevant conservation principle is applied. The system may be of finite size, in which case the 'integral' balance equations are obtained. Nonetheless, a particularly useful form of the equations is obtained when the system is chosen as a differential volume surrounding the point considered. The balance equations are then obtained in differential form.

The principle of conservation of mass, as applied to any specified system, can be written as

$$\begin{pmatrix} \text{Net flow of} \\ \text{mass entering} \\ \text{the system} \end{pmatrix} = \begin{pmatrix} \text{Increase of} \\ \text{mass within} \\ \text{the system} \end{pmatrix} \qquad (1\text{-}1.2)$$

It is obvious that both the density and the velocity will appear in a mathematical formulation of eq. (1-1.2). Density is a scalar quantity, and velocity is a vector; all terms appearing in eq. (1-1.2) are scalars, because the quantity whose conservation principle is being considered (mass) is a scalar. Even if eq. (1-1.1) is assumed to hold, i.e., if constant density fluids are considered, eq. (1-1.2) cannot be solved for the velocity, because a scalar equation is not sufficient to determine an unknown vector.

† Figures in square brackets refer to the Bibliography at the end of each chapter.

The principle of conservation of momentum is therefore always required. It may be written as

$$
\begin{pmatrix} \text{Net flow of} \\ \text{momentum} \\ \text{into the} \\ \text{system} \end{pmatrix} + \begin{pmatrix} \text{Sum of all} \\ \text{surface forces} \\ \text{acting on the} \\ \text{system} \end{pmatrix} + \begin{pmatrix} \text{Sum of all} \\ \text{body forces} \\ \text{acting on} \\ \text{the system} \end{pmatrix} = \begin{pmatrix} \text{Rate of} \\ \text{increase of} \\ \text{momentum of} \\ \text{the system} \end{pmatrix}
$$

$$(1\text{-}1.3)$$

In eq. (1-1.3), the second term on the left-hand side represents all the forces acting across the surfaces which are the boundaries of the system, while the third term represents the forces, such as gravity, acting throughout the system. The variables appearing in eq. (1-1.3) are again density and velocity, but two new variables appear: *pressure*, which acts across boundaries and therefore appears in the second term, and *stress*. In fact, in order to calculate the second term in eq. (1-1.3) it is necessary to be able to calculate the forces acting across any arbitrary surface in the material, so that the system to which eq. (1-1.3) is applied may be chosen at will. The force acting across any given surface is not simply due to pressure, because it need not be orthogonal to the surface; nor does its value need to be independent of the surface orientation in space. Stress is a 'tensor' (to be defined exactly in section 1-3), which relates the force vector to the surface vector. A surface is a vector in the sense that not only its magnitude, but also its orientation in space are required in order to identify it.

The system of eqs. (1-1.2) and (1-1.3) is not determined because the four unknowns (density, pressure, velocity, and stress) cannot be determined from the two equations. Hence the need for equations representing the physical behavior of the material, i.e., a constitutive equation and a thermodynamic equation of state. The latter may take the simplified form of eq. (1-1.1).

The constitutive equation is a relationship which allows stress to be calculated as a function of the kinematic variables and ultimately as a function of the possibly time-dependent velocity field. If attention is restricted to constant density fluids, the system of eqs. (1-1.1)–(1-1.3), together with the constitutive equation, can be solved in principle, as shown in Table 1-1.

This book is concerned mainly with the solution of problems based on the system of equations reported in Table 1-1, and in particular with materials whose behavior requires to be considered by rather complex forms of the constitutive equation.

Note that classical hydrodynamics is concerned with the case where the constitutive equation simply states that the state of stress is always isotropic, i.e., completely determined by the value of pressure, while classical Newtonian fluid mechanics is concerned with the case where the constitutive equation has the simple linear form embodied in Newton's law. More complex forms of the constitutive equation give rise to the discipline of non-Newtonian fluid mechanics.

If the restriction to constant density fluids is relaxed, the equation of state takes the form of a relationship among density, pressure, and temperature. The appearance of a temperature variable requires the simultaneous solution of the energy conservation equation (the first law of thermodynamics), which in turn introduces two new variables—heat flux and internal energy. Fourier's law (relating heat flux to temperature distribution) and an 'energetic' equation of state then make the complete set of equations given in Table 1-2.

The entire system of seven equations must be solved simultaneously for non-isothermal systems. Isothermal systems can be dealt with on the basis of the equations listed in Table 1-1.

The 'energetic' equation of state relates internal energy to temperature, density, and state of deformation (in some sense to be defined). For simple Newtonian fluids, the dependence on the state of deformation can be neglected, so that the energetic equation of state reduces to a dependence of specific heat on temperature. For isothermal systems, the energy balance equation can then be solved independently for the energy dissipation.

For more complex materials, which possess some degree of elasticity, internal energy may be stored reversibly in consequence of deformation, and the energetic equation of state necessarily includes kinematic independent

Table 1-1   EQUATIONS TO BE SOLVED FOR
CONSTANT DENSITY FLUIDS

Equations	Variables
State (scalar)	Density (scalar)
Mass balance (scalar)	Pressure† (scalar)
Momentum balance (vectorial)	Velocity (vector)
Constitutive† (tensorial)	Stress† (tensor)

† A 'total' stress, including isotropic pressure, could be regarded as a single tensor variable. The constitutive equation determines the total stress within an arbitrary additive isotropic tensor. The scalar by which the unit tensor is multiplied to obtain this isotropic tensor is then the scalar variable to be introduced in place of the pressure. This will be further clarified in section 1-8.

variables. Very little is known about the form of the energetic equation of state for real elastic fluids, i.e., on reasonable constitutive assumptions for the internal energy. This point raises some problems to be discussed in detail in later chapters; in general, it may be stated that non-Newtonian fluid mechanics is analyzed basically in terms of momentum considerations, and little information can be obtained from the energy conservation principle at present.

1-2 VECTORS

Before proceeding to a discussion on the basic equations of fluid mechanics, a certain amount of vector and tensor algebra is required. This, and the following three sections, are dedicated to some basic mathematical concepts that form a necessary introduction to the material given in later sections.

We shall use mainly abstract tensor notation [4, 5, 6] and, in a subordinate way, indicial notation. The difference is more than a simple question of notation, and readers who already have a working knowledge of tensor analysis may find it useful to read these sections. Many proofs are not given, because it is felt that fluid mechanicists may use mathematics simply as a tool for studying physical problems.

It is assumed that the reader is familiar with the visualization of vectors as arrows in Euclidean three-dimensional space. The vectors considered here are 'free' in the sense that two parallel arrows having the same length and direction are considered as one vector. It is also assumed that the rules of addition and multiplication by a scalar and inner product are well known. These rules are easily visualized by thinking intuitively of vectors as arrows, although formally they are just the basic rules that *define* a vector space. The geometric interpretation of vectors (and tensors) is advantageous in

Table 1-2 SYSTEM OF EQUATIONS TO BE SOLVED IN GENERAL

Equations	Tensorial character	Variables	Tensorial character
State	Scalar	Pressure	Scalar
Mass balance	Scalar	Density	Scalar
Momentum balance	Vector	Velocity	Vector
Constitutive	Tensor	Stress	Tensor
Energy balance	Scalar	Internal energy	Scalar
Fourier's law	Vector	Heat flux	Vector
'Energetic' equation	Scalar	Temperature	Scalar

keeping track of the physical meaning of the operations involving these quantities, and will therefore be used extensively in this text.

The alternative representation of vectors (and tensors) as ordered sets of numbers called the components is also often encountered. With respect to the geometrical representation, the representation by means of components has the disadvantage of being dependent upon the choice of a vector basis and frequently, therefore, of a coordinate system; that is, upon a change of vector basis, a given vector (an arrow in space) will change its components.

A vector basis is a set of three vectors that are not all parallel to the same plane. If the basis vectors are mutually orthogonal and of unit length, the basis is called orthonormal. Given a vector basis e_1, e_2, e_3, any vector a can be constructed from the basis vectors through the operations of multiplication by a scalar and addition:

$$a = a^1 e_1 + a^2 e_2 + a^3 e_3 \qquad (1\text{-}2.1)$$

The ordered set of numbers a^1, a^2, a^3 is uniquely related to the vector a and constitutes the set of the *components* of vector a with respect to the chosen basis.

It is of practical importance to associate to the space of points a coordinate system. This is a relation between ordered triplets of numbers, called the coordinates, and points of the space. The Cartesian coordinate system is a familiar example. Smoothness conditions for this relation are tacitly assumed.

If a coordinate system has been introduced, it is customary to choose a vector basis in each point in space, called the *natural basis*, defined as

$$e_1 = \frac{\partial X}{\partial x^1}$$

$$e_2 = \frac{\partial X}{\partial x^2} \qquad (1\text{-}2.2)$$

$$e_3 = \frac{\partial X}{\partial x^3}$$

Let us now consider the meaning of the 'derivatives' appearing in eq. (1-2.2). Given a point X with coordinates x^1, x^2, x^3, a point Y with coordinates $x^1 + \Delta x^1$, x^2, x^3 is chosen and the vector formed:[†]

$$\frac{Y - X}{\Delta x^1} \qquad (1\text{-}2.3)$$

[†] The 'difference' of two points, $Y - X$, is interpreted as the vector joining the two points and oriented from X to Y. Similarly, the 'sum' of a point X and a vector a is the point Y, endpoint of the arrow a whose initial point is taken in X.

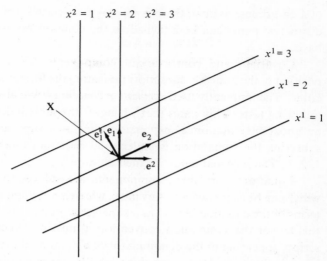

FIGURE 1-1
Natural basis and its dual.

The limit of this ratio for **Y** approaching **X**, i.e., for $\Delta x^1 \to 0$, is the vector $\mathbf{e}_1$. Vectors $\mathbf{e}_2$ and $\mathbf{e}_3$ are similarly obtained. The vectors of the natural basis are therefore at any point tangent to the three curves through **X** along which only one of the coordinates varies (see Fig. 1-1). The basis is, in general, not orthonormal and is not the same at two different points. For a Cartesian coordinate system the natural basis is orthonormal and does not change from point to point. The components of a vector with respect to the natural basis are called *contravariant*.

It is also convenient at this point to introduce the *dual* of the natural basis whose vectors are indicated by $\mathbf{e}^1$, $\mathbf{e}^2$, $\mathbf{e}^3$. Two bases are dual if the following equalities hold:

$$\mathbf{e}^i \cdot \mathbf{e}_j \begin{cases} =0 & \text{if} \quad i \neq j \\ =1 & \text{if} \quad i = j \end{cases} \tag{1-2.4}$$

As shown in Fig. 1-1, the dual of the natural basis at any point is made of vectors that are orthogonal to the three surfaces through **X**, along which one of the coordinates is constant. If the coordinate system is Cartesian, the vectors of the natural basis coincide with the corresponding vectors of its dual in all points, as for any orthonormal basis.

The components of a vector with respect to the dual of the natural basis are called *covariant*. The distinction between covariant and contra-variant components is meaningful only in relation to the existence of a coordinate system. When two mutually dual bases are chosen independently of a coordinate system, there is no way of choosing between them, and no distinctive name can be attached to the components with respect to each basis.

Covariant and contravariant components are distinguished by the position of the indices—subscripts indicating the former and superscripts the latter. The frequently used *summation convention* will also be adopted in the following tests. This states that whenever the same index is encountered in an apparently monomial expression once as a superscript and once as a subscript, the summation is understood over that index taking the values 1, 2, 3. The repeated index is called 'dummy'.

Equations containing components are subject to a consistency test which can be expressed as: 'Any index which is not dummy must appear in all terms of the equation always as a superscript or as a subscript.' In applying this test or the summation convention it must be remembered that superscripts appearing in the denominator of a monomial expression are equivalent to subscripts of the expression. Finally, any equation containing indices other than dummy must be interpreted as a set of equations, one for each index other than dummy taking the values 1, 2, 3.

Summarizing, the definitions of the contravariant and covariant components of a vector **a** are, respectively,

$$\mathbf{a} = a^i \mathbf{e}_i \qquad (1\text{-}2.5)$$

$$\mathbf{a} = a_i \mathbf{e}^i \qquad (1\text{-}2.6)$$

where $\mathbf{e}_i$ are the vectors of the natural basis defined in eq. (1-2.2) and $\mathbf{e}^i$ those of the dual basis defined in eq. (1-2.4). If a Cartesian coordinate system is used, covariant and contravariant components coincide.

The contravariant and covariant components, defined by eqs (1-2.5) and (1-2.6), can also be obtained as scalar products of the vector **a** with the basis vectors:

$$a^i = \mathbf{a} \cdot \mathbf{e}^i \qquad (1\text{-}2.7)$$

$$a_i = \mathbf{a} \cdot \mathbf{e}_i \qquad (1\text{-}2.8)$$

Equations (1-2.7) and (1-2.8) are easily obtained from eqs (1-2.5) and (1-2.6) respectively by making use of eqs (1-2.4).

Whenever the coordinate system is changed, the covariant and contravariant components of a vector change. A change of coordinates is a set of three relationships of the form

$$x'^i = x'^i(x^j) \qquad (1\text{-}2.9)$$

where x^j are the 'old' coordinates and x'^i the corresponding 'new' coordinates of the same point. It is easily shown that the 'new' contravariant components are obtained from the 'old' contravariant components of the same vector by the relationship

$$a'^i = \frac{\partial x'^i}{\partial x^j} a^j \qquad (1\text{-}2.10)$$

while the covariant components transform according to

$$a'_i = \frac{\partial x^j}{\partial x'^i} a_j \qquad (1\text{-}2.11)$$

On the other hand, if a set of three numbers is known to transform upon a change of coordinate system according to eq. (1-2.10) or (1-2.11), then some vector exists whose contravariant or covariant components are given by the set.

Following a different approach, many texts of vector and tensor analysis (linear algebra) use the transformation properties expressed in eqs (1-2.10) and (1-2.11) to define ordered sets of numbers called contravariant and covariant vectors, respectively.

The operations among vectors, which are defined independently of the introduction of the components, have their counterparts in terms of components. For example, the sum of two vectors, which is visualized through the rule of the parallelogram, gives, in terms of components (covariant or contravariant),

$$(\mathbf{a} + \mathbf{b})_i = a_i + b_i \qquad (1\text{-}2.12)$$

The rule used to calculate the scalar product of two vectors via components is of some importance:

$$\mathbf{a} \cdot \mathbf{b} = a^i b_i = a_i b^i \qquad (1\text{-}2.13)$$

The scalar product of a vector multiplied by itself is also indicated by

$$\mathbf{a} \cdot \mathbf{a} = \mathbf{a}^2 = a^i a_i \qquad (1\text{-}2.14)$$

The algebraic square root of $\mathbf{a}^2$ is called the magnitude of $\mathbf{a}$:

$$|\mathbf{a}| = \sqrt{\mathbf{a}^2} \qquad (1\text{-}2.15)$$

1-3 TENSORS

A tensor is best visualized as a machine which transforms vectors into vectors. Defining a tensor means giving the rules by which the machine works; that is, given any vector as input to the machine, we have to know the vector that emerges as the output. The machine at work is indicated in the following way:

$$\mathbf{b} = \mathbf{A} \cdot \mathbf{a} \qquad (1\text{-}3.1)$$

where $\mathbf{A}$ is the machine, $\mathbf{a}$ the input vector, and $\mathbf{b}$ the output. The 'dot' connecting $\mathbf{A}$ and $\mathbf{a}$ in eq. (1-3.1) is the same symbol used to indicate the scalar product of vectors, although the 'meaning' of the dot is, of course, different. Its use in both cases is justified by the similarity of the component form of eqs (1-3.1) and (1-2.13) which will be shown below (eq. (1-3.30)).

Actually, if $\mathbf{A}$ is to be a tensor an important restriction must be made on the kind of machine. It has in fact to be *linear*; that is, the following two properties must be satisfied for any vectors and scalars:

$$\alpha(\mathbf{A} \cdot \mathbf{a}) = \mathbf{A} \cdot (\alpha\mathbf{a}) \qquad (1\text{-}3.2)$$

$$\mathbf{A} \cdot \mathbf{a} + \mathbf{A} \cdot \mathbf{b} = \mathbf{A} \cdot (\mathbf{a} + \mathbf{b}) \qquad (1\text{-}3.3)$$

Thus, a tensor is a linear operator which transforms vectors into vectors.

A familiar example of a tensor is the stress tensor which can be introduced in the following way. One method of revealing the presence of a stress in a point of a body is by making a cut through that point (which is, of course, a mental experiment) and by observing that the two parts of the body exert a force on each other. (The force is uniquely determined as that force to be applied to the surface of the cut so as to preserve the conditions which existed before the cut had been made.) In order that the situation *at* the point be examined, the surface of the cut has to be infinitesimal and can therefore be considered plane; consequently, it can be individuated by an infinitesimal vector normal to it and proportional to its area. The force acting on that surface will also be an infinitesimal vector. If a different cut is considered (represented by a different vector) the exchanged force will also be different. The stress tensor is defined as that operator which associates a corresponding force to any vector representing the cut. By using momentum balance arguments, it can be shown that the operator is linear and is therefore a tensor.

Other examples of tensors of important physical significance will be encountered in later sections.

The unit tensor is that tensor which transforms any vector into itself:

$$\mathbf{a} = \mathbf{1} \cdot \mathbf{a} \qquad (1\text{-}3.4)$$

The zero tensor is that tensor which transforms all vectors into the zero vector:

$$\mathbf{0} = \mathbf{0} \cdot \mathbf{a} \qquad (1\text{-}3.5)$$

A special set of tensors is that of the *dyads* or *dyadic products* of two vectors. The dyadic product of vectors **c** and **d**, indicated by **cd**, is the tensor defined by

$$\mathbf{cd} \cdot \mathbf{a} = \mathbf{c}(\mathbf{d} \cdot \mathbf{a}) \qquad (1\text{-}3.6)$$

that is, the result of **cd** operating on any vector **a** is always a vector multiple of vector **c**. The factor depends linearly on **d** and the given vector **a**. Because of the properties of the scalar product, it is immediately shown that the definition of eq. (1-3.6) satisfies the linearity conditions, eqs (1-3.2) and (1-3.3). It is also apparent that, in general, the dyads **cd** and **dc** are different. The following tensor operations are considered:

(a)   Multiplication by a scalar   Given a scalar α and a tensor **A**, the tensor $\alpha\mathbf{A}$ is such as to give, for all vectors,

$$(\alpha\mathbf{A}) \cdot \mathbf{a} = \alpha(\mathbf{A} \cdot \mathbf{a}) \qquad (1\text{-}3.7)$$

Tensors which are multiples of the unit tensor are called *isotropic*.

(b)   Addition of tensors   Given the tensors **A** and **B**, the tensor $\mathbf{A} + \mathbf{B}$ is defined by

$$(\mathbf{A} + \mathbf{B}) \cdot \mathbf{a} = \mathbf{A} \cdot \mathbf{a} + \mathbf{B} \cdot \mathbf{a} \qquad (1\text{-}3.8)$$

valid for all vectors **a**.

(c)   Product of tensors   Given the tensors **A** and **B**, the tensor $\mathbf{A} \cdot \mathbf{B}$ is defined as the application in series of the tensors **B** and **A**. If, for example, $\mathbf{b} = \mathbf{B} \cdot \mathbf{a}$ and $\mathbf{c} = \mathbf{A} \cdot \mathbf{b}$, then

$$\mathbf{c} = (\mathbf{A} \cdot \mathbf{B}) \cdot \mathbf{a} \qquad (1\text{-}3.9)$$

In general, therefore, $\mathbf{A} \cdot \mathbf{B} \neq \mathbf{B} \cdot \mathbf{A}$.

Such expressions as $\mathbf{A} \cdot \mathbf{A}$, $\mathbf{A} \cdot \mathbf{A} \cdot \mathbf{A}$, etc. are usually shortened to $\mathbf{A}^2$, $\mathbf{A}^3$, etc. (It may be noted that $\mathbf{A}^2$, $\mathbf{A}^3$, etc. are tensors, while $\mathbf{a}^2$ is a scalar.)

(d)   Inverse of a tensor   The inverse of a tensor **A** is the machine working in the opposite direction. If, for example, $\mathbf{b} = \mathbf{A} \cdot \mathbf{a}$, then the inverse of **A**, indicated by $\mathbf{A}^{-1}$, transforms **b** into **a**:

$$\mathbf{b} = \mathbf{A} \cdot \mathbf{a} \quad \rightarrow \quad \mathbf{a} = \mathbf{A}^{-1} \cdot \mathbf{b} \qquad (1\text{-}3.10)$$

Not all tensors have an inverse; for instance, the inverses of the dyads and of the zero tensor are not defined. If the inverse of tensor **A** exists, then

$$\mathbf{A} \cdot \mathbf{A}^{-1} = \mathbf{A}^{-1} \cdot \mathbf{A} = \mathbf{1} \qquad (1\text{-}3.11)$$

(e)   Transpose of a tensor   The transpose of a tensor **A**, indicated by $\mathbf{A}^{\mathrm{T}}$, is such that the following equality holds for all pairs of vectors:

$$\mathbf{b} \cdot (\mathbf{A} \cdot \mathbf{a}) = \mathbf{a} \cdot (\mathbf{A}^{\mathrm{T}} \cdot \mathbf{b}) \qquad (1\text{-}3.12)$$

It is immediately recognized that the transpose of a dyad **cd** is the dyad **dc**.

A tensor is said to be *symmetric* if it coincides with its transpose. The unit tensor and the zero tensor are symmetric. A dyad is usually not.

A tensor is *skew-* or *antisymmetric* if it coincides with the opposite of its transpose (the opposite is obtained by multiplying by -1):

$$\mathbf{A} = -\mathbf{A}^\mathrm{T} \qquad (1\text{-}3.13)$$

A tensor is said to be *orthogonal* if its inverse exists and coincides with its transpose:

$$\mathbf{Q}^{-1} = \mathbf{Q}^\mathrm{T} \qquad (1\text{-}3.14)$$

If a tensor is orthogonal, the following property holds from the definition and eq. (1-3.11):

$$\mathbf{Q} \cdot \mathbf{Q}^\mathrm{T} = \mathbf{Q}^\mathrm{T} \cdot \mathbf{Q} = \mathbf{1} \qquad (1\text{-}3.15)$$

So far we have considered tensors and tensor operations without introducing the concept of components; that is, until now the situation is the same as that encountered with vectors when visualized as arrows in space. On the introduction of a vector basis, $\mathbf{e}_1, \mathbf{e}_2, \mathbf{e}_3$, the components of a tensor with respect to the chosen basis are defined as

$$A_{ij} = \mathbf{e}_i \cdot (\mathbf{A} \cdot \mathbf{e}_j) \qquad (1\text{-}3.16)$$

that is, the tensor is made to operate upon one of the basis vectors. The scalar product between the resulting vectors and another basis vector is then calculated. Clearly, nine components are obtained from eq. (1-3.16), which are usually arranged in a 3×3 matrix, the first index identifying the row and the second one the column.

As we have seen in discussing vector components, the vector basis is generally associated with a coordinate system. By using the natural basis $\mathbf{e}_i$ and its dual $\mathbf{e}^i$ the following types of tensor component are defined:

Covariant

$$A_{ij} = \mathbf{e}_i \cdot \mathbf{A} \cdot \mathbf{e}_j \qquad (1\text{-}3.17)$$

Contravariant

$$A^{ij} = \mathbf{e}^i \cdot \mathbf{A} \cdot \mathbf{e}^j \qquad (1\text{-}3.18)$$

Mixed

$$A^i{}_j = \mathbf{e}^i \cdot \mathbf{A} \cdot \mathbf{e}_j \qquad (1\text{-}3.19)$$

$$A_i{}^j = \mathbf{e}_i \cdot \mathbf{A} \cdot \mathbf{e}^j \qquad (1\text{-}3.20)$$

(Here and in the following text the brackets in such expressions as $\mathbf{a} \cdot (\mathbf{A} \cdot \mathbf{b})$, or analogous expressions, are dropped when no confusion may arise.) One may notice the analogy between eqs (1-3.17)–(1-3.20) and the corresponding equations, (1-2.7) and (1-2.8), for vector components. It has also to be noted that two types of mixed components are defined which, in general, are different. They are distinguished by the position of the indices as shown in eqs (1-3.19) and (1-3.20). The two sets of mixed components coincide only for a symmetric tensor, as can be shown by applying eq. (1-3.12). In such cases the indices can be set in any order, and the 'vertical' arrangement is preferred:

$$A^i_{\ j} = A^{\ i}_j = A^i_j \qquad (1\text{-}3.21)$$

Of course, if a Cartesian coordinate system is used, all kinds of components coincide. For the case of the stress tensor, the Cartesian components assume the familiar meaning of components of the forces per unit area acting on the faces of a cube with the sides parallel to the coordinate axes. For instance, T_{11} is given by

$$T_{11} = \mathbf{e}_1 \cdot \mathbf{T} \cdot \mathbf{e}_1 \qquad (1\text{-}3.22)$$

Because $\mathbf{e}_1$ is a unit vector in the direction of the x^1 axis, $\mathbf{T} \cdot \mathbf{e}_1$ is the force acting per unit area on the surface normal to $\mathbf{e}_1$. T_{11} is then, by eq. (1-3.22), the component of such force in the 1 direction.

We have now to remove an indeterminateness that was left in the definition of the stress tensor. Defining the stress tensor as the operator which associates forces to surfaces, we have not yet specified which of the two possible ways the vector representing the surface has to point, nor which of the two opposite forces the two parts of the body exert on one another.

We complete the definition by specifying that, choosing arbitrarily one of the two parts of the body, the normal to the surface has to point outwards from it and the corresponding force is that which the other part exerts on the part we have chosen (see Fig. 1-2). With this convention, it is immediately seen that the normal components of the stress tensor (such as T_{11}) are positive when a traction exists along that direction, and are negative when there is a compression.

The 'sign' convention adopted here is the classical one in the theory of elasticity, and it is also frequently used by rheologists when dealing with fluids, especially if the elastic aspect of their behavior is the one under study. The opposite convention, that of a positive compression and a negative traction, is classical of fluid mechanics and is most frequently encountered when dealing with 'ideal' or Newtonian fluids.

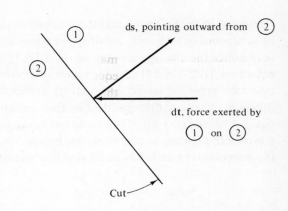

FIGURE 1-2
Definition of stress tensor.

The equations of transformation of tensor components from one system of coordinates to another are analogous to those for vectors (see eqs (1-2.10) and (1-2.11)).

For contravariant components:

$$A'^{ij} = \frac{\partial x'^i}{\partial x^m} \frac{\partial x'^j}{\partial x^n} A^{mn} \qquad (1\text{-}3.23)$$

For covariant components:

$$A'_{ij} = \frac{\partial x^m}{\partial x'^j} \frac{\partial x^n}{\partial x'^j} A_{mn} \qquad (1\text{-}3.24)$$

For mixed components:

$$A'^i_{\ j} = \frac{\partial x'^i}{\partial x^m} \frac{\partial x^n}{\partial x'^j} A^m_{\ n} \qquad (1\text{-}3.25)$$

The operations among tensors have their counterpart in component form. For example:

(a) Multiplication by a scalar:

$$(\alpha \mathbf{A})_{ij} = \alpha A_{ij} \qquad (1\text{-}3.26)$$

(b) Addition:

$$(\mathbf{A} + \mathbf{B})_{ij} = A_{ij} + B_{ij} \qquad (1\text{-}3.27)$$

(c) Product of tensors:

$$(\mathbf{A} \cdot \mathbf{B})_{ij} = A_{im} B^m_{\ j} = A_i^{\ m} B_{mj} \qquad (1\text{-}3.28)$$

(d) Transpose of a tensor:

$$(\mathbf{A}^\mathrm{T})_{ij} = A_{ji} \qquad (1\text{-}3.29)$$

The rule for the inverse is more complex and will be given in a later chapter. The equivalent of eq. (1-3.1) in component form is

$$b^i = A^{ij}a_j = A^i{}_j a^j \qquad (1\text{-}3.30)$$
$$b_i = A_{ij}a^j = A_i{}^j a_j$$

Equations (1-3.23) to (1-3.30) can be easily demonstrated on the basis of the already established properties of tensors and vectors and the definition of components. We give here one example by showing how one of eqs (1-3.30) is obtained from eq. (1-3.1).

From eq. (1-2.6):

$$\mathbf{a} = a_j \mathbf{e}^j$$

Substitution into eq. (1-3.1) gives

$$\mathbf{b} = a_j \mathbf{A} \cdot \mathbf{e}^j$$

Taking the scalar product of this equation with $\mathbf{e}^i$ gives

$$\mathbf{b} \cdot \mathbf{e}^i = a_j \mathbf{e}^i \cdot \mathbf{A} \cdot \mathbf{e}^j$$

Hence, by eqs (1-2.7) and (1-3.18),

$$b^i = A^{ij}a_j$$

Equations (1-3.30) show that any of the sets of nine components of a tensor completely determines the tensor. In fact, from the components of any vector, those of the resulting vector are calculated from the appropriate equation of (1-3.30).

A special mention has to be made of the components of the unit tensor, **1**. From eqs (1-3.17)–(1-3.20), these are the scalar products of the vectors of the natural basis and its dual.

The contravariant components of the unit tensor, called *contravariant metric*, are given by

$$g^{ij} = \mathbf{e}^i \cdot \mathbf{e}^j \qquad (1\text{-}3.31)$$

The *covariant metric* is

$$g_{ij} = \mathbf{e}_i \cdot \mathbf{e}_j \qquad (1\text{-}3.32)$$

The mixed components are always either 0 or 1, according to eqs (1-2.4). This is frequently indicated by a symbol called the Kronecker delta:

$$\delta^i_j = \mathbf{e}^i \cdot \mathbf{e}_j = \mathbf{e}_j \cdot \mathbf{e}^i \begin{cases} =0 & \text{if} \quad i \neq j \\ =1 & \text{if} \quad i = j \end{cases} \qquad (1\text{-}3.33)$$

The matrix of the mixed components of the unit tensor is therefore a 3×3 unit matrix:

$$[\delta_j^i] = \begin{Vmatrix} 1 & 0 & 0 \\ 0 & 1 & 0 \\ 0 & 0 & 1 \end{Vmatrix} \qquad (1\text{-}3.34)$$

In a Cartesian coordinate system, the contravariant and covariant metric coincide with the unit matrix.

The metric is very useful for obtaining one kind of component of vectors or tensors from another kind; for example,

$$a^i = g^{ij}a_j$$
$$A^{ij} = g^{im}A_m{}^j = g^{im}g^{jn}A_{mn} \qquad (1\text{-}3.35)$$
$$A_i{}^j = g_{im}A^{mj}$$

Operations, such as those of eqs (1-3.35), are called 'raising and lowering of indices.' In abstract notation they always correspond to $\mathbf{a} = \mathbf{1} \cdot \mathbf{a}$ or $\mathbf{A} = \mathbf{1} \cdot \mathbf{A}$.

Let us now briefly consider the subject of scalar functions of tensor arguments. These are relationships which associate to any given tensor a scalar result:

$$\alpha = f(\mathbf{A})$$

There are two categories of scalar functions of tensor arguments: those for which the relationship is dependent on the choice of some other quantity, and those for which the relationship is unique. These are called *invariants* or isotropic functions. For example, the relationship which associates to any given tensor one of its components is a scalar function which depends on the choice of the vector basis. The relationship

$$A_{12} = f(\mathbf{A})$$

is defined only if a basis is assigned, and it changes upon a change of basis. It is therefore non-invariant. We are particularly interested in the other category, that of the invariant relationships. One of the most important of these is the *trace* of a tensor. The trace is defined by the following properties:

$$\text{tr}\,(\mathbf{A} + \mathbf{B}) = \text{tr}\,\mathbf{A} + \text{tr}\,\mathbf{B}$$
$$\text{tr}\,(\alpha\mathbf{A}) = \alpha\,\text{tr}\,\mathbf{A} \qquad (1\text{-}3.36)$$
$$\text{tr}\,(\mathbf{ab}) = \mathbf{a} \cdot \mathbf{b}$$

The first two express the fact that the trace is a linear function. The third gives explicitly the value of the trace if the tensor is a dyad. It can be shown that the above definition uniquely determines the function for all values of the argument, that is, for all tensors.

The practical method of calculating the trace of a tensor is by means of the mixed components:

$$\text{tr } \mathbf{A} = A^i{}_i = A_i{}^i \qquad (1\text{-}3.37)$$

It is worthwhile stressing that despite the fact that the trace is usually calculated by means of eq. (1-3.37), its value is independent of the basis used to calculate the components. This justifies the name of the first invariant of $\mathbf{A}$, indicated by $I_\mathbf{A}$, which is used interchangeably with that of the trace of $\mathbf{A}$.

The following properties of the trace are easily demonstrated:

$$\text{tr } \mathbf{A}^\mathsf{T} = \text{tr } \mathbf{A} \qquad (1\text{-}3.38)$$

$$\text{tr } (\mathbf{A} \cdot \mathbf{B}) = \text{tr } (\mathbf{B} \cdot \mathbf{A}) \qquad (1\text{-}3.39)$$

The last quantity is also indicated by

$$\text{tr } (\mathbf{A} \cdot \mathbf{B}) \equiv \mathbf{A} : \mathbf{B} \qquad (1\text{-}3.40)$$

The *magnitude* of a tensor is defined as

$$|\mathbf{A}| = \sqrt{\text{tr } (\mathbf{A} \cdot \mathbf{A}^\mathsf{T})} = \sqrt{(\mathbf{A} : \mathbf{A}^\mathsf{T})} \qquad (1\text{-}3.41)$$

The *second invariant* of a tensor $\mathbf{A}$, indicated by $II_\mathbf{A}$, is also based on the operation of trace. Its definition is

$$II_\mathbf{A} = \tfrac{1}{2}[I_\mathbf{A}^2 - \text{tr } (\mathbf{A}^2)] \qquad (1\text{-}3.42)$$

The *third invariant*, $III_\mathbf{A}$, or *determinant* of a tensor, is another example of isotropic scalar function. It can be defined as follows: Given any three vectors, not all parallel to the same plane, consider the volume of the parallelepiped spanned by the three vectors. Consider then the three vectors obtained by letting the tensor $\mathbf{A}$ operate on the given vectors, and again calculate the volume of the parallelepiped spanned by these three vectors. The ratio of the latter volume to the former is the determinant of $\mathbf{A}$. The sign is established by assuming that it is positive if the handedness of the three vectors is preserved after the tensor has operated, and it is negative in the other case.† It can be shown that the value of the determinant so obtained does not depend upon the choice of the three vectors but only upon the tensor $\mathbf{A}$.

---

† The handedness of three ordered vectors not all parallel to the same plane is determined in the following way: apply the three vectors to the same point and consider the plane α of the first and second vector. If, from the endpoint of the third vector, the rotation of the first vector towards the second along the smallest angle in the plane α is seen as a clockwise rotation, then the set of vectors is left-handed; if the rotation is counterclockwise, it is right-handed.

It can also be shown that the value of III_A can be calculated as the determinant of the matrix of the components of $\mathbf{A}$ with respect to an orthonormal basis.

The following properties of the determinant are of importance:

$$\det (\mathbf{A} \cdot \mathbf{B}) = (\det \mathbf{A})(\det \mathbf{B}) \qquad (1\text{-}3.43)$$

$$\det (\mathbf{A}^{-1}) = (\det \mathbf{A})^{-1} \qquad (1\text{-}3.44)$$

$$\det (\mathbf{A}^T) = \det \mathbf{A} \qquad (1\text{-}3.45)$$

In relation to eq. (1-3.44) we note, incidentally, that the necessary and sufficient condition for the existence of $\mathbf{A}^{-1}$ is that $\det \mathbf{A}$ be different from zero.

The three invariants defined above, called collectively *principal invariants*, are important because of the following *representation theorem for symmetric tensors*.

Any isotropic scalar function (i.e., invariant) of a symmetric tensor argument can be expressed as a function of the three principal invariants of the argument:

$$\alpha = f(\mathbf{A}) = g(I_A, II_A, III_A) \qquad (1\text{-}3.46)$$

Other invariants commonly encountered are the *moments*:

$$\bar{I}_A = \text{tr } \mathbf{A}$$
$$\bar{II}_A = \text{tr } (\mathbf{A}^2) \qquad (1\text{-}3.47)$$
$$\bar{III}_A = \text{tr } (\mathbf{A}^3)$$

The representation theorem, eq. (1-3.46), can also be written in terms of moments:

$$\alpha = f(\mathbf{A}) = \bar{g}(\bar{I}_A, \bar{II}_A, \bar{III}_A) \qquad (1\text{-}3.48)$$

The principal invariants are also encountered in the following useful identity, known as the Hamilton–Cayley theorem:

$$\mathbf{A}^3 - I_A \mathbf{A}^2 + II_A \mathbf{A} - III_A \mathbf{1} = 0 \qquad (1\text{-}3.49)$$

1-4 DIFFERENTIATION OF SCALARS, VECTORS, AND TENSORS

Let us assume that, at a given time, we associate to each point in space, or at least in a continuous part of it, a scalar value. This 'function of the point' is called a *scalar field*. Assumptions of continuity are usually made which, in non-rigorous terms, mean that the function varies smoothly from point to point. The temperature distribution in a body is an example of a scalar field. The field is, in general, also a function of time; that is, for different values of time, different fields are present in space. In the relation between the scalar values and the points in space, time plays the role of a parameter.

Likewise, we may associate to each point in space a vector value or a tensor value, and we then speak of a *vector field* or a *tensor field* respectively. The velocity and stress in a fluid are examples of these types of field.

In this section we are concerned with the definition and properties of quantities that are related to the concepts of differential operations in *space* of fields. All the values involved are those at a given time. Differentiation with respect to time will be dealt with in a later chapter.

In the following text, all the conditions of continuity and differentiability —that is, all the conditions for the existence of the quantities to be defined— are assumed to be fulfilled by the fields we are considering and will not be specified.

(a) The gradient of a scalar

Let the scalar field at a given time be represented by the function $f(\mathbf{X})$ where $\mathbf{X}$ is any point in the region of space where the field is defined. The *gradient* of f at $\mathbf{X}$ is a *vector*, designated by the symbol ∇f such that

$$d f = f(\mathbf{X} + d\mathbf{X}) - f(\mathbf{X}) = \nabla f \cdot d\mathbf{X} \qquad (1\text{-}4.1)$$

Here $d\mathbf{X}$ is any infinitesimal vector and $\mathbf{X} + d\mathbf{X}$ is the point obtained by 'summing' to $\mathbf{X}$ the vector $d\mathbf{X}$ in the sense specified in section 1-2. It can be shown that the vector ∇f defined by eq. (1-4.1) is unique, i.e., it does not depend on $d\mathbf{X}$. Needless to say, as ∇f is defined in all points where the field $f(\mathbf{X})$ is defined, it is in turn a field and precisely a vector field.

If a coordinate system is chosen, the scalar field $f(\mathbf{X})$ can be represented by the function of three variables $f(x^i)$, where x^i are the coordinates of $\mathbf{X}$. We may then see how the components of vector ∇f are related to the function $f(x^i)$.

In order to do so we must digress briefly and show that the differentials of the coordinates dx^i are the contravariant components of vector $d\mathbf{X}$, the difference between the point whose coordinates are $x^i + dx^i$, and the point whose coordinates are x^i.

If $x'^i(x^j)$ are the equations of a change of coordinate system, we have

$$dx'^i = \frac{\partial x'^i}{\partial x^j} dx^j \qquad (1\text{-}4.2)$$

Comparing eq. (1-4.2) with eq. (1-2.10) we observe that the quantities dx^i must be the contravariant components of some vector. To discover the vector in question we may choose a Cartesian system, which shows immediately that the vector is $d\mathbf{X}$.

Because the dx^i are contravariant components of $d\mathbf{X}$, we write

$$d\mathbf{X} = \mathbf{e}_i \, dx^i \qquad (1\text{-}4.3)$$

(Equation (1-4.3) can also be demonstrated directly on the basis of the definitions of the basis vectors $\mathbf{e}_i$, eqs (1-2.2).)

We now substitute eq. (1-4.3) into eq. (1-4.1) and express df through the derivatives with respect to the coordinates:

$$df = \frac{\partial f}{\partial x^i} \, dx^i = \nabla f \cdot \mathbf{e}_i \, dx^i \qquad (1\text{-}4.4)$$

Because of the arbitrariness of dx^i, eq. (1-4.4) is equivalent to

$$\nabla f \cdot \mathbf{e}_i = \frac{\partial f}{\partial x^i} \qquad (1\text{-}4.5)$$

The left-hand side of eq. (1-4.5) can be compared with eq. (1-2.8) and is recognized as the set of covariant components of ∇f. Thus: the *covariant components of the gradient of a scalar field $f(\mathbf{X})$ are the partial derivatives of $f(x^i)$ with respect to the coordinates.*

The contravariant components can be obtained by raising the index, that is, through the use of the metric. The covariant and contravariant components of ∇f are sometimes indicated by $D_i f$ and $D^i f$, respectively.

Each of the three coordinates of a coordinate system is a scalar field, because to any point $\mathbf{X}$ a scalar, x^i, is associated. One may then ask 'What is the corresponding gradient ∇x^i?' From the definition (eq. (1-4.1)) we have

$$dx^i = \nabla x^i \cdot d\mathbf{X} \qquad (1\text{-}4.6)$$

Recalling that dx^i are the contravariant components of $d\mathbf{X}$, and comparing eq. (1-4.6) with eq. (1-2.7), we find

$$\nabla x^i = \mathbf{e}^i \qquad (1\text{-}4.7)$$

that is, the gradients of the coordinate fields are the vectors of the basis dual of the natural basis.

(b) The gradient of a vector

Let the vector field be represented by the function $\mathbf{a}(\mathbf{X})$.

The gradient of **a** at **X** is a *tensor*, indicated by ∇**a** and defined as

$$d\mathbf{a} = \mathbf{a}(\mathbf{X} + d\mathbf{X}) - \mathbf{a}(\mathbf{X}) = \nabla\mathbf{a} \cdot d\mathbf{X} \qquad (1\text{-}4.8)$$

Again, if a coordinate system is chosen, we may find the relation between the components of ∇**a** and the components of **a**. This relation turns out to be more complex than in the previous case of the gradient of a scalar. In fact the components of ∇**a** are not the derivatives of the components of **a** with respect to the coordinates as the analogy between eqs (1-4.8) and (1-4.1) might suggest. This simple result is obtained only if the coordinate system is Cartesian.

Through a lengthy procedure it can be shown that for the general case the components of ∇**a** are given by:

Mixed components

$$(\nabla\mathbf{a})^i{}_j = \frac{\partial a^i}{\partial x^j} + \left\{ \begin{matrix} i \\ j \ \ k \end{matrix} \right\} a^k \qquad (1\text{-}4.9)$$

where

$$\left\{ \begin{matrix} i \\ j \ \ k \end{matrix} \right\} = \left\{ \begin{matrix} i \\ k \ \ j \end{matrix} \right\} = g^{im}[jk, m] \qquad (1\text{-}4.10)$$

$$[jk, m] = [kj, m] = \frac{1}{2}\left(\frac{\partial g_{jm}}{\partial x^k} + \frac{\partial g_{km}}{\partial x^j} - \frac{\partial g_{kj}}{\partial x^m} \right) \qquad (1\text{-}4.11)$$

The symbols defined in eqs (1-4.11) and (1-4.10) are called Christoffel symbols of the first and second kind, respectively. As shown, they are a combination of the derivatives of the metric with respect to the coordinates and are all zero when the metric is constant, as in a Cartesian system. The summation convention also applies to these symbols. The indices in Christoffel symbols of the first kind must be interpreted as subscripts; the Christoffel symbol of the second kind has one superscript and two subscripts.

Because the tensor ∇**a** is in general not symmetric, the position of the indices is significant. The mixed components so far considered are usually indicated by one of the following notations:

$$(\nabla\mathbf{a})^i{}_j = a^i{}_{,j} = D_j a^i \qquad (1\text{-}4.12)$$

In both notations, i must be interpreted as the first index and j as the second; that is,†

$$a^i{}_{,j} = D_j a^i = \mathbf{e}^i \cdot \nabla\mathbf{a} \cdot \mathbf{e}_j \qquad (1\text{-}4.13)$$

---

† In some texts the opposite interpretation prevails, i.e., in eq. (1-4.12) j is interpreted as the first index. This corresponds to either a different definition of tensor components

$$A_{ij} = \mathbf{e}_j \cdot \mathbf{A} \cdot \mathbf{e}_i$$

or to a different definition of the gradient than that of eq. (1-4.8)

$$d\mathbf{a} = (\nabla\mathbf{a})^\mathsf{T} \cdot d\mathbf{X}$$

The current name for $a^i{}_{,j}$ —for the expression given by eq. (1-4.9)—is the covariant derivative of a contravariant vector.

Covariant components

$$(\mathbf{\nabla a})_{ij} = a_{i,j} = D_j a_i = \frac{\partial a_i}{\partial x^j} - \left\{ \begin{matrix} k \\ i \quad j \end{matrix} \right\} a_k \qquad (1\text{-}4.14)$$

This expression is called the covariant derivative of a covariant vector. Instead of using eq. (1-4.14), $a_{i,j}$ can be obtained from $a^i{}_{,j}$ by a lowering index procedure:

$$a_{i,j} = g_{im}a^m{}_{,j} \qquad (1\text{-}4.15)$$

The contravariant components and the other kind of mixed components of $\mathbf{\nabla a}$ are obtained by raising the second index in eqs (1-4.9) and (1-4.14), respectively. The symbols are $a^{i,j}$ and $a_i{}^{,j}$ and take the names of contravariant derivative of a contravariant and covariant vector, respectively.

The divergence of a vector

The divergence of a vector field is a *scalar* quantity defined by

$$\operatorname{div} \mathbf{a} = \operatorname{tr}(\mathbf{\nabla a}) \qquad (1\text{-}4.16)$$

By recalling eq. (1-3.37) and the previous section on the gradient of **a** we have

$$\operatorname{div} \mathbf{a} = a^i{}_{,i} = a_m{}^{,m} \qquad (1\text{-}4.17)$$

Another symbol used to indicate the divergence of vector **a** is

$$\operatorname{div} \mathbf{a} = \mathbf{\nabla} \cdot \mathbf{a} \qquad (1\text{-}4.18)$$

It has to be stressed that $\mathbf{\nabla} \cdot \mathbf{a}$, in spite of the notation, cannot be interpreted as a scalar product, and $\mathbf{\nabla a}$ cannot be considered a dyad.

The gradient of a tensor

The gradient of a tensor is a *third-order tensor*. (In a generalization of tensor analysis, or linear algebra, scalars are considered zero-order tensors, vectors first-order tensors, 'tensors' second-order tensors, and higher order tensors are also considered. These have components with more than two indices; the components transform, upon a change of a coordinate system, according to rules analogous to eqs (1-2.10)–(1-2.11) and (1-3.23)–(1-3.25).)

The components of the gradient of a tensor **A** are given by

$$A^{ij}{}_{,k} = \frac{\partial A^{ij}}{\partial x^k} + \left\{ \begin{matrix} i \\ k \quad m \end{matrix} \right\} A^{mj} + \left\{ \begin{matrix} j \\ k \quad m \end{matrix} \right\} A^{im} \qquad (1\text{-}4.19)$$

Equation (1-4.19) can be considered as the extension of eq. (1-4.9) for the gradient of a vector.

Other components may be obtained by means of similar expressions or by raising and lowering index procedures.

The divergence of a tensor

The divergence of a tensor field is a *vector* indicated by the notation div **A**, or **V** · **A**. It has rather an involved definition: Consider the field of the transpose of tensor **A** and a fixed vector **a**. $\mathbf{A}^T \cdot \mathbf{a}$ is a vector field of which the divergence can be calculated. The divergence of **A** is then the vector of which the following equality holds true:

$$\left. \begin{matrix} (\text{div } \mathbf{A}) \cdot \mathbf{a} = \text{div}\,(\mathbf{A}^T \cdot \mathbf{a}) \\ (\mathbf{V} \cdot \mathbf{A}) \cdot \mathbf{a} = \mathbf{V} \cdot (\mathbf{A}^T \cdot \mathbf{a}) \end{matrix} \right\} \qquad (1\text{-}4.20)$$

It can be shown that the vector so defined is independent of **a** and only depends on the tensor field **A**.

The components of this vector are easily obtained from those of the gradient of **A**.

Covariant

$$(\mathbf{V} \cdot \mathbf{A})_i = A_{im}{}^{,m} = A_i{}^m{}_{,m} \qquad (1\text{-}4.21)$$

Contravariant

$$(\mathbf{V} \cdot \mathbf{A})^i = A^{im}{}_{,m} = A^i{}_m{}^{,m} \qquad (1\text{-}4.22)$$

Quantities involving second derivatives

The quantities considered here are based on the concepts already introduced in the previous sections. They are actually combinations in series of the 'operations' of divergence and gradient already considered.

The Laplacian of a scalar is the divergence of the gradient of a scalar field, $f(\mathbf{X})$. It is therefore a *scalar* quantity, indicated by $\nabla^2 f$ for $\mathbf{V} \cdot \mathbf{V} f$. We have

$$\nabla^2 f = (\nabla f)^i{}_{,i} = (\nabla f)_i{}^{,i} \qquad (1\text{-}4.23)$$

The *Laplacian of a vector* is the divergence of the gradient of a vector field, $\mathbf{a}(\mathbf{X})$. It is therefore a *vector*, indicated by $\nabla^2\mathbf{a}$ or $\nabla \cdot \nabla\mathbf{a}$. We have:

Contravariant components

$$(\nabla^2\mathbf{a})^i = a^i_{,m}{}^{,m} = a^{i,m}{}_{,m} \qquad (1\text{-}4.24)$$

Covariant components

$$(\nabla^2\mathbf{a})_i = a_{i,m}{}^{,m} = a_i{}^{,m}{}_{,m} \qquad (1\text{-}4.25)$$

It is apparent from the above presentation that great use is made of the symbol ∇ in introducing different quantities. The symbol ∇ has also received the name of *Nabla operator*. As these circumstances may lead to some confusion, it is important to recognize that the operation implied by the symbol depends on the nature of the quantity upon which it operates; in this respect it is different for scalars, vectors, and tensors. On the other hand, in component form, the operation can receive a general formulation by defining the covariant derivative of an nth-order tensor. Also, the difference between the operators ∇ and $\nabla \cdot$, which indicate the gradient and divergence respectively, should be noted.

1-5 FRAMES OF REFERENCE AND COORDINATE SYSTEMS

In dealing with problems of motion it is essential to introduce a 'frame of reference' or 'observer.' The concept of motion is in fact relative, and only such expressions as 'the motion of something *with respect* to something else' have a physical significance.

In studying the motion of some body, reference is made to other bodies which are fixed, each with respect to all others, and constitute a *frame of reference*. The fact that the bodies constituting the frame are fixed is recognized, observing that their mutual distances do not change appreciably with time, at least over the time scale of the experiment. An Euclidean geometrical space is then 'fixed' to these bodies and the motion of the body under study is then imagined to occur 'through' this space, in the sense that the 'particles' or 'material points' of the body occupy different points of the space of the frame of reference as time continues.

The *velocity* of a particle is defined in the following way: Let $\mathbf{X}(t)$ be the point occupied by the particle at time t. With passing time $\mathbf{X}(t)$ describes the 'trajectory' of the particle. The limit

$$\lim_{\Delta t \to 0} \frac{\mathbf{X}(t + \Delta t) - \mathbf{X}(t)}{\Delta t} = \mathbf{v} = \dot{\mathbf{X}} \qquad (1\text{-}5.1)$$

is the velocity vector of the particle at time t. It is indicated either as **v** or as $\dot{\mathbf{X}}$, the time derivative of the 'point' function $\mathbf{X}(t)$.

If we now consider a second frame of reference (another set of mutually fixed objects and an Euclidean space joint to them) which is in motion with respect to the first frame, the motion of the same body will appear different to the two frames. In particular the velocity of a particle will be a different vector.

Further, consider two material points of one of the bodies which individuate a frame. These particles are fixed with respect to that frame; that is, they occupy fixed points of the space contained within the frame. The difference between the two points is a vector which is constant in time. If we consider a second frame in motion with respect to the first, the same two particles will move, and the difference between the two points occupied by these particles will be a variable vector in the second frame. Even if the relative motion of the two frames is stopped from a certain point in time, the two vectors will in general be different; they will be 'rotated' one with respect to the other.´

In general, therefore, upon a change of frame vectors and tensors (which are but relations among vectors) change.

It is important to distinguish carefully between frames and coordinate systems. In section 1-2 we introduced the concept of a coordinate system as a relationship which associates ordered triplets of numbers to points in space. It is clearly possible to assign this relationship in an infinite number of different ways within the same space, that is, within the same frame. If there is a change of coordinate system within the same frame, vectors and tensors do not change, only components do.

For example, in studying the process of a fiber spinning and twisting out of a spinneret, the space fixed to the walls of the laboratory might be chosen as a frame of reference. The velocity of the particles and all other vectors and tensors of interest are thus individuated. For the purpose of performing certain calculations, it might then be convenient to choose a coordinate system, say a Cartesian system. Because of the cylindrical symmetry of the fiber, one might instead choose a cylindrical coordinate system, or for some other reason one might choose yet another coordinate system, but in each of these choices only the components of vectors and tensors are affected, not the vectors and tensors themselves.

On the other hand, one might choose to observe the motion from a frame of reference which is fixed to a small piece of fiber and which, therefore, translates and rotates with respect to the walls of the laboratory following the advancing and twisting motion of the fiber. In this type of frame the tensors and vectors (e.g., the velocity) are changed. Again, within the new frame, one may choose any coordinate system that seems convenient, and this choice only affects the components.

From the above statements it follows that there are many Euclidean spaces, one for each frame of reference, that one may choose. They occupy the same 'space we live in' but are nevertheless distinct. They are in relative motion, one 'through' the other or even simply 'rotated' one with respect to the other.

The considerations developed so far are, of course, largely intuitive. The concept which can receive a precise mathematical formulation is that of change of frame. However, before presenting this concept two further intuitive observations will be made.

Because the set of objects and the associated space constituting a frame are mutually fixed, the relative motion of two frames can only be a rigid motion. It is therefore, at any instant of time, the superposition of a translation and a rotation.

An orthogonal tensor $\mathbf{Q}$ has the following property. Let $\mathbf{a}$ and $\mathbf{b}$ be any two vectors; we may write (see eqs (1-3.9), (1-3.12), and (1-3.15)):

$$(\mathbf{Q} \cdot \mathbf{a}) \cdot (\mathbf{Q} \cdot \mathbf{b}) = \mathbf{b} \cdot \mathbf{Q}^{\mathrm{T}} \cdot (\mathbf{Q} \cdot \mathbf{a}) = \mathbf{b} \cdot \mathbf{a} \qquad (1\text{-}5.2)$$

that is, the resulting vectors $\mathbf{Q} \cdot \mathbf{a}$ and $\mathbf{Q} \cdot \mathbf{b}$ have the same scalar product as $\mathbf{b}$ and $\mathbf{a}$. Therefore, $\mathbf{Q}$ transforms vectors into vectors which have the same length and form the same angles among themselves. $\mathbf{Q}$ can thus be interpreted as a tensor which rotates rigidly the space of vectors.

Change of frame

Following the work of Noll [7], a change of frame is represented by

$$\mathbf{X}^* - \mathbf{Y}(t) = \mathbf{Q}(t) \cdot (\mathbf{X} - \mathbf{Z}) \qquad (1\text{-}5.3)$$

$$t^* = t + a \qquad (1\text{-}5.4)$$

Here t is the time in the 'old' frame, t^* that in the new frame, and a is a constant; $\mathbf{Q}(t)$ is an orthogonal tensor; $\mathbf{X}$, $\mathbf{Y}(t)$, and $\mathbf{Z}$ are points in the 'old' frame, and $\mathbf{X}^*$ is the 'transform' of $\mathbf{X}$ in the 'new' frame.

The interpretation of eq. (1-5.4) is obvious: it is a change in the time origin. Equation (1-5.3) is the equation of transformation of points which describes the relative motion of the two frames; $\mathbf{Q}(t)$ is representative of the rigid rotation, and the vector $\mathbf{Y}(t) - \mathbf{Z}$ is representative of the relative displacement of the two frames at any instant of time; i.e., it describes the translation. If $\mathbf{Q}(t) = \mathbf{1}$, the relative motion is only a translation; if $\mathbf{Y}(t) - \mathbf{Z}$

is a constant vector, the relative motion is only a rotation.†

The equation of transformation of 'geometrical' vectors, such as vectors obtained as differences of points, is immediately obtained from eq. (1-5.4). Let X_1 and X_2 be two points and X_1^*, X_2^* their transforms. We have

$$X_1^* - X_2^* = Q(t) \cdot (X_1 - X_2) \qquad (1\text{-}5.5)$$

A vector **a** which transforms like a geometrical vector is called *indifferent*:

$$a^* = Q(t) \cdot a \qquad (1\text{-}5.6)$$

Let **A** be a tensor such that

$$b = A \cdot a \qquad (1\text{-}5.7)$$

where both the vectors **a**, upon which the tensor operates, and **b**, those obtained as result, are indifferent.

$$\begin{aligned} a^* &= Q(t) \cdot a \\ b^* &= Q(t) \cdot b \end{aligned} \qquad (1\text{-}5.8)$$

We wish to find the relation between **b*** and **a***. Premultiplying eqs (1-5.8) by Q^T we have

$$\begin{aligned} a &= Q^T(t) \cdot a^* \\ b &= Q^T(t) \cdot b^* \end{aligned} \qquad (1\text{-}5.9)$$

Substituting into eq. (1-5.7) and premultiplying by $Q(t)$ we have

$$b^* = (Q(t) \cdot A \cdot Q^T(t)) \cdot a^* \qquad (1\text{-}5.10)$$

Therefore, the relation between **b*** and **a*** is the tensor **A** given by

$$A^* = Q(t) \cdot A \cdot Q^T(t) \qquad (1\text{-}5.11)$$

A* can be considered the transform of **A**. A tensor which transforms according to eq. (1-5.11) is called *indifferent*.

Not all vectors and tensors are indifferent. We shall encounter many examples of tensors which have different rules of transformation than eq. (1-5.11). A typical non-indifferent vector is the velocity **v**. Because of its use in the following, we shall derive the equation of transformation of **v**.

In eq. (1-5.3), which gives the transformation rule of a point, let us interpret $X(t)$ as the position, variable with time, of a particle and thus $X^*(t)$ as the corresponding position of the same particle in a different frame. Differentiation of eq. (1-5.3) with respect to time gives:

$$\dot{X}^* - \dot{Y} = Q \cdot \dot{X} + \dot{Q} \cdot (X - Z) \qquad (1\text{-}5.12)$$

---

† **Z** can also be a function of time. However, taking **Z** as a constant point does not reduce generality. Only the difference, $Y(t) - Z$, is representative of the translatory motion.

Equation (1-5.12) has been obtained using the rule:

$$\frac{d}{dt}(\mathbf{A} \cdot \mathbf{a}) = \dot{\mathbf{A}} \cdot \mathbf{a} + \mathbf{A} \cdot \dot{\mathbf{a}} \qquad (1\text{-}5.13)$$

which is proved by using the linearity properties of a tensor together with eq. (1-3.8).

Equation (1-5.12) can be written as

$$\mathbf{v}^* = \mathbf{Q} \cdot \mathbf{v} + \dot{\mathbf{Y}} + \dot{\mathbf{Q}} \cdot (\mathbf{X} - \mathbf{Z}) \qquad (1\text{-}5.14)$$

This shows that eq. (1-5.6) is not fulfilled, namely that the velocity is not an indifferent vector.

So far, we have not mentioned scalar quantities and their behavior in a change of frame. Leaving aside those scalars which may be changed even within a frame of reference (e.g., the components of vectors and tensors), scalar quantities may be considered which are again of two categories with respect to a change of frame; that is, indifferent and non-indifferent.

A scalar quantity is said to be indifferent if it remains unchanged after a change of frame; that is, if

$$\alpha^* = \alpha \qquad (1\text{-}5.15)$$

Some examples of indifferent scalars are density, temperature, internal energy, etc. Other examples of indifferent scalars are those uniquely related to indifferent vectors and tensors; for example, the 'length' or magnitude of an indifferent vector is indifferent. In fact, if $\mathbf{a}$ is such a vector, we have

$$\mathbf{a}^* = \mathbf{Q} \cdot \mathbf{a} \qquad (1\text{-}5.16)$$

$$|\mathbf{a}^*| = \sqrt{(\mathbf{a}^* \cdot \mathbf{a}^*)} = \sqrt{[(\mathbf{Q} \cdot \mathbf{a}) \cdot (\mathbf{Q} \cdot \mathbf{a})]} = \sqrt{(\mathbf{a} \cdot \mathbf{a})} = |\mathbf{a}| \qquad (1\text{-}5.17)$$

It can be shown that the invariants of an indifferent tensor are indifferent; that is, if

$$\alpha = f(\mathbf{A}) \qquad (1\text{-}5.18)$$

$$\mathbf{A}^* = \mathbf{Q} \cdot \mathbf{A} \cdot \mathbf{Q}^{\mathrm{T}} \qquad (1\text{-}5.19)$$

then

$$\alpha = f(\mathbf{A}^*) \qquad (1\text{-}5.20)$$

Actually the condition that, for any orthogonal tensor $\mathbf{Q}$,

$$f(\mathbf{A}) = f(\mathbf{Q} \cdot \mathbf{A} \cdot \mathbf{Q}^{\mathrm{T}}) \qquad (1\text{-}5.21)$$

can be taken as the *definition* of isotropic scalar functions of a tensor argument, or invariants of a tensor, such as those encountered in section 1-3.

Scalars related to non-indifferent vectors and tensors are non-indifferent; for example, the magnitude of the velocity vector changes upon a change of frame.

The invariants of a non-indifferent tensor are non-indifferent. If **B** is non-indifferent, then

$$f(\mathbf{B}^*) \neq f(\mathbf{B}) \qquad (1\text{-}5.22)$$

This fact must not be confused with the definition of 'invariant' given above. It remains true that

$$f(\mathbf{B}) = f(\mathbf{Q} \cdot \mathbf{B} \cdot \mathbf{Q}^{\mathrm{T}}) \qquad (1\text{-}5.23)$$

only that $\mathbf{Q} \cdot \mathbf{B} \cdot \mathbf{Q}^{\mathrm{T}}$ is not $\mathbf{B}^*$, if **B** is not indifferent.

In conclusion, upon a change of frame, an indifferent tensor **A** gives rise to a tensor **A*** which has the same invariants as **A**, while a non-indifferent tensor **B** gives rise to a tensor **B*** which has different invariants.

A last example of a non-indifferent scalar is time, as shown by eq. (1-5.4). Conversely, a time interval between two events is indifferent.

1-6   THE EQUATION OF CONTINUITY

Consider a material undergoing flow in the Euclidean space of a frame of reference. Let **v** be the velocity vector, ρ the density, **X** any point in space, and t time. Both **v** and ρ are in general functions of both position and time (i.e., time-dependent fields):

$$\mathbf{v} = \mathbf{v}(\mathbf{X}, t) \qquad (1\text{-}6.1)$$

$$\rho = \rho(\mathbf{X}, t) \qquad (1\text{-}6.2)$$

The vector $\rho\mathbf{v}$ represents the mass flux (measured in units of grammes per square centimeter and second, or equivalent) crossing a differential surface orthogonal to the vector **v**. Now, consider the following identity known as the Gauss–Ostrogradskii theorem:

$$\iiint_V (\boldsymbol{\nabla} \cdot \mathbf{a}) \, dv = \iint_S \mathbf{a} \cdot d\mathbf{s} \qquad (1\text{-}6.3)$$

where **a** is a vector field, V is the volume of an arbitrary region of space enclosed by the surface S, and d**s** is a differential vector representing a differential surface element, directed outwards from the region of space considered. If **a** can be interpreted as the flux of some scalar property, then $\mathbf{a} \cdot d\mathbf{s}$ is the rate of flow of that property through the elementary surface represented by ds, and the integral on the right-hand side of eq. (1-6.3) is the overall efflux of the property out of the region of space considered. Consequently, $\mathbf{V} \cdot \mathbf{a}$ can be interpreted as the local value of such efflux per unit volume of space.

Applying these concepts to the case at hand, the divergence of $\rho\mathbf{v}$ is seen to represent the net outlet mass flow per unit volume from a differential volume in the neighborhood of the arbitrary point $\mathbf{X}$. If such a differential volume is chosen as the system, eq. (1-1.2) takes the form

$$\mathbf{V} \cdot (\rho\mathbf{v}) = -\frac{\partial\rho}{\partial t} \qquad (1\text{-}6.4)$$

Let us now transform eq. (1-6.4) so that the rate of change of density at the material point itself appears explicitly. The trajectory of the material point is the function $\mathbf{X}(t)$; one has:

$$\rho = \rho(\mathbf{X}, t) = \rho[\mathbf{X}(t), t] = \bar{\rho}(t) \qquad (1\text{-}6.5)$$

where the *function* $\rho(\ )$ is different from the *function* $\bar{\rho}(\ )$ though their values are equal when $\mathbf{X}$ in $\rho(\ )$ is taken as $\mathbf{X}(t)$. We have

$$\frac{d\bar{\rho}}{dt} = \dot{\bar{\rho}} = \frac{\partial\rho}{\partial t} + \mathbf{V}\rho \cdot \dot{\mathbf{X}} \qquad (1\text{-}6.6)$$

The quantity $\dot{\mathbf{X}}$ is the velocity $\mathbf{v}$. The derivative $\dot{\bar{\rho}}$ is the rate of change of density 'following the particle'; it is indicated by the symbol $D\rho/Dt$, and is called the *substantial derivative*. We have then

$$\frac{D\rho}{Dt} = \frac{\partial\rho}{\partial t} + \mathbf{V}\rho \cdot \mathbf{v} \qquad (1\text{-}6.7)$$

Expanding the left-hand side of eq. (1-6.4) we obtain (see problem 1-6):

$$\mathbf{V} \cdot (\rho\mathbf{v}) = \rho(\mathbf{V} \cdot \mathbf{v}) + \mathbf{v} \cdot \mathbf{V}\rho = -\frac{\partial\rho}{\partial t} \qquad (1\text{-}6.8)$$

Taking into account eq. (1-6.7), the following form of the mass balance equation is obtained:

$$\frac{D\rho}{Dt} = -\rho\mathbf{V} \cdot \mathbf{v} \qquad (1\text{-}6.9)$$

For constant density fluids, both forms of the differential mass balance equation simplify to

$$\mathbf{V} \cdot \mathbf{v} = 0 \qquad (1\text{-}6.10)$$

Equations (1-6.4) and (1-6.9) are called the 'Eulerian' and 'Lagrangian' forms of the continuity equation, respectively. The Lagrangian form may be regarded as written with reference to a frame with respect to which the material point is fixed. In fact, consider any change of frame which makes the velocity v^* at the material point X^* zero for all times. Equation (1-6.4) then transforms to

$$\mathbf{V} \cdot (\rho^* \mathbf{v}^*) = -\frac{\partial \rho^*}{\partial t} \qquad (1\text{-}6.11)$$

In eq. (1-6.11), ρ^* is the density field $\rho^*(X^*, t)$, which is *different* from the field $\rho(X, t)$ though the *values* of ρ and ρ^* are equal when X^* is taken as the transform of X (in this sense, the density is indifferent). Inasmuch as we have considered a frame such that the position X^* of the material point is fixed, $\partial \rho^* / \partial t$ is identified with $\partial \bar{\rho} / \partial t$ and hence with $D\rho / Dt$.

Equation (1-6.11) can be written as

$$-\frac{D\rho}{Dt} = \rho \mathbf{V} \cdot \mathbf{v}^* + \mathbf{v}^* \cdot \mathbf{V}\rho^* \qquad (1\text{-}6.12)$$

In the first term on the right-hand side of eq. (1-6.12) the density is written as ρ rather than ρ^*, because its *value* is considered (ρ appears as a factor) which is indifferent. The second term is zero because the value of v^* at X^* is zero by choice of frame. Thus,

$$-\frac{D\rho}{Dt} = \rho \mathbf{V} \cdot \mathbf{v}^* \qquad (1\text{-}6.13)$$

Equation (1-6.13) is still different formally from eq. (1-6.9), because the divergence of the field v^* appears rather than the divergence of the field v. But it may be proved (see section 2-2) that

$$\mathbf{V} \cdot \mathbf{v}^* = \mathbf{V} \cdot \mathbf{v} \qquad (1\text{-}6.14)$$

and hence eqs (1-6.9) and (1-6.13) are seen to be identical.

In the development given above, it was implicitly assumed that eq. (1-6.4) could be rewritten in the new frame in exactly the same form, i.e., that the equation was *frame-indifferent*. Frame-indifference is a property of some physical laws, but not necessarily of all physical laws: the dynamical equation, to be discussed in the next section, is not frame-indifferent. Frame-indifference will be discussed in detail in chapter 2.

1-7 THE DYNAMICAL EQUATION

The principle of conservation of momentum holds only in a so-called 'inertial' frame, which we assume exists in the Euclidian space of classical physics. If one inertial frame exists, any other frame translating with a constant velocity with respect to it is also inertial. The dynamical equation is written on the assumption that the frame is inertial. Indeed, the validity of the dynamical equation may be regarded as a definition of an inertial frame.

The Eulerian form of the dynamical equation is obtained by writing out explicitly the terms appearing in eq. (1-1.3) as applied to a differential neighborhood of the point $\mathbf{X}$.

In order to express the net inlet flow of momentum, consider the dyad $\rho\mathbf{vv}$. By definition of a dyad (see eq. (1-3.6)), we have

$$\rho\mathbf{vv} \cdot \mathbf{ds} = \mathbf{v}(\rho\mathbf{v} \cdot \mathbf{ds}) \qquad (1\text{-}7.1)$$

where $\mathbf{ds}$ is a surface element on which the considered point lies. The term in brackets on the right-hand side is the mass flow rate through the surface element considered; thus $\rho\mathbf{vv} \cdot \mathbf{ds}$ is seen to be the momentum flow through the same surface element. This identifies the dyad $\rho\mathbf{vv}$ as the momentum flux.

An identity analogous to eq. (1-6.3) for a tensor field $\mathbf{A}(\mathbf{X})$:

$$\iiint_V (\nabla \cdot \mathbf{A}) \, dV = \iint_S \mathbf{A} \cdot \mathbf{ds} \qquad (1\text{-}7.2)$$

allows us to identify the divergence of $\rho\mathbf{vv}$ as the net outlet flow of momentum per unit volume:

$$\nabla \cdot (\rho\mathbf{vv}) = \begin{pmatrix} \text{Net outlet flow} \\ \text{of momentum per} \\ \text{unit volume} \end{pmatrix} \qquad (1\text{-}7.3)$$

Now consider the second term in eq. (1-1.3), i.e., the sum of all surface forces. The stress force $\mathbf{dt}$ acting across any surface element $\mathbf{ds}$ is, by definition of the total stress tensor,

$$\mathbf{dt} = \mathbf{T} \cdot \mathbf{ds} \qquad (1\text{-}7.4)$$

By reconsidering eq. (1-7.2), the right-hand side is thus seen to be the resultant of the surface forces, and the divergence of $\mathbf{T}$ is the resultant per unit volume:

$$\nabla \cdot \mathbf{T} = \begin{pmatrix} \text{Resultant of} \\ \text{surface forces} \\ \text{per unit volume} \end{pmatrix} \qquad (1\text{-}7.5)$$

As will be discussed in chapter 4, for constant density fluids the constitutive equation determines the total stress **T** within an arbitrary additive isotropic tensor. It is therefore useful to decompose the total stress as follows:

$$\mathbf{T} = -p\mathbf{1} + \boldsymbol{\tau}' \qquad (1\text{-}7.6)$$

where p is called the pressure (not in the thermodynamic sense, see section 1-8), and $\boldsymbol{\tau}'$ is a *deviatoric* tensor, i.e.,

$$\text{tr } \boldsymbol{\tau}' = 0 \qquad (1\text{-}7.7)$$

The decomposition called for in eqs (1-7.6) and (1-7.7) is unique: in fact, if the trace of both sides is taken, we obtain

$$\text{tr } \mathbf{T} = -3p \qquad (1\text{-}7.8)$$

which is the operational definition of pressure.

The divergence of the total stress can now be decomposed as

$$\mathbf{V} \cdot \mathbf{T} = \mathbf{V} \cdot \boldsymbol{\tau}' - \mathbf{V}p \qquad (1\text{-}7.9)\dagger$$

The third term in eq. (1-1.3), when expressed per unit volume of the system, is simply $\rho\mathbf{g}$, where **g** is the field acceleration (gravity in a majority of cases).

The right-hand side of eq. (1-1.3), when expressed per unit volume of the system, is the partial derivative with respect to time of the vector $\rho\mathbf{v}$. Thus, finally, considering eqs (1-7.3), (1-7.5) and (1-7.9), the Eulerian form of the dynamical equation is obtained:

$$-\mathbf{V} \cdot (\rho\mathbf{v}\mathbf{v}) - \mathbf{V}p + \mathbf{V} \cdot \boldsymbol{\tau}' + \rho\mathbf{g} = \frac{\partial}{\partial t}(\rho\mathbf{v}) \qquad (1\text{-}7.10)$$

The Lagrangian form of the dynamical equation is obtained by algebraic manipulation of eq. (1-7.9). By using the result of problem 1-7, we have

$$\frac{\partial}{\partial t}(\rho\mathbf{v}) + \mathbf{V} \cdot (\rho\mathbf{v}\mathbf{v}) = \rho\frac{\partial\mathbf{v}}{\partial t} + \mathbf{v}\frac{\partial\rho}{\partial t} + \rho\mathbf{V}\mathbf{v} \cdot \mathbf{v} + \mathbf{v}[\mathbf{V} \cdot (\rho\mathbf{v})] \qquad (1\text{-}7.11)$$

---

† Equation (1-7.9) is obtained as follows: In general (see problem 1-9), we have for any tensor **A**:

$$\mathbf{V} \cdot (\alpha\mathbf{A}) = \alpha\mathbf{V} \cdot \mathbf{A} + \mathbf{A} \cdot \mathbf{V}\alpha$$

For an isotropic tensor $\alpha\mathbf{1}$:

$$\mathbf{V} \cdot (\alpha\mathbf{1}) = \alpha\mathbf{V} \cdot \mathbf{1} + \mathbf{V}\alpha = \mathbf{V}\alpha$$

In fact $\mathbf{V} \cdot \mathbf{1} = 0$, as can be seen directly from the definition of the divergence of a tensor.

and substituting eq. (1-6.4)† :

$$\frac{\partial}{\partial t}(\rho \mathbf{v}) + \mathbf{\nabla} \cdot (\rho \mathbf{v} \mathbf{v}) = \rho \left[\frac{\partial \mathbf{v}}{\partial t} + \mathbf{\nabla} \mathbf{v} \cdot \mathbf{v} \right] \equiv \rho \frac{D\mathbf{v}}{Dt} \qquad (1\text{-}7.12)$$

The quantity in square brackets in eq. (1-7.12) is recognized to be the substantial derivative of the velocity; the algebra leading to eq. (1-6.7) is simply duplicated by substituting the velocity field $\mathbf{v}$ for the density field ρ. Substituting eq. (1-7.12) into eq. (1-7.10) we get the Lagrangian form of the dynamical equation:

$$\rho \frac{D\mathbf{v}}{Dt} = -\mathbf{\nabla}p + \mathbf{\nabla} \cdot \mathbf{\tau}' + \rho \mathbf{g} \qquad (1\text{-}7.13)$$

The left-hand side of eq. (1-7.13) is the inertial force due to the acceleration of the particle, which equals the surface and body forces acting on it and appearing on the right-hand side. Thus eq. (1-7.13) is recognized as a direct formulation of Newton's law.

It is customary in classical fluid mechanics to consider a so-called 'mechanical energy equation.' Of course, there is no principle of conservation of mechanical energy; the mechanical energy equation is obtained by taking the scalar product of the dynamical equation with the velocity vector [8]. The mechanical energy equation contains no additional information to that contained in the dynamical equation, and in fact contains less: it is a scalar equation, while the dynamical equation is vectorial. Nonetheless, the mechanical energy equation is very useful in classical hydrodynamics, where the deviatoric stress $\mathbf{\tau}'$ is assumed to be zero; it is of limited usefulness in Newtonian fluid mechanics, and almost useless in non-Newtonian fluid mechanics.

It may be demonstrated that the principle of conservation of the moment of momentum implies that the stress is symmetric, i.e., $\mathbf{T} = \mathbf{T}^\mathsf{T}$. This is true in the so-called non-polar case, i.e., in the absence of body couples and couple stresses.

This book only deals with the non-polar case. For the non-polar case, the principle of conservation of moment of momentum poses no further restriction than that of the symmetry of the stress tensor; thus, this principle will not be referred to in the following, and the stress tensor will always be assumed to be symmetric.

----

† Equivalent notations for $D\mathbf{v}/Dt$ are $\ddot{\mathbf{X}}$ and $\dot{\mathbf{v}}$.

1-8 PRESSURE

The physical meaning of pressure in the case of constant density fluids needs to be clarified. In fact, pressure as appearing in eqs (1-7.10) and (1-7.13) cannot simply be identified with thermodynamic pressure (i.e., the independent variable entering the thermodynamic equation of state) when the density is independent of pressure. Indeed, the thermodynamic pressure is undetermined for constant density fluids, because the thermodynamic equation of state cannot be solved for pressure.†

The constitutive equation for constant density fluids determines the stress tensor only within an arbitrary additive isotropic tensor. The total stress $\mathbf{T}$ can be decomposed as follows:

$$\mathbf{T} = -\alpha\mathbf{1} + \boldsymbol{\tau} \qquad (1\text{-}8.1)$$

with α an *arbitrary* scalar and $\boldsymbol{\tau}$ the so called 'extra stress,' which is given by the constitutive equation. The scalar α should not be regarded as the pressure, since the trace of $\boldsymbol{\tau}$ need not be zero.

Comparison of eq. (1-7.6) and eq. (1-8.1) gives the following relationships among $\boldsymbol{\tau}$, $\boldsymbol{\tau}'$, α, and p:

$$-\alpha\mathbf{1} + \boldsymbol{\tau} = -p\mathbf{1} + \boldsymbol{\tau}' \qquad (1\text{-}8.2)$$

and hence

$$\operatorname{tr}\boldsymbol{\tau} = 3(\alpha - p) \qquad (1\text{-}8.3)$$

Substituting eq. (1-8.3) into eq. (1-8.2) and solving for the deviatoric stress:

$$\boldsymbol{\tau}' = \boldsymbol{\tau} - \tfrac{1}{3}(\operatorname{tr}\boldsymbol{\tau})\mathbf{1} \qquad (1\text{-}8.4)$$

The role of the dynamical equation with regards to either α or p needs to be discussed. Assume that the velocity field is assigned and that a constitutive equation for the fluid is available. If the constitutive equation is of a type which gives a deviatoric stress, then $\boldsymbol{\tau}'$ is calculated from the known kinematics, and from the dynamical equation (eq. (1-7.13)) ∇p is determined. Hence, the pressure field is finally calculated within an arbitrary additive constant. If, as is more frequently the case, the constitutive equation is of the form which gives a non-deviatoric extra stress, then from the calculated $\boldsymbol{\tau}$, $\boldsymbol{\tau}'$ is obtained from eq. (1-8.4) and ∇p from eq. (1-7.13) as before.

---

† The thermodynamic pressure can be defined in terms of the 'energetic' equation of state as minus the partial derivative of internal energy with respect of specific volume. Partial derivation of energy implies holding constant all other independent variables, among them the kinematic variables describing deformation. This poses an inherent difficulty to an operational definition of thermodynamic pressure in fluids possessing some elasticity (see the discussion in section 1-10). For constant density fluids, the thermodynamic pressure is simply not defined.

By using eqs (1-8.3) and (1-8.4), the dynamical equation can be written in terms of extra stress τ and α or p as desired:

$$\rho \frac{D\mathbf{v}}{Dt} = -\nabla\alpha + \nabla \cdot \tau + \rho\mathbf{g}$$

$$= -\nabla(p + \tfrac{1}{3}\operatorname{tr}\tau) + \nabla \cdot \tau + \rho\mathbf{g} \tag{1-8.5}$$

1-9 SPECIAL FORMS OF THE DYNAMICAL EQUATION

In classical hydrodynamics, one defines an *ideal fluid* as a material which is incapable of sustaining deviatoric stresses, so that the total stress tensor is always isotropic. This is tantamount to considering a very special constitutive equation, i.e.,

$$\text{Definition of an ideal fluid:} \quad \tau' = 0 \tag{1-9.1}$$

The dynamical equation then takes the following form, known as the Euler equation:

$$\rho \frac{D\mathbf{v}}{Dt} = -\nabla p + \rho\mathbf{g} \tag{1-9.2}$$

A further simplification arises by limiting attention to gravity as the only body force acting; in that case, $\mathbf{g} = -g\nabla z$, where z is the vertical position and g the value of the gravity acceleration, and the Euler equation reduces to

$$\rho \frac{D\mathbf{v}}{Dt} = -\nabla(p + \rho g z) \tag{1-9.3}$$

The scalar product of eq. (1-9.3) with the velocity vector is known as the differential Bernoulli equation; it is a form of the mechanical energy equation for the special case where $\tau' = 0$.

In textbooks on classical hydrodynamics, the Bernoulli equation is often derived from the energy conservation principle alone, following a procedure which will be discussed in the next section. There is a logical flaw in such an approach; while the dynamical equation is not used at all, the Bernoulli equation is obtained by making use of *two* constitutive assumptions, i.e., the one embodied in eq. (1-9.1) and the additional one that no mechanical energy is converted irreversibly to internal energy, i.e., that there is no energy dissipation.

The two constitutive assumptions are reduced to only one, i.e., eq. (1-9.1), when the Bernoulli equation is derived from the dynamical equation, as shown above. Following this procedure, it can then be proved that the second assumption is a consequence of the first.

In Newtonian incompressible fluid mechanics, Newton's law defining the viscosity μ is generalized to the following form:[†]

$$\boldsymbol{\tau} = 2\mu\mathbf{D} = \boldsymbol{\tau}' \qquad (1\text{-}9.4)$$

where $\mathbf{D}$ is the 'stretching' tensor (also called 'rate of stretch' or 'rate of strain' tensor), defined by

$$\mathbf{D} = \tfrac{1}{2}(\nabla\mathbf{v} + \nabla\mathbf{v}^{\mathrm{T}}) \qquad (1\text{-}9.5)$$

Clearly, $\mathbf{D}$ is symmetric. In general, any tensor can be uniquely divided into the sum of a symmetric and an antisymmetric tensor. For the velocity gradient we have

$$\nabla\mathbf{v} = \mathbf{D} + \mathbf{W} \qquad (1\text{-}9.6)$$

where

$$\mathbf{W} = \tfrac{1}{2}(\nabla\mathbf{v} - \nabla\mathbf{v}^{\mathrm{T}}) \qquad (1\text{-}9.7)$$

is the 'vorticity' or 'spin' tensor.

Substitution of eq. (1-9.4) into eq. (1-7.13) gives the Navier–Stokes equation:

$$\rho\frac{D\mathbf{v}}{Dt} = -\nabla p + \rho\mathbf{g} + \mu\nabla^2\mathbf{v} \qquad (1\text{-}9.8)$$

which is the starting point of classical Newtonian fluid mechanics.

The last term in eq. (1-9.8) is obtained as follows:

$$\begin{aligned}\nabla\cdot\boldsymbol{\tau}' = \nabla\cdot(2\mu\mathbf{D}) &= \nabla\cdot[\mu(\nabla\mathbf{v} + \nabla\mathbf{v}^{\mathrm{T}})] \\ &= \mu\nabla^2\mathbf{v} + \mu\nabla\cdot\nabla\mathbf{v}^{\mathrm{T}}\end{aligned} \qquad (1\text{-}9.9)$$

The last term in eq. (1-9.9) is calculated from the definition of the divergence of a tensor:

$$\begin{aligned}(\nabla\cdot\nabla\mathbf{v}^{\mathrm{T}})\cdot d\mathbf{X} = \nabla\cdot(\nabla\mathbf{v}\cdot d\mathbf{X}) &= \nabla\cdot d\mathbf{v} \\ &= \nabla\cdot[(\mathbf{v} + d\mathbf{v}) - \mathbf{v}] = \nabla\cdot(\mathbf{v} + d\mathbf{v}) - \nabla\cdot\mathbf{v}\end{aligned} \qquad (1\text{-}9.10)$$

Considering the equation of continuity (1-6.10), it is seen that $\nabla\cdot\nabla\mathbf{v}^{\mathrm{T}} = \mathbf{0}$. In general, the divergence of the transpose of the gradient of a vector with zero divergence is zero.

---

† For compressible fluids, one may write

$$\boldsymbol{\tau}' = 2\mu\mathbf{D} - \tfrac{2}{3}\mu(\nabla\cdot\mathbf{v})\mathbf{1}$$

so that eq. (1-7.7) is satisfied. It is evident that, for constant density fluids, (1-7.7) is satisfied by eq. (1-9.4) because of eq. (1-6.10).

1-10 ENERGY EQUATIONS

The 'mechanical energy' equation is obtained by taking the scalar product of the dynamical equation with the velocity vector.

Starting from the result in problem 1-5, we have

$$
\begin{aligned}
\mathbf{V}(\tfrac{1}{2}v^2) = \tfrac{1}{2}\mathbf{V}(\mathbf{v}\cdot\mathbf{v}) &= \mathbf{V}\mathbf{v}^T\cdot\mathbf{v} \\
&= (\mathbf{D}+\mathbf{W})^T\cdot\mathbf{v}
\end{aligned}
\tag{1-10.1}
$$

Inasmuch as $\mathbf{D}$ is a symmetric tensor and $\mathbf{W}$ an antisymmetric one, eq. (1-10.1) reduces to

$$
\mathbf{V}(\tfrac{1}{2}v^2) = \mathbf{D}\cdot\mathbf{v} - \mathbf{W}\cdot\mathbf{v} = \mathbf{V}\mathbf{v}\cdot\mathbf{v} - 2\mathbf{W}\cdot\mathbf{v}
\tag{1-10.2}
$$

The scalar product of the left-hand side of eq. (1-7.13) with the velocity vector can now be calculated:

$$
\begin{aligned}
\mathbf{v}\cdot\rho\frac{D\mathbf{v}}{Dt} &= \mathbf{v}\cdot\rho\frac{\partial\mathbf{v}}{\partial t} + \mathbf{v}\cdot\rho(\mathbf{V}\mathbf{v}\cdot\mathbf{v}) \\
&= \rho\frac{\partial}{\partial t}(\tfrac{1}{2}v^2) + \mathbf{v}\cdot\rho\mathbf{V}(\tfrac{1}{2}v^2) + 2\mathbf{v}\cdot\rho\mathbf{W}\cdot\mathbf{v}
\end{aligned}
\tag{1-10.3}
$$

The last term in eq. (1-10.3) is

$$
2\mathbf{v}\cdot\mathbf{W}\cdot\mathbf{v} = \mathbf{v}\cdot\mathbf{V}\mathbf{v}\cdot\mathbf{v} - \mathbf{v}\cdot\mathbf{V}\mathbf{v}^T\cdot\mathbf{v} = 0
\tag{1-10.4}
$$

where the last equality is simply obtained from the definition of the transpose of a tensor. Thus, finally,

$$
\mathbf{v}\cdot\rho\frac{D\mathbf{v}}{Dt} = \rho\frac{D}{Dt}(\tfrac{1}{2}v^2)
\tag{1-10.5}
$$

The mechanical energy equation is then obtained, in the Lagrangian form, as

$$
\rho\frac{D}{Dt}(\tfrac{1}{2}v^2) = -\mathbf{v}\cdot\mathbf{V}p + \mathbf{v}\cdot\mathbf{V}\cdot\mathbf{\tau}' + \mathbf{v}\cdot\rho\mathbf{g}
\tag{1-10.6}
$$

So far, only the dynamical equation has been used. It cannot be over-emphasized that eq. (1-10.6) is *not* a conservation principle, but only a scalar form of the dynamical equation.

The principle of conservation of energy, i.e., the first law of thermodynamics, can be written as follows. Let U be the internal energy per unit mass, and gz the potential energy per unit mass ($g\mathbf{V}z = -\mathbf{g}$). We then have

(a) Accumulation of energy in the volume element:

$$
\frac{\partial}{\partial t}[\rho(U + \tfrac{1}{2}v^2 + gz)] = gz\frac{\partial\rho}{\partial t} + \frac{\partial}{\partial t}[\rho(U + \tfrac{1}{2}v^2)]
\tag{1-10.7}
$$

(b) Net inlet flow of energy due to flow:

$$-\mathbf{V} \cdot [\rho(U + \tfrac{1}{2}v^2 + gz)\mathbf{v}]$$
$$= -\mathbf{V} \cdot [\rho(U + \tfrac{1}{2}v^2)\mathbf{v}] - gz\mathbf{V} \cdot \rho\mathbf{v} - \rho\mathbf{v} \cdot g\mathbf{V}z \qquad (1\text{-}10.8)$$

where the second term on the right-hand side is, because of eq. (1-6.4), equal to the term $gz(\partial\rho/\partial t)$ appearing in eq. (1-10.7).

(c) Net inlet flow of energy due to the heat flux $\mathbf{q}$:

$$-\mathbf{V} \cdot \mathbf{q} \qquad (1\text{-}10.9)$$

(d) Net inlet flow of energy as work of the surface forces. If $d\mathbf{t}$ is the stress force acting across the surface $d\mathbf{s}$, the work done, i.e., the non-thermal energy flux, is $\mathbf{v} \cdot d\mathbf{t}$ (the velocity is the displacement per unit time). Thus, the flow of energy through the surface $d\mathbf{s}$ is

$$\mathbf{v} \cdot d\mathbf{t} = \mathbf{v} \cdot (\mathbf{T} \cdot d\mathbf{s}) = d\mathbf{s} \cdot \mathbf{T}^{\mathrm{T}} \cdot \mathbf{v} = d\mathbf{s} \cdot \mathbf{T} \cdot \mathbf{v} \qquad (1\text{-}10.10)$$

where use has been made of the symmetry of $\mathbf{T}$. The vector $\mathbf{T} \cdot \mathbf{v}$ is now recognized as the flux of energy, and its divergence is the net inflow of energy as work of the surface forces:

$$\mathbf{V} \cdot (\mathbf{T} \cdot \mathbf{v}) \qquad (1\text{-}10.11)$$

Considering eq. (1-7.6), we have

$$\mathbf{V} \cdot (\mathbf{T} \cdot \mathbf{v}) = -\mathbf{V} \cdot (p\mathbf{v}) + \mathbf{V} \cdot (\boldsymbol{\tau}' \cdot \mathbf{v}) \qquad (1\text{-}10.12)$$

No work of the body forces needs to be considered, because the latter contribute to the flow of energy only through the term $-\rho\mathbf{v} \cdot g\mathbf{V}z$ appearing in eq. (1-10.8).

Simple algebra yields then the Lagrangian form of the energy balance equation:

$$\rho\frac{\mathbf{D}}{\mathbf{D}t}[U + \tfrac{1}{2}v^2] = -\mathbf{V} \cdot \mathbf{q} - \mathbf{V} \cdot (p\mathbf{v}) + \mathbf{V} \cdot (\boldsymbol{\tau}' \cdot \mathbf{v}) - \rho\mathbf{v} \cdot g\mathbf{V}z \qquad (1\text{-}10.13)$$

A very interesting relationship, sometimes called the 'thermal energy equation,' is obtained by subtracting eq. (1-10.6) from eq. (1-10.13):[†]

$$\rho\frac{\mathbf{D}U}{\mathbf{D}t} = -\mathbf{V} \cdot \mathbf{q} + \boldsymbol{\tau}':\mathbf{V}\mathbf{v} \qquad (1\text{-}10.14)$$

[†] In obtaining eq. (1-10.14) use has been made of the constant density hypothesis, so that $\mathbf{V} \cdot (p\mathbf{v}) = \mathbf{v} \cdot \mathbf{V}p$, and of the relation (see problem 1-8)

$$\mathbf{V} \cdot (\boldsymbol{\tau}' \cdot \mathbf{v}) = \mathbf{v} \cdot \mathbf{V} \cdot \boldsymbol{\tau}' + \boldsymbol{\tau}':\mathbf{V}\mathbf{v}$$

Equation (1-10.14) shows that the term $\tau':\nabla\mathbf{v}$ represents the conversion into internal energy of work done by the deviatoric stresses. In classical fluid mechanics, one assumes that constant density fluids can increase their internal energy only through an increase of entropy. In fact, one makes use of the Maxwell relation:

$$\frac{DU}{Dt} = -p\frac{D(1/\rho)}{Dt} + T\frac{DS}{Dt} \qquad (1\text{-}10.15)$$

where the first term on the right-hand side is zero for constant density fluids. Equation (1-10.15) is based on the assumption that U is entirely determined by the density and the entropy, i.e., the assumption that no kinematic variables enter into the energetic equation of state.

Substituting eq. (1-10.15) into (1-10.14), one has

$$\frac{DS}{Dt} + \frac{\nabla\cdot\mathbf{q}}{\rho T} = \frac{\tau':\nabla\mathbf{v}}{\rho T} \qquad (1\text{-}10.16)$$

The second term on the left-hand side is recognized to be the entropy increase of the universe surrounding the volume element considered, per unit volume of the latter. Thus, the left-hand side is the *total* entropy increase, and $\tau':\nabla\mathbf{v}$ represents the 'dissipation' of energy, i.e., its rate of irreversible conversion to internal energy.

These considerations further clarify the point raised in section 1-9 concerning the derivation of the Bernoulli equation from the first law of thermodynamics that is encountered in hydrodynamics textbooks. In fact, when the constitutive assumption (1-9.1) is made, the dissipation $\tau':\nabla\mathbf{v}$ is zero; i.e., ideal fluids do not dissipate energy. If this concept is initially accepted as intuitive, eq. (1-10.14), with the last term on the right-hand side set equal to zero, can be written directly and subtracted from the energy balance equation (eq. (1-10.13)). Of course, one obtains eq. (1-10.6) (with $\mathbf{v}\cdot\nabla\cdot\tau' = 0$), which is the Bernoulli equation. It is evident that this approach includes the constitutive assumption that there is no dissipation at some point along the line. In spite of this, the approach has such a long tradition that it has spilled over to Newtonian fluid mechanics, where it is not only logically unsatisfactory, but even leads to incorrect results.[†]

In this book we shall frequently deal with fluids possessing some degree of elasticity. Such fluids may accumulate internal energy in elastic form, so that eq. (1-10.15) should be written (for constant density fluids) as

$$\frac{DU}{Dt} = T\frac{DS}{Dt} + \frac{DU_{\text{el}}}{Dt} \qquad (1\text{-}10.17)$$

[†] An example of this is the familiar assumption that a Pitot tube reads the 'kinetic head,' which cannot be proved unless ideal fluids are considered, but is commonly applied to any Newtonian fluid.

Equation (1-10.17) can indeed be regarded as the definition of U_{el}, because U and S are defined independently. Of course, unless the unwarranted assumption is made that $U_{el} = 0$, the term $\boldsymbol{\tau}' : \nabla\mathbf{v}$ is recognized as the sum of the dissipation and the accumulation of elastic energy:

$$\boldsymbol{\tau}' : \nabla\mathbf{v} = \left(\rho T \frac{DS}{Dt} + \nabla \cdot \mathbf{q} \right) + \rho \frac{DU_{el}}{Dt} \quad (1\text{-}10.18)$$

It is clear that no use can be made of the energy equation unless the dependence of U_{el} on kinematic variables is known. This dependence is embodied in the 'energetic equation of state' referred to in section 1-1; such an equation is independent of the constitutive equation. As a consequence of this difficulty, very little use of energy considerations is made in non-Newtonian fluid mechanics; the relationship of the latter with thermodynamics will be discussed in detail in chapter 4.

In conclusion, it may be worth while to mention certain problems concerning the logical foundation of the conservation principles. The classical approach is that the four principles of conservation of mass, momentum, moment of momentum, and energy are logically independent of each other. Some recent work [9, 10, 11] on the foundations of continuum mechanics replaces this classical assumption with a postulate of frame-indifference for the 'mechanical power'; i.e., one of the terms appearing in the equation of energy is assumed to have a form that is independent of the frame of reference. With this postulate, the dynamical equation and the conservation of the moment of momentum can be deduced from the energy equation. It is clear that this new approach uses three postulates as starting points, and obtains exactly the same final equations as the classical approach, which starts from *four* postulates rather than three.

PROBLEMS

1-1 Consider the operator $\mathbf{a}+$, where $\mathbf{a}$ is a constant vector. When operating on any vector $\mathbf{b}$, it yields a vector $\mathbf{c} = \mathbf{a} + \mathbf{b}$. Is $\mathbf{a}+$ a tensor?

1-2 Calculate $\operatorname{tr}(\mathbf{1})$.

1-3 Calculate $\operatorname{tr}\mathbf{D}$ and $\operatorname{tr}\mathbf{W}$.

1-4 Calculate the trace of any antisymmetric tensor $\mathbf{A} = -\mathbf{A}^{\mathrm{T}}$.

1-5 Prove the following identity:

$$\nabla(\mathbf{a} \cdot \mathbf{b}) = \nabla\mathbf{b}^{\mathrm{T}} \cdot \mathbf{a} + \nabla\mathbf{a}^{\mathrm{T}} \cdot \mathbf{b}$$

for any two vector fields $\mathbf{a}$ and $\mathbf{b}$.

1-6 Prove the following identity:

$$\mathbf{V} \cdot (\alpha \mathbf{a}) = \alpha \mathbf{V} \cdot \mathbf{a} + \mathbf{a} \cdot \mathbf{V}\alpha$$

for any scalar field α and vector field **a**. *Hint*: first prove the analogous equation for $\mathbf{V}(\alpha \mathbf{a})$.

1-7 Prove the following identity:

$$\mathbf{V} \cdot (\mathbf{ab}) = \mathbf{a}(\mathbf{V} \cdot \mathbf{b}) + \mathbf{V}\mathbf{a} \cdot \mathbf{b}$$

for any dyad field **ab**. *Hint*: use the definition of divergence of a tensor and the results of problems 1-5 and 1-6.

1-8 Prove the following identity:

$$\mathbf{V} \cdot (\mathbf{A} \cdot \mathbf{a}) = \mathbf{a} \cdot \mathbf{V} \cdot \mathbf{A} + \mathbf{A} \colon \mathbf{V}\mathbf{a}$$

where **A** is any symmetric tensor field and **a** is any vector field. *Hint*: work with components, because the gradient of **A** is involved. Note carefully where the symmetry condition is required.

1-9 Prove the following identity:

$$\mathbf{V} \cdot (\alpha \mathbf{A}) = \alpha \mathbf{V} \cdot \mathbf{A} + \mathbf{A} \cdot \mathbf{V}\alpha$$

for any scalar field α and tensor field **A**.

1-10 Obtain eq. (1-7.10) by writing down the dynamical equation on a cubical element $dx^1 \, dx^2 \, dx^3$ (where x^i are Cartesian coordinates).

BIBLIOGRAPHY

1. COLEMAN, B. D.: *Arch. Ratl Mech. Anal.*, **17**, 230 (1964).
2. COLEMAN, B. D.: *Arch. Ratl Mech. Anal.*, **17**, 1 (1964).
3. COLEMAN, B. D., and MIZEL, W. J.: *Arch. Ratl Mech. Anal.*, **27**, 255 (1968).
4. COLEMAN, B. D., MARKOWITZ, H., and NOLL, W.: *Viscometric Flows of Non-Newtonian Fluids*, pp. 91–108. Springer-Verlag (1966).
5. GREUB, W.: *Linear Algebra* (2nd edn). Springer, Berlin (1963).
6. HALMOS, P. R.: *Finite Dimensional Vector Spaces* (2nd edn). Princeton University Press, Princeton (1958).
7. NOLL, W.: *Am. Math. Monthly*, **71**, 129 (1964).
8. BIRD, R. B.: *Chem. Engng Sci.*, **6**, 123 (1957).
9. BEATTY, M. F.: 'On the foundation principles of continuum mechanics,' University of Delaware, Tech. Rep. No. 44 (1965).
10. GREEN, A. E., and RIVLIN, R. S.: *Arch. Ratl Mech. Anal.*, **16**, 325 (1964).
11. GREEN, A. E., and RIVLIN, R. S.: *Zeit. angew. Math. Phys.*, **15**, 290 (1964).

2

PURELY VISCOUS NON-NEWTONIAN
CONSTITUTIVE EQUATIONS

2-1 VISCOSITY OF REAL FLUIDS

$$\tau = 2\mu D = \tau'$$

The viscosity of Newtonian fluids is defined by eq. (1-9.4) as one half the proportionality constant relating the stress tensor τ to the stretching tensor **D**. Equation (1-9.4) implies that the components of the stress tensor should be proportional to those of the stretching tensor for any given flow pattern. One of the well-known consequences of the Navier–Stokes equation (eq. (1-9.8)) is the Hagen–Poiseuille law relating the volumetric flow rate Q to the pressure gradient in the axial direction for steady rectilinear flow down a long circular pipe:

$$Q = \frac{\pi R^4 \, \Delta p}{16 \mu L} \qquad (2\text{-}1.1)$$

where Δp is the pressure drop, L the pipe length, and R the pipe radius. Equation (2-1.1) will be obtained as a special case of the generalized treatment in section 2-5.

The direct proportionality between volumetric flow rate Q and pressure drop Δp predicted by eq. (2-1.1) is observed experimentally, under laminar flow conditions, for a great variety of ordinary, low molecular weight liquids.

At the same time, many real materials fail to exhibit the predicted behavior, and a non-linear dependence of Q on Δp is observed experimentally. Thick suspensions, paints, polymer melts and solutions are typical examples of materials which exhibit this 'non-Newtonian' behavior.

The flow down a circular pipe is an example of a class of flows, called viscometric flows, which will be discussed in detail in chapter 5 and will be shown to be equivalent to each other. The simplest example of viscometric flow is lineal Couette flow, which takes place between two parallel flat plates sliding with respect to each other. In a Cartesian coordinate system x^i, the lineal Couette flow (sometimes called simple shear in the literature) can be described by the following equations for the components of the velocity vector:

$$v^1 = \gamma x^2 \qquad (2\text{-}1.2)$$

$$v^2 = v^3 = 0 \qquad (2\text{-}1.3)$$

Notice that the distinction between contravariant and covariant components is irrelevant because a Cartesian coordinate system has been chosen.

The components of the stretching tensor **D** for lineal Couette flow are

$$[\mathbf{D}] = \frac{\gamma}{2} \begin{Vmatrix} 0 & 1 & 0 \\ 1 & 0 & 0 \\ 0 & 0 & 0 \end{Vmatrix} \qquad (2\text{-}1.4)$$

and, according to the Newtonian constitutive equation (eq. (1-9.4)), the components of the stress tensor should be

$$[\tau] = [\tau'] = \mu\gamma \begin{Vmatrix} 0 & 1 & 0 \\ 1 & 0 & 0 \\ 0 & 0 & 0 \end{Vmatrix} \qquad (2\text{-}1.5)$$

Fluids which fail to obey the Hagen–Poiseuille law also fail to exhibit the linear dependence of τ_{12} on γ predicted by eq. (2-1.5). For such fluids, an 'apparent viscometric viscosity,' η, can be defined as

$$\eta = \frac{\tau_{12}}{\gamma} \qquad (2\text{-}1.6)$$

as measured in a viscometric flow experiment. The quantity η depends on the value of the shear rate γ.

Basically two types of fluid behavior have been observed as far as the apparent viscometric viscosity is concerned: *dilatant* fluids, for which η is an increasing function of γ:

$$\frac{\mathrm{d}\eta}{\mathrm{d}\gamma} > 0 \quad \text{(dilatant)} \qquad (2\text{-}1.7)$$

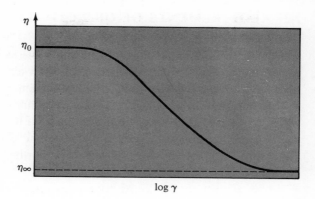

FIGURE 2-1
Viscosity curve for pseudoplastic fluid.

and *pseudoplastic* fluids, for which the opposite is true:

$$\frac{d\eta}{d\gamma} < 0 \quad \text{(pseudoplastic)} \qquad (2\text{-}1.8)$$

Typical examples of dilatant fluids are concentrated suspensions of solids; on the other hand, polymer melts and solutions are almost invariably pseudoplastic.

When the apparent viscometric viscosity of a real fluid is measured over a range of shear rate values covering several orders of magnitude, the behavior reported in Fig. 2-1 is generally observed. Newtonian behavior (i.e., a constant value of η) is observed both at very low and at very high shear rates. The limiting values η_0 and η_∞, called the lower and upper limiting viscometric viscosities, are often several orders of magnitude different from each other.

It is clear that the Newtonian constitutive equation (eq. (1-9.4)) is inadequate to represent the behavior of real fluids exhibiting a variable η. This is one of the basic physical justifications for analyzing more complex forms of the constitutive equation which may be able to predict the observed behavior.

It is obvious that the first step in this direction is to assume a non-linear dependence of the stress on the stretching tensor. However, to do so an analysis of the requirements of invariance for constitutive equations is in order, so that physically impossible forms of the constitutive equation may be avoided. The next section deals with this analysis.

2-2 REQUIREMENTS OF OBJECTIVITY FOR CONSTITUTIVE EQUATIONS

A constitutive equation is the mathematical statement of some assumptions concerning the mechanical behavior of a material, or, more generally, of a class of materials. Such a mathematical statement must conform to the requirements that the behavior of materials cannot depend on the artificial conventions that have been chosen to describe it mathematically; i.e., the *same* behavior must be described whatever the convention. This implies that the constitutive equation must be *invariant* when a change of convention is considered.

It is evident that a constitutive equation must be invariant under a change of coordinate system: the choice of the latter, in fact, is a convention used to assign components to vectors and tensors. When a constitutive equation is written in tensor form, it is always invariant under a change of coordinate system. In fact, given a frame chosen for observation, tensors remain unchanged under a change of coordinate system although their components may change. This is immediately obvious when tensors are defined as linear operators, because their definition is independent of the choice of any coordinate system.

A more subtle, yet equally basic, requirement of invariance for constitutive equations is that they remain unchanged under a change of frame, even a time-dependent frame. That this must be so may either be regarded as a postulate, or accepted intuitively. A good example of intuitive acceptance of this 'principle of material objectivity' is given by Truesdell and Noll [1]:

'A body of known weight, say one pound, when suspended by a given spring, is observed to extend it by a given amount, say one inch. The spring and weight, still connected, are then laid upon a horizontal disc, to the center of which the spring is attached. The disc is then caused to spin at a steady speed such as to extend the spring again by one inch. The spectators *are expected to agree* [our italics] that the centripetal force required to hold the weight from flying off is exactly one pound. That is, the *response of the spring* is unaffected by a rigid motion.'

It is important to observe that frame-indifference is not required for all physical laws; e.g., the dynamical equation *is not* frame-indifferent.† In fact, the dynamical equation defines an 'inertial' frame, and its validity breaks

---

† This is true unless the acceleration Dv/Dt is substituted with the acceleration relative to the 'fixed stars'. In this case, one is really choosing a 'preferred' frame, identified by the fixed stars, and frame-indifference is obtained only formally.

down if a reference frame is chosen which is accelerated relative to an inertial frame.

In contrast, physical laws exist which are required to be frame-indifferent. In section 1-6 we have adopted the viewpoint that the mass balance equation is frame-indifferent. Analogously, the response of a material to its deformation is required to be frame-indifferent.

Note that the principle of material objectivity does not imply an assumption of isotropy of the materials; anisotropic materials must obey the principle of material objectivity. If any, the principle of material objectivity implies an assumption of isotropy of space: a change of observer (i.e., of reference frame) must leave the behavior of the material unaffected. Note also that the principle of material objectivity is stronger than an assumption of indifference to rotations because frame-indifference is also required under *improper* (i.e., non-preserving handedness) 'rotations' [2].

In spite of its apparent simplicity, the application of the principle may be difficult if one considers how recently it has been stated in rigorous form [3]. This may partly be due to the fact that the requirement of frame-indifference does not apply to the dynamical equation, which is used in conjunction with the constitutive equation in the solution of practical problems.

The literature contains more than one constitutive equation which fails to obey the principle of material objectivity. In particular, some of the literature on linear viscoelasticity suffers from this drawback. This is an unfortunate situation, because available experimental data are occasionally useless, inasmuch as the results have been published in a form obtained after manipulation based on a non-invariant (and hence physically impossible) constitutive equation. In particular, we shall see in chapter 6 that in the case of constitutive equations involving time derivatives of the stress tensor, the principle must be carefully obeyed.

A third requirement of invariance is, of course, dimensional invariance. This requirement does not impose any restriction on the *form* of the constitutive equation, but only implies that the latter must contain a certain minimum number of dimensional parameters. It can be shown that in the most general case three parameters are required, with dimensions of stress, time, and length, respectively.

The requirement of dimensional invariance gives, through dimensional analysis, certain scale-up rules for a variety of problems in engineering. Unfortunately, this is true only when linearized forms of the constitutive assumptions are chosen; when non-linear forms are used (and such is the case in non-Newtonian fluid mechanics) scale-up rules can be found only when the *same* material is used both in the model and in the prototype. Indeed, the asymptotic validity of linear (i.e., Newtonian) theory is demonstrated mainly by the successful use of scale-up rules as applied to different materials, rather than by direct experimental confirmation of the fundamental assumptions [4].

Finally, a requirement which is not one of invariance should be considered; namely, that the second law of thermodynamics must not be violated. This requirement for a Newtonian fluid is simply met by assigning a non-negative value to the viscosity, so that eq. (1-10.16) always gives a positive dissipation. For more complex constitutive assumptions, the matter may not be equally straightforward; the second law of thermodynamics imposes restrictions on both the constitutive and the energetic equations of state. Recent work by Coleman [5] attacks this very important problem, and will be discussed in chapter 4.

The requirement that constitutive equations remain invariant under a change of frame clearly imposes some restrictions on constitutive equations: when all tensors appearing in it are transformed to the new frame, the constitutive equation must remain the same.

First consider the stress tensor $\mathbf{T}$. Let us use asterisks to indicate vectors and tensors in the 'new' frame, as we did in section 1-5. We then have, from the definition of $\mathbf{T}$,

$$d\mathbf{t} = \mathbf{T} \cdot d\mathbf{s} \qquad (2\text{-}2.1)$$

$$d\mathbf{t}^* = \mathbf{T}^* \cdot d\mathbf{s}^* \qquad (2\text{-}2.2)$$

The area vector, $d\mathbf{s}$, is obviously indifferent because it is a geometrical vector. By *postulating* that the stress force vector is also indifferent, we have

$$d\mathbf{s}^* = \mathbf{Q} \cdot d\mathbf{s} \qquad (2\text{-}2.3)$$

$$d\mathbf{t}^* = \mathbf{Q} \cdot d\mathbf{t} \qquad (2\text{-}2.4)$$

By combining eqs (2-2.1)–(2-2.4) one obtains

$$\mathbf{Q} \cdot \mathbf{T} \cdot d\mathbf{s} = \mathbf{T}^* \cdot \mathbf{Q} \cdot d\mathbf{s} \qquad (2\text{-}2.5)$$

for *any* surface $d\mathbf{s}$; thus,

$$\mathbf{Q} \cdot \mathbf{T} \cdot \mathbf{Q}^{-1} = \mathbf{T}^* = \mathbf{Q} \cdot \mathbf{T} \cdot \mathbf{Q}^{\mathrm{T}} \qquad (2\text{-}2.6)$$

Equation (2-2.6) shows that the total stress tensor $\mathbf{T}$ is indifferent.

At this point let us state the logical process by which this concept has been proved. Equations (2-2.1) and (2-2.2) have the same form, and both are the *definition* of tensor $\mathbf{T}$. In general, we assume that the *definition* of a tensor is frame-indifferent; thus, we shall write

$$\mathbf{b} = \mathbf{A} \cdot \mathbf{a} \qquad (2\text{-}2.7)$$

$$\mathbf{b}^* = \mathbf{A}^* \cdot \mathbf{a}^* \qquad (2\text{-}2.8)$$

where both equations define $\mathbf{A}$ as that linear operator which assigns to the vector $\mathbf{a}$ the vector $\mathbf{b}$—both vectors, of course, having been previously defined. If $\mathbf{b}$ and $\mathbf{a}$ are indifferent, then $\mathbf{A}$ is also indifferent.

In particular, let us consider the unit tensor $\mathbf{1}$, which is defined as

$$\mathbf{a} = \mathbf{1} \cdot \mathbf{a} \qquad (2\text{-}2.9)$$

for *any* vector $\mathbf{a}$. Of course,

$$\mathbf{1} = \mathbf{1}^* = \mathbf{Q} \cdot \mathbf{Q}^T = \mathbf{Q} \cdot \mathbf{1} \cdot \mathbf{Q}^T \qquad (2\text{-}2.10)$$

which shows that the unit tensor, and any isotropic tensor, are indifferent.

When this result is combined with eq. (2-2.6) it is seen that both the deviatoric stress τ' and the stress τ are indifferent (the sum of two indifferent tensors is an indifferent tensor):

$$\tau'^* = \mathbf{Q} \cdot \tau' \cdot \mathbf{Q}^T \qquad (2\text{-}2.11)$$

$$\tau^* = \mathbf{Q} \cdot \tau \cdot \mathbf{Q}^T \qquad (2\text{-}2.12)$$

Let us now consider kinematic tensors such as the velocity gradient $\nabla\mathbf{v}$ and the stretching tensor $\mathbf{D}$. From the definition of velocity gradient, we have

$$\nabla\mathbf{v} \cdot d\mathbf{X} = d\mathbf{v} \qquad (2\text{-}2.13)$$

$$\nabla\mathbf{v}^* \cdot d\mathbf{X}^* = d\mathbf{v}^* \qquad (2\text{-}2.14)$$

$d\mathbf{X}$ is a geometrical vector, and hence is indifferent:

$$d\mathbf{X}^* = \mathbf{Q} \cdot d\mathbf{X} \qquad (2\text{-}2.15)$$

From eq. (1-5.14), the rule of transformation for $d\mathbf{v}$ is obtained as

$$d\mathbf{v}^* = \mathbf{Q} \cdot d\mathbf{v} + \dot{\mathbf{Q}} \cdot d\mathbf{X} \qquad (2\text{-}2.16)$$

where use has been made of the fact that $\mathbf{Q}$ and $\dot{\mathbf{Y}}$ are not fields, but are a fixed tensor and a fixed vector, respectively (the time-dependent change of frame is a rigid-body motion of the frame). The translation of the 'new' frame, $\mathbf{Y}(t)$, gives no contribution to $d\mathbf{v}^*$ though it does to $\mathbf{v}^*$: in fact, velocity differences are clearly indifferent to a superimposed rigid-translation.

Combining eqs (2-2.13)–(2-2.16), one obtains

$$\nabla\mathbf{v}^* \cdot \mathbf{Q} \cdot d\mathbf{X} = \mathbf{Q} \cdot \nabla\mathbf{v} \cdot d\mathbf{X} + \dot{\mathbf{Q}} \cdot d\mathbf{X} \qquad (2\text{-}2.17)$$

which holds for *any* d**X**. Thus,

$$\nabla v^* = Q \cdot \nabla v \cdot Q^T + \dot{Q} \cdot Q^T \qquad (2\text{-}2.18)$$

Equation (2-2.18) shows that the velocity gradient *is not* indifferent.

The transformation of the stretching tensor **D** is obtained from its definition:†

$$
\begin{aligned}
D^* &= \tfrac{1}{2}(\nabla v^* + \nabla v^{T*}) \\
&= \tfrac{1}{2}(Q \cdot \nabla v \cdot Q^T + Q \cdot \nabla v^T \cdot Q^T + \dot{Q} \cdot Q^T + Q \cdot \dot{Q}^T) \qquad (2\text{-}2.19) \\
&= Q \cdot D \cdot Q^T + \tfrac{1}{2}\overline{Q \cdot Q^T}
\end{aligned}
$$

The last term in eq. (2-2.19) is zero, being the time derivative of the unit tensor; thus,

$$D^* = Q \cdot D \cdot Q^T \qquad (2\text{-}2.20)$$

which shows that **D** is indifferent.

Equations (2-2.11), (2-2.12), and (2-2.20) show immediately that the Newtonian constitutive equation (eq. (1-9.4)) satisfies the principle of material objectivity. The continuity equation for constant density fluids, eq. (1-6.10), which also embodies a constitutive assumption, satisfies the principle. In fact,

$$\nabla \cdot v = \operatorname{tr} \nabla v = \operatorname{tr}\left[\tfrac{1}{2}(\nabla v + \nabla v^T)\right] = \operatorname{tr} D \qquad (2\text{-}2.21)$$

$$\operatorname{tr} D^* = \operatorname{tr}(Q \cdot D \cdot Q^T) = \operatorname{tr} D \qquad (2\text{-}2.22)$$

The equality of $(\nabla \cdot v)^*$ and $\nabla \cdot v$ was made use of in section 1-6, when the Lagrangian form of the continuity equation was obtained by considering a change of frame.

2-3 REINER–RIVLIN FLUIDS

After having laid down the principle of material objectivity, a non-linear constitutive equation relating the stress tensor τ to the stretching tensor **D** can be analyzed:‡

$$\tau = g(D) \qquad (2\text{-}3.1)$$

---

† Use is made of results to be given in section 2-7.

‡ Equation (2-3.1) can be regarded as a rigorous formulation (for incompressible fluids) of the basic hypothesis of Stokes, laid down in 1845, that the stress is determined by the rate of deformation. A suggestion of Boussinesq, that the stress may depend both on **D** and on the vorticity **W**, can be shown [6] to disobey the principle of material objectivity unless it degenerates to eq. (2-3.1).

Equation (2-3.1) is still very restricted. In fact, it assumes that the stress at a point at a given time is entirely determined by the stretching rate *at that point and at that time*. No restrictions of linearity are imposed, but deformations taking place at any other point and/or at any other time are assumed to be irrelevant. In dealing with more complex equations, we will remove the time restriction but will keep the space limitation ; this point will be discussed in detail in chapter 4.

Let us now apply the principle of material objectivity to eq. (2-3.1). The principle imposes the following restriction on the form of the function **g** :

$$\mathbf{Q} \cdot \mathbf{g}(\mathbf{D}) \cdot \mathbf{Q}^\mathrm{T} = \mathbf{g}(\mathbf{Q} \cdot \mathbf{D} \cdot \mathbf{Q}^\mathrm{T}) \qquad (2\text{-}3.2)$$

for *any* time-dependent orthogonal tensor **Q**. Symmetric tensor functions satisfying eq. (2-3.2) are called isotropic. The principle of material objectivity imposes the condition that *any* material obeying eq. (2-3.1) be isotropic [6] ; of course, non-isotropic materials will still obey the principle, but cannot be described by eq. (2-3.1).

A theorem of tensor analysis states that *any* isotropic symmetric tensor function **g(D)** can only be of the form

$$\mathbf{g}(\mathbf{D}) = \phi_0 \mathbf{1} + \phi_1 \mathbf{D} + \phi_2 \mathbf{D}^2 \qquad (2\text{-}3.3)$$

where the ϕ_i are scalar functions of the three principal invariants of **D**.

In substituting eq. (2-3.3) into (2-3.1) we take further advantage from the condition of constant density and write

$$\tau = \phi_1(\mathrm{II}_\mathbf{D}, \mathrm{III}_\mathbf{D})\mathbf{D} + \phi_2(\mathrm{II}_\mathbf{D}, \mathrm{III}_\mathbf{D})\mathbf{D}^2 \qquad (2\text{-}3.4)$$

In fact, the term $\phi_0 \mathbf{1}$ is unnecessary when determining the extra stress. Moreover, the first invariant, $\mathrm{I}_\mathbf{D}$, is identically zero for constant density fluids (see eq. (2-2.21)), and therefore the dependence of ϕ_1 and ϕ_2 on $\mathrm{I}_\mathbf{D}$ can be dropped.

Equation (2-3.4) is the equation defining a Reiner–Rivlin fluid. It is as general as eq. (2-3.1), the steps leading to the simplified form of eq. (2-3.4) being *dictated* by the principle of material objectivity. Consequently, if the behavior of a real fluid fails to be adequately described by eq. (2-3.4), we can conclude that in such a fluid the stress *is not* uniquely determined by the stretching tensor.

The quantities ϕ_1 and ϕ_2 are *material functions*, in the sense that any specific Reiner–Rivlin fluid is identified by assigning these two functions. The Newtonian fluids are a very special case of Reiner–Rivlin fluids, for which $\phi_1 = 2\mu$ and $\phi_2 = 0$.

It is important to point out that eq. (2-3.4) cannot hold with ϕ_1 and ϕ_2 both constant, unless $\phi_2 = 0$, in which case the Newtonian constitutive equation is obtained [7]. The argument now to be developed is thermodynamic, and is worth considering in some detail in order to show the pitfalls of developing a purely mechanical theory without due regard to thermodynamics.

Assume that ϕ_1 and ϕ_2 are both constant, and calculate the value of the term $\tau' : \nabla v$ appearing in the equation of thermal energy, eq. (1-10.14). We have, from eq. (2-3.4),

$$\text{tr } \tau = \phi_1 \text{ tr } \mathbf{D} + \phi_2 \text{ tr } (\mathbf{D}^2) = -2\phi_2 \text{II}_{\mathbf{D}} \qquad (2\text{-}3.5)$$

where use has been made of eqs (1-3.42), (1-6.10), and (2-2.21). From eq. (1-8.4) the deviatoric stress τ' can be calculated:

$$\tau' = \tau - \tfrac{1}{3}(\text{tr } \tau)\mathbf{1}$$
$$= \tfrac{2}{3}\phi_2 \text{II}_{\mathbf{D}}\mathbf{1} + \phi_1\mathbf{D} + \phi_2\mathbf{D}^2 \qquad (2\text{-}3.6)$$

Thus,

$$\tau' \cdot \mathbf{D} = \tfrac{2}{3}\phi_2 \text{II}_{\mathbf{D}}\mathbf{D} + \phi_1\mathbf{D}^2 + \phi_2\mathbf{D}^3 \qquad (2\text{-}3.7)$$

The term $\mathbf{D}^3$ can be calculated from the Cayley–Hamilton theorem, eq. (1-3.49), as

$$\mathbf{D}^3 = -\text{II}_{\mathbf{D}}\mathbf{D} + \text{III}_{\mathbf{D}}\mathbf{1} \qquad (2\text{-}3.8)$$

Substituting eq. (2-3.8) into eq. (2-3.7), we have

$$\tau' \cdot \mathbf{D} = -\tfrac{1}{3}\phi_2 \text{II}_{\mathbf{D}}\mathbf{D} + \phi_1\mathbf{D}^2 + \text{III}_{\mathbf{D}}\phi_2\mathbf{1} \qquad (2\text{-}3.9)$$

The quantity $\tau' : \nabla v$ can now be calculated:

$$\tau' : \nabla v = \tau' : (\mathbf{D} + \mathbf{W}) = \tau' : \mathbf{D}$$
$$= \text{tr } (\tau' \cdot \mathbf{D}) = -2\phi_1 \text{II}_{\mathbf{D}} + 3\phi_2 \text{III}_{\mathbf{D}} \qquad (2\text{-}3.10)$$

The second invariant, $\text{II}_{\mathbf{D}}$, of the stretching tensor is an intrinsically negative quantity, but $\text{III}_{\mathbf{D}}$ may have any sign. Thus, eq. (2-3.10) shows that, if ϕ_1 and ϕ_2 are constants, there exist some flow patterns for which $\tau' : \nabla v$ is negative, in contrast with the interpretation of the latter as the rate of energy dissipation discussed in section 1-10 (it may safely be assumed that $U_{\text{el}} = 0$ for Reiner–Rivlin fluids, as will become clear from the discussion in section 2-6). Of course, if $\phi_2 = 0$, eq. (2-3.10) always gives a positive value for $\tau' : \nabla v$, but in that case the Newtonian constitutive equation is obtained.

If ϕ_1 and ϕ_2 are taken as variables, functional forms which always yield a positive value of $\tau' : \nabla v$ are conceivable. As an example, if ϕ_1 is always positive and ϕ_2 has always the opposite sign of III_D, eq. (2-3.10) again yields positive values for $\tau' : \mathbf{Vv}$. The requirement of a positive dissipation imposes the following restriction on the form of the functions ϕ_1 and ϕ_2:

$$-2II_D\phi_1(II_D, III_D) > -3III_D\phi_2(II_D, III_D) \qquad (2\text{-}3.11)$$

Let us now consider the lineal Couette flow of a Reiner–Rivlin fluid. From eq. (2-3.4) the following equations for the components of the stress tensor are obtained (see Illustration 2A):

$$\tau_{12} = \tau_{21} = \phi_1\gamma/2 \qquad (2\text{-}3.12)$$

$$\tau_{13} = \tau_{31} = \tau_{23} = \tau_{32} = 0 \qquad (2\text{-}3.13)$$

$$\tau_{11} - \tau_{22} = 0 \qquad (2\text{-}3.14)$$

$$\tau_{22} - \tau_{33} = \phi_2\gamma^2/4 \qquad (2\text{-}3.15)$$

The functions ϕ_1 and ϕ_2 are to be taken at $II_D = -\gamma^2/4$ and $III_D = 0$, and are therefore even functions of the shear rate.

Equation (2-3.12) shows that the apparent viscometric viscosity η equals one half the value of ϕ_1. Equation (2-3.12) can fit *any* apparent viscometric viscosity curve, because no restriction is imposed on the form of the function ϕ_1 and thus none on the form of the $\eta(\gamma)$ curve except that it must be an even function. The evenness of $\eta(\gamma)$ is required also on thermodynamic grounds. Equation (2-3.13) is not a restriction of the predictive adequacy of eq. (2-3.4), because it can be shown in general that for *any* isotropic material eq. (2-3.13) holds for lineal Couette flow (see, in particular, problem 2-2).

Equation (2-3.15) predicts that, in lineal Couette flow, the three normal stresses are not all equal, in contrast with the predictions of the Newtonian equation (eq. (1-9.4)). Normal stress differences have indeed been measured for a variety of fluids in viscometric flow (such data will be discussed in chapter 5), but the equality of τ_{11} and τ_{22} indicated by eq. (2-3.14) *has not* been confirmed for any real material exhibiting a non-zero value of $\tau_{22} - \tau_{33}$.

This is a definite drawback of eq. (2-3.4), and one which cannot be resolved without making use of constitutive assumptions that are more complex than eq. (2-3.1). In other words, the behavior of real materials which exhibit a non-zero first normal stress difference $(\tau_{11} - \tau_{22} \neq 0)$ in viscometric flow cannot be explained by assuming that the stress tensor is uniquely determined by the stretching tensor.

Two different lines of attack of non-Newtonian fluid mechanics are therefore possible at this stage. On one side, one may wish to focus attention on flow problems for which, in some sense to be defined, only the apparent viscometric viscosity is relevant, so that the inadequacy of eq. (2-3.4) is regarded as irrelevant. This line of thought is pursued in what we shall call the discipline of generalized Newtonian fluid mechanics. The approach may be justified either because the flow pattern considered is one where only the viscometric viscosity is relevant (laminar flows are in this category, at least as a first approximation), or because the material one is considering exhibits a shear-dependent viscometric viscosity but no other non-Newtonian mechanical properties. (Suspensions of solids are often of this type, but unfortunately the pragmatically more important polymer melts and solutions are generally not.)

On the other hand, one may wish to explore the possibilities of constitutive equations more complex than eq. (2-3.1), which are required in order to describe adequately the behavior of real materials even in the simplest conceivable flow pattern, namely lineal Couette flow. This second line of thought is pursued in the discipline we shall call 'memory fluids mechanics.'

A question of nomenclature is involved here. In much of the technical literature, the terms 'non-Newtonian' and 'viscoelastic' are used rather loosely, with meanings that are different according to different authors. We use here the term 'non-Newtonian' for any material which fails to obey eq. (1-9.4); the term 'Reiner–Rivlin fluid' for any material obeying eq. (2-3.1)—Newtonian fluids are a special case; and 'memory fluids' for materials which fail to obey eq. (2-3.1), the stress being influenced also by deformations taking place before the instant of observation.

2-4 GENERALIZED NEWTONIAN FLUIDS

If attention is focused on the apparent viscometric viscosity of real fluids, there is no need to carry on the last term on the right-hand side of eq. (2-3.4), because that only leads to the prediction of normal stresses (a prediction which in any event fails to be verified for any known real fluid except those for which $\tau_{11} = \tau_{22} = \tau_{33}$), and does not contribute to the value of η, because $\mathbf{D}^2$ has a zero value of the off-diagonal components in the case of viscometric flow.

Moreover, the second independent variable on which ϕ_1 depends, $III_\mathbf{D}$, is identically zero in viscometric flows, and therefore the dependence of ϕ_1 on $III_\mathbf{D}$ cannot be detected in a viscometric flow experiment. As a tentative hypothesis it may be assumed that ϕ_1 *does not* depend on $III_\mathbf{D}$. There are indeed some experimental indications that such is the case, although the evidence is not strong [8].

The considerations above suggest the utility of the following constitutive definition of the generalized Newtonian fluid:

$$\tau' = \tau = 2\eta(S)\mathbf{D} \qquad (2\text{-}4.1)$$

where the parameter S, having dimensions of the square of a frequency, is defined for constant density fluids as

$$S = -4\text{II}_\mathbf{D} = 2\,\text{tr}\,(\mathbf{D}^2) = 2\mathbf{D}{:}\mathbf{D} \qquad (2\text{-}4.2)$$

The factor 2 is introduced in the definition of S so that, for lineal Couette flow, $S = \gamma^2$.

Several forms of the $\eta(S)$ function in eq. (2-4.1) have been proposed in the literature, and widely used in flow calculations. The difference among the Newtonian constitutive equation (eq. (1-9.4)) and eq. (2-4.1) is, conceptually, only slight; substitution of eq. (2-4.1) into the dynamical equation (1-7.13) leads to a generalized form of the Navier–Stokes equation:

$$\rho\frac{D\mathbf{v}}{Dt} = -\nabla p + \rho\mathbf{g} + \eta(S)\nabla^2\mathbf{v} + 2\mathbf{D}\cdot\nabla\eta \qquad (2\text{-}4.3)$$

The integration of eq. (2-4.3) for specified sets of boundary conditions is often more cumbersome than, but not conceptually different from, the integration of eq. (1-9.8). Flow calculations based on eq. (2-4.3) constitute the discipline of generalized Newtonian fluid mechanics.

A form of the $\eta(S)$ function which has been used quite extensively in the literature is the so-called 'power law':

$$\eta(S) = KS^{(n-1)/2} \qquad (2\text{-}4.4)$$

where n and K are constant parameters, which are called the power-law index and the consistency, respectively. The power-law index is dimensionless, while the consistency has units which depend on the value of n:

$$K\,[=]\,\text{g/s}^{2-n}\,\text{cm} \qquad (2\text{-}4.5)$$

The Newtonian constitutive equation is obtained as a special case when $n = 1$. Pseudoplastic fluid behavior corresponds to $n < 1$, dilatant behavior to $n > 1$. Although eq. (2-4.4) often describes accurately the viscometric viscosity curve for real materials over a range of S values of one or even a few orders of magnitude, it fails to predict an upper and a lower limiting viscosity. In particular, for pseudoplastic fluids ($n < 1$), eq. (2-4.4) predicts an infinite viscosity in the limit of diminishingly small shear rates. In spite of this difficulty, flow calculations based on eq. (2-4.4) have been successful in engineering analyses of a variety of laminar flow problems; the book of Skelland [9] gives a survey of such type of calculations.

Another form for the function $\eta(S)$ that has received attention is the Prandtl–Eyring model [10], which is at least partly based on molecular considerations. The $\eta(S)$ function is assumed to be

$$\eta = \eta_0 \frac{\text{arcsinh} (\sqrt{S}\,\lambda)}{\sqrt{S}\,\lambda} \qquad (2\text{-}4.6)$$

where η_0 and λ are constants. Here η_0 has the dimensions of a viscosity, and is indeed equal to the lower limiting viscosity:

$$S \to 0, \qquad \eta \to \eta_0 \qquad (2\text{-}4.7)$$

The parameter λ has dimensions of time. Equation (2-4.6) always predicts pseudoplastic behavior, but it fails to predict an upper limiting viscosity. This drawback is eliminated by a slightly more complicated form of the $\eta(S)$ function, usually referred to as the Powell–Eyring model [11]:

$$\eta = \eta_\infty + (\eta_0 - \eta_\infty)\frac{\text{arcsinh} (\sqrt{S}\,\lambda)}{\sqrt{S}\,\lambda} \qquad (2\text{-}4.8)$$

2-5 LAMINAR FLOW OF GENERALIZED NEWTONIAN FLUID

The theory of generalized Newtonian fluids is useful in particular in the analysis of steady laminar flows through constant section ducts, where the approximations involved in eq. (2-4.1) are best justified.

A typical example is that of steady laminar flow down a circular pipe. If a cylindrical coordinate system with the z axis along the center-line of the pipe is chosen, the flow can be described as

$$v_z = u(r) \qquad (2\text{-}5.1)$$

$$v_r = v_\theta = 0 \qquad (2\text{-}5.2)$$

and only the r–z component of the stretching tensor is different from zero, viz.,

$$\frac{du}{dr} = -\gamma(r) \qquad (2\text{-}5.3)$$

(for the r–z component, distinction between component types is irrelevant).

The volumetric flow rate Q is given by

$$Q \equiv \pi R^2 V = \int_0^R 2\pi r u \, dr \qquad (2\text{-}5.4)$$

where V is the average velocity. Integrating eq. (2-5.4) by parts, we obtain

$$V = u_s + \frac{1}{R^2} \int_0^R r^2 \gamma \, dr \qquad (2\text{-}5.5)$$

where u_s is the slip velocity at the tube wall. The usual assumption in classical fluid mechanics is that the fluid adheres to any solid boundary, and hence u_s would be taken as zero. In the case of non-Newtonian fluids, u_s has been shown [12, 13] to be different from zero under a variety of conditions.

Equation (2-4.1) can, for the problem at hand, be written as

$$\tau = \eta(\gamma^2)\gamma \qquad (2\text{-}5.6)$$

where τ is the r–z component of the stress tensor. Equation (2-5.6) can in principle be solved for γ,† i.e., it can be written in the form

$$\gamma = f(\tau) \qquad (2\text{-}5.7)$$

where $f(\ )$ is as yet an unspecified function. Application of the dynamical equation to the problem at hand yields the well-known linear distribution of stresses:

$$\tau = \tau_w \frac{r}{R} = \frac{\Delta p}{2L} r \qquad (2\text{-}5.8)$$

where τ_w is the stress at the wall, and Δp is the pressure drop over a length L of pipe. The stress τ_w is a measurable quantity.

Substituting eqs (2-5.7) and (2-5.8) into (2-5.5) gives

$$\gamma_a(1 - A) = \frac{4}{\tau_w^3} \int_0^{\tau_w} \tau^2 f(\tau) \, d\tau \qquad (2\text{-}5.9)$$

where

$$\gamma_a = \frac{4V}{R} \qquad (2\text{-}5.10)$$

$$A = \frac{u_s}{V} \qquad (2\text{-}5.11)$$

Equation (2-5.9) shows that the quantity $\gamma_a(1 - A)$ is a unique function of the wall stress τ_w.

Differentiating eq. (2-5.9) with respect to τ_w gives, after some algebraic manipulation,

$$\frac{1}{n'} + 3(1 - A) - A\beta = 4\frac{\gamma_w}{\gamma_a} \qquad (2\text{-}5.12)$$

---

† The assumption is made here that the $\tau(\gamma)$ curve is monotonous and thus invertible. This is not necessarily true, except in the limit of diminishingly small values of γ.

where

$$n' = \frac{\partial \ln \tau_w}{\partial \ln \gamma_a}\bigg|_R = \frac{\partial \ln \Delta p}{\partial \ln \gamma_a}\bigg|_R \qquad (2\text{-}5.13)$$

$$\beta = \frac{d \ln u_s}{d \ln \tau_w} \qquad (2\text{-}5.14)$$

$$\gamma_w = f(\tau_w) \qquad (2\text{-}5.15)$$

The quantity γ_w is recognized to be the shear rate at the wall, see eq. (2-5.7). u_s has been assumed to depend only on τ_w.

Let us first consider the case where there is no slip velocity, i.e., $A = 0$. Equation (2-5.12) degenerates to

$$\gamma_w = \gamma_a \frac{3n' + 1}{4n'} \qquad (2\text{-}5.16)$$

Equation (2-5.16), which is known as the Mooney–Rabinowitsch equation, is the starting point for the determination of the $\eta(S)$ curve from laminar flow pressure drop data. In fact, both τ_w and γ_a are directly measurable quantities; a log–log plot of τ_w versus γ_a yields the value of n'. Of course, n' is in general a function of γ_a, but in most cases it shows a very weak dependence on γ_a. Equation (2-5.16) can then be used to calculate the actual shear rate at the wall, γ_w. The apparent viscometric viscosity and the corresponding value of S are then obtained as

$$\eta = \frac{\tau_w}{\gamma_w}; \qquad S = \gamma_w^2 \qquad (2\text{-}5.17)$$

We have here illustrated the procedure by which the viscometric viscosity can be obtained from pressure drop data in laminar flow through circular pipes. Flows in other apparatuses from which η can be obtained will be discussed in later chapters.

In spite of a superficial analogy, the flow behavior index, n', should not be confused with the power-law index n defined in eq. (2-4.4). An apparent consistency K' can be defined as

$$\tau_w = K'\gamma_a^{n'} \qquad (2\text{-}5.18)$$

When n' is constant, it coincides with the power-law index n, while the relationship between K' and K is

$$K = K'\left[\frac{4n}{3n + 1}\right]^n \qquad (2\text{-}5.19)$$

Let us now go back to the general case where there is slip at the wall, i.e., $A \neq 0$. Differentiation of eq. (2-5.9) with respect to τ_w at constant Δp yields

$$q = (2 + \beta)A - 3 + 4\frac{\gamma_w}{\gamma_a} \qquad (2\text{-}5.20)$$

where

$$q = \frac{\partial \ln \gamma_a}{\partial \ln R}\bigg|_{\Delta p} = \frac{\partial \ln \gamma_a}{\partial \ln \tau_w}\bigg|_{\Delta p} \qquad (2\text{-}5.21)$$

Substitution of eq. (2-5.12) into eq. (2-5.20) gives

$$q = \frac{1}{n'} - A \qquad (2\text{-}5.22)$$

When $A = 0$, eq. (2-5.9) shows that γ_a is an unique function of τ_w, and therefore the τ_w–γ_a curves obtained with different pipe radii superimpose on each other. When there is slip at the wall, a shift with pipe radius will be observed: in fact, it is physically unrealistic that A be an unique function of τ_w. When such a shift is observed, eq. (2-5.22) can be used to calculate the value of A. If, now, the *assumption* is made that $\beta = 1$, eq. (2-5.12) or (2-5.20) can be used to calculate γ_w, and therefore the apparent viscometric viscosity η can be obtained even in the presence of wall slip.

In the very special case of Newtonian fluids with no slip at the wall, $n' = n = 1$, $K' = K = \mu$, and eq. (2-5.18) degenerates to:

$$\tau_w = \mu\gamma_a \qquad (2\text{-}5.23)$$

which is but another form of the Hagen–Poiseuille law, eq. (2-1.1). Equation (2-5.23) shows that, for Newtonian fluids, γ_a equals the wall shear rate; by extension, γ_a may be regarded as the apparent wall shear rate.

No mention has been made of possible end effects in the discussion above; the results are therefore asymptotically valid for very long tubes. End effects can, of course, be investigated by experiments with different tube lengths.

In classical fluid mechanics, it is customary to recast eq. (2-1.1), or its equivalent, eq. (2-5.23), in the form of a relationship between the friction factor and the Reynolds number:

$$f \equiv \frac{\Delta p R}{\rho V^2 L} = \frac{16}{Re} \qquad (2\text{-}5.24)$$

The validity of eq. (2-5.24) can be extended to generalized Newtonian fluids by a proper definition of the generalized Reynolds number. In fact, substituting eqs (2-5.8) and (2-5.18) into (2-5.24) and solving for the Reynolds number, one obtains

$$Re = \frac{\rho V^{2-n'} D^{n'}}{K' \cdot 8^{n'-1}} \qquad (2\text{-}5.25)$$

The definition of the generalized Reynolds number in eq. (2-5.25) implies, in pipe flow calculations, the use of the K' and n' values corresponding to the shear stress at the wall. When an extension is made to different laminar or creeping flow problems, either a characteristic shear rate or a characteristic stress needs to be defined, so that the values of n' and K' to be used are determined.

The procedure outlined above for the problem of laminar flow down a circular pipe has been extended to other laminar flow problems, such as, for example, flow down an inclined flat plate [12]. Several creeping flow problems have also been discussed in the literature [14, 15]. Generalized Newtonian fluid mechanics are discussed at length in the book by Skelland [9].

2-6 THE CONCEPT OF MEMORY FOR FLUID-LIKE MATERIALS

The problem of describing the mechanical behavior of real materials can be approached both from an entirely axiomatic and an entirely phenomenological viewpoint. Both approaches have advantages and disadvantages. The axiomatic approach, typical of rational mechanics, has the advantages of rigor and generality, the drawback of solving only the problems which *can* be solved rather than those which *need* to be solved. The phenomenological approach has the advantage of a higher level of pragmatic relevance to engineering problems; it sometimes justifies and motivates an axiomatic approach to a certain class of problems. In this section, a strictly phenomenological viewpoint is taken, and some concepts are discussed which are largely intuitive, and are not clearly defined in mathematical terms. The reader is asked not to look for a rigorous treatment, but to grasp some intuitive ideas which may motivate him to study the considerable amount of theory required for the axiomatic approach to be given in chapter 4.

The inadequacy of eq. (2-3.1) to predict correctly the behavior of real materials in even such a simple flow pattern as lineal Couette flow poses the problem of obtaining a more general form of the constitutive equation, where the stress tensor τ is not uniquely determined by the stretching tensor.

The behavior that is actually observed in real materials in lineal Couette flow is the existence of a rather large first normal stress difference, $\tau_{11} - \tau_{22}$, and only a much smaller second normal stress difference, $\tau_{22} - \tau_{33}$. This behavior is reminiscent of the Poynting effect which is obtained in the theory of isotropically elastic solids: a solid block subjected to a shear deformation exhibits a non-zero first normal stress difference.

There are other elements suggesting that real fluid-like materials whose behavior is not representable through eq. (2-3.1) possess some degree of elasticity. In fact, phenomena such as recoil, which are highly suggestive of elasticity, are commonly observed in such materials.

When trying to include the idea of elasticity in the constitutive equation, one is faced with a basic problem of definition of elasticity and of 'fluid'. Elasticity is, intuitively, that property of materials which implies that the internal stresses are determined by deformations. Deformation in its turn can only be defined in terms of a reference 'configuration,' say, in non-rigorous terms of a preferred shape of the material being considered. Deformation is intended as a distortion from the preferred shape.

In contrast with this, 'fluid' materials are intended as those materials which have *no* preferred shape, so that bringing together the intuitive concepts of elasticity and fluidity leads, at least superficially, to an inner contradiction. Indeed, the idea that a fluid-like material is insensitive to deformation leads to the concept that internal stresses are determined by the deformation *rate*, a concept which is embodied in eq. (2-3.1). (The stretching tensor **D**, as will be shown in the next chapter, describes the instantaneous rate of deformation.)

Truesdell [16] has proposed a model of constitutive equation which, while satisfying the principle of material objectivity, brings the two concepts of elasticity and fluidity within a common framework. The 'fluid of convected elasticity' is defined as a material for which the stress depends on the deformation (i.e., an 'elastic' material); but the deformation is not defined in terms of a preferred shape, but in terms of the distortion between the configuration of the material at the instant of observation (when stress is measured), and the configuration that the material had some fixed time before the instant of observation.

Though Truesdell's fluid of convected elasticity is no more satisfactory than the Reiner–Rivlin fluid in portraying the behavior of real fluids (instead of eq. (2-3.14), it predicts that

$$\tau_{11} - \tau_{22} \propto \gamma \tau_{12} \qquad (2\text{-}6.1)$$

which is an equally restrictive prediction), it does show that the concepts of elasticity and fluidity are not mutually exclusive.

Furthermore, Truesdell's model may lead us to introduce a concept which is very useful in analyzing elastic fluid mechanics, namely the concept of memory. This concept needs to be dealt with in some detail.

The concept of elasticity as having stress-dependence upon deformation from a preferred shape or reference configuration implies that a material is sensitive to distortion from its preferred shape, no matter how long ago that shape had actually been assumed (indeed, it may *never* have been assumed, as is demonstrated by the residual stresses existing in solid metals when solidified from a melt). At the other extreme, the concept of viscosity having stress-dependence on the deformation *rate* (embodied in eq. (2-3.1)) implies that a material is sensitive only to the instantaneous rate of change of its shape, while configurations assumed at any time in the past, except at the instant of observation and immediately before, are irrelevant.

One may, in non-rigorous terms, regard elastic solids as being materials which possess a perfect memory, and thus remember a preferred shape for ever; while viscous (i.e., in general, Reiner–Rivlin) fluids have no memory, and are only aware of the instantaneous rate of distortion. There is clearly space between these two concepts; that is, materials may be envisaged which, although devoid of a reference configuration of special physical significance— incapable of remembering a preferred shape for ever and thus essentially 'fluid'—may have some memory of past deformation. A concept of 'fading memory,' to be formalized, is clearly involved here. One may wish to observe that while solids remember *one* shape for ever, memory fluids remember *all* shapes assumed in the past, but not for ever. The concepts discussed here are clearly linked to the discussion in the paragraph following eq. (2-3.1). One may envisage a theory of memory fluids, which will degenerate into the theory of purely viscous fluids in the limit of a very short memory span. The remainder of this book will be mainly dedicated to such a theory. Purely viscous, and in particular, generalized Newtonian fluid mechanics is then recognized to be a discipline yielding results which are asymptotically valid under conditions such that the memory of the material considered can be neglected.

It may be noted that we have so far only considered kinematic variables, such as velocity, rate of stretching, etc., representing instantaneous rates of change. Clearly these variables are not adequate for a theory of memory fluids where a description of the 'deformation history' is required in order to formalize the intuitive concepts introduced in this section. The next chapter is dedicated to differential kinematics, a discipline which is required in the treatment of memory fluid behavior. A few mathematical concepts required in differential kinematics are discussed in the next section.

The concepts of elasticity in fluid-like materials and of memory for past deformation, though intimately connected to each other, are not to be regarded as equivalent. Phenomena such as recoil are clearly in the realm of what is intuitively regarded as elasticity; but there exist some phenomena observed in real materials which, although suggestive of a memory of the material for past deformation, do not conform to our intuitive notion of elasticity. Typical phenomena in this category are those known as 'rheopexy' and 'thixotropy.' A rheopectic or thixotropic material, when subjected to shear such as, for example, in a lineal Couette flow, exhibits a time-dependence of the apparent viscometric viscosity, the value of which depends on the time of duration of shear, reaching an asymptote after a very long period. Yet such materials, upon instantaneous cessation of the deformation, do not necessarily exhibit the recoil phenomenon.

Clearly, a generalized theory of the behavior of materials with memory should include both elastic fluids and rheopectic and thixotropic materials; such is indeed the case of the 'simple fluid' theory to be discussed in chapter 4. Yet the behavior of rheopectic and thixotropic materials are rather special cases, which may deserve specialized treatment, although very little theoretical research has been carried out in this direction. Finally, one should observe that while the concept of memory in a fluid material can be rigorously formalized, the intuitive concept of the elasticity of fluid materials cannot. For this reason, we will use the word 'viscoelastic' only in the sense of fluids endowed with memory.

2-7   FURTHER USEFUL RELATIONS OF TENSOR ALGEBRA

In this section we shall consider some further relations of vector and tensor analysis which were not given in chapter 1. These will be used in the next and subsequent chapters and are collected here for ready reference. The content of this section is somewhat miscellaneous and the topics are often unrelated to each other. The reader should remember that this book is not intended to present a complete and ordered analysis of vectors and tensors, but just that part of it which is of use in the mechanics of complex fluids.

(a) The unit tensor as sum of three dyads

The unit tensor can be written as

$$\mathbf{1} = \mathbf{e}_k\mathbf{e}^k = \mathbf{e}^k\mathbf{e}_k \qquad (2\text{-}7.1)$$

Here $\mathbf{e}_k$ are three vectors which form a basis and $\mathbf{e}^k$ those of the corresponding dual. To prove the first of these identities, let us apply the sum of the three dyads $\mathbf{e}_k\mathbf{e}^k$ to any vector $\mathbf{a}$. From eqs (1-2.7) and (1-2.5), we have

$$\mathbf{e}_k\mathbf{e}^k \cdot \mathbf{a} = \mathbf{e}_k(\mathbf{e}^k \cdot \mathbf{a}) = \mathbf{e}_k a^k = \mathbf{a} \qquad (2\text{-}7.2)$$

Thus $\mathbf{e}_k\mathbf{e}^k$ equals the unit tensor. The second identity is similarly proved.

Before showing an example of the application of eq. (2-7.1), let us prove the following identity:

$$(\mathbf{a} \cdot \mathbf{A} \cdot \mathbf{b})(\mathbf{c} \cdot \mathbf{B} \cdot \mathbf{d}) = \mathbf{a} \cdot \mathbf{A} \cdot \mathbf{b}\mathbf{c} \cdot \mathbf{B} \cdot \mathbf{d} \qquad (2\text{-}7.3)$$

This follows immediately from the definition of a dyad. In fact,

$$\mathbf{b}\mathbf{c} \cdot (\mathbf{B} \cdot \mathbf{d}) = \mathbf{b}(\mathbf{c} \cdot \mathbf{B} \cdot \mathbf{d}) \qquad (2\text{-}7.4)$$

from which eq. (2-7.3) is easily obtained.

The use of eq. (2-7.1) is shown by the following example. We wish to prove eq. (1-3.28); that is,

$$(\mathbf{A} \cdot \mathbf{B})_{ij} = A_{im}B^m{}_j$$

We write

$$(\mathbf{A} \cdot \mathbf{B})_{ij} = \mathbf{e}_i \cdot \mathbf{A} \cdot \mathbf{B} \cdot \mathbf{e}_j = \mathbf{e}_i \cdot \mathbf{A} \cdot \mathbf{1} \cdot \mathbf{B} \cdot \mathbf{e}_j$$
$$= \mathbf{e}_i \cdot \mathbf{A} \cdot \mathbf{e}_m\mathbf{e}^m \cdot \mathbf{B} \cdot \mathbf{e}_j$$
$$= (\mathbf{e}_i \cdot \mathbf{A} \cdot \mathbf{e}_m)(\mathbf{e}^m \cdot \mathbf{B} \cdot \mathbf{e}_j) = A_{im}B^m{}_j$$

Here we have also used eq. (2-7.3) in conjunction with the summation convention. This is possible because eq. (2-7.3) can be applied to every term of the summation, as can easily be verified in the expanded form.

(b) Time derivatives

Consider a function $\psi(\tau)$ of a single scalar argument, τ, which in particular is interpreted as a time. The value of ψ may be a scalar, a vector, a point, or a tensor.

The derivative of ψ with respect to τ, indicated by $\dot{\psi}$, is defined by

$$\dot{\psi} = \lim_{\Delta\tau \to 0} \frac{\psi(\tau + \Delta\tau) - \psi(\tau)}{\Delta\tau} \qquad (2\text{-}7.5)$$

The definition, already familiar for the case of a scalar-valued function, is thus extended to vector-, point-, and tensor-valued functions. The derivative $\dot{\psi}$, which is a scalar for the case of scalar-valued functions, is a vector in the cases of vector- and point-valued functions, and a tensor for tensor-valued functions. We have already encountered examples of these derivatives in chapter 1.

Most rules of ordinary differential calculus extend to the differentiation of vector and tensor functions. The only differences arise from the fact that the commutative law is not valid in general (i.e., $\mathbf{A} \cdot \mathbf{B} \neq \mathbf{B} \cdot \mathbf{A}$). For example,

$$\overline{\mathbf{A} \cdot \mathbf{B}} = \dot{\mathbf{A}} \cdot \mathbf{B} + \mathbf{A} \cdot \dot{\mathbf{B}} \qquad (2\text{-}7.6)$$

and thus

$$\overline{\mathbf{A}^2} = \dot{\mathbf{A}} \cdot \mathbf{A} + \mathbf{A} \cdot \dot{\mathbf{A}} \qquad (2\text{-}7.7)$$

Notice that eq. (2-7.7) cannot be written as $\overline{\mathbf{A}^2} = 2\mathbf{A} \cdot \dot{\mathbf{A}}$.

Similarly, by differentiating the identity $\mathbf{A}^{-1} \cdot \mathbf{A} = \mathbf{1}$ we have

$$\overline{\mathbf{A}^{-1}} = -\mathbf{A}^{-1} \cdot \dot{\mathbf{A}} \cdot \mathbf{A}^{-1} \qquad (2\text{-}7.8)$$

which can be compared with the familiar expression for scalar-valued functions:

$$\overline{f^{-1}} = -f^{-2}\dot{f}$$

It can easily be shown that transposition and differentiation commute:

$$\overline{\mathbf{A}^{\mathrm{T}}} = (\dot{\mathbf{A}})^{\mathrm{T}} = \dot{\mathbf{A}}^{\mathrm{T}} \qquad (2\text{-}7.9)$$

It must be stressed that the definition given by eq. (2-7.5) makes no use of concepts of coordinate system and components. Thus, for instance, the tensor $\dot{\mathbf{A}}$ is the operator defined on the basis of the operators $\mathbf{A}(\tau)$ following the operations indicated by eq. (2-7.5).

In general, the operations of calculating the components and of differentiation do not commute; for example,

$$(\dot{\mathbf{A}})_{ij} \neq \dot{A}_{ij} \qquad (2\text{-}7.10)$$

In fact, by differentiating the equality

$$A_{ij} = \mathbf{e}_i \cdot \mathbf{A} \cdot \mathbf{e}_j \qquad (2\text{-}7.11)$$

the equality $\dot{A}_{ij} = (\dot{\mathbf{A}})_{ij}$ would follow only in the special case that the basis vectors are independent of τ. Further considerations on time derivatives of tensors will be made in chapter 3.

(c) Physical components

Among the possible coordinate systems, those of special importance for practical purposes are the orthogonal coordinate systems. These are such that, in all points, the vectors of the natural basis are mutually orthogonal (although not usually of unit length). Familiar examples of orthogonal systems are, together with the Cartesian, the cylindrical and the spherical coordinate systems.

When dealing with orthogonal coordinate systems it is often useful to consider the *physical* components of vectors and tensors. These are the components, with respect to an orthonormal basis, formed of vectors that have the same directions as those of the natural basis (which also coincide with those of its dual).

The vectors of the orthonormal basis associated to the natural basis (or its dual) of an orthogonal coordinate system will be indicated by $\mathbf{e}\langle i \rangle$. Being of unit length, they are given by

$$\mathbf{e}\langle i \rangle = \frac{\mathbf{e}_i}{|\mathbf{e}_i|} = \frac{\mathbf{e}^i}{|\mathbf{e}^i|} \qquad (2\text{-}7.12)$$

In eq. (2-7.12), and in all equations of this subsection, no summation is intended on the repeated index.

Through eq. (2-7.12), the physical components of vectors and tensors are easily related to the correponding covariant, contravariant, or mixed components. Indicating physical components by enclosing the indices in the special brackets $\langle \ \rangle$, we have, for example,

$$a\langle i \rangle = \mathbf{a} \cdot \mathbf{e}\langle i \rangle = \frac{1}{|\mathbf{e}_i|}\mathbf{a} \cdot \mathbf{e}_i$$

$$= \frac{1}{|\mathbf{e}_i|}a_i \qquad (2\text{-}7.13)$$

$$A\langle ij \rangle = \mathbf{e}\langle i \rangle \cdot \mathbf{A} \cdot \mathbf{e}\langle j \rangle$$

$$= \frac{1}{|\mathbf{e}^i|\,|\mathbf{e}^j|}\mathbf{e}^i \cdot \mathbf{A} \cdot \mathbf{e}^j = \frac{1}{|\mathbf{e}^i|\,|\mathbf{e}^j|}A^{ij} \qquad (2\text{-}7.14)$$

The magnitude of the basis vectors, $|\mathbf{e}^i|$, $|\mathbf{e}_i|$, can be expressed through the metric:

$$|\mathbf{e}_i| = \sqrt{(\mathbf{e}_i \cdot \mathbf{e}_i)} = \sqrt{g_{ii}}$$
$$|\mathbf{e}^i| = \sqrt{(\mathbf{e}^i \cdot \mathbf{e}^i)} = \sqrt{g^{ii}} \qquad (2\text{-}7.15)$$

so that eqs (2-7.13) and (2-7.14) can also be written:

$$a\langle i\rangle = \frac{a_i}{\sqrt{g_{ii}}} \qquad (2\text{-}7.16)$$

$$A\langle ij\rangle = \frac{A^{ij}}{\sqrt{(g^{ii}g^{jj})}} \qquad (2\text{-}7.17)$$

For orthogonal systems the following relation holds true, as will be shown below:

$$g_{ii} = \frac{1}{g^{ii}} \qquad (2\text{-}7.18)$$

Using eq. (2-7.18) and developing the other possible relations like those of eqs (2-7.13) and (2-7.14), we obtain the full set of possible relations between physical components and the other types of components of vectors and tensors:

$$a\langle i\rangle = \frac{a_i}{\sqrt{g_{ii}}} = a_i\sqrt{g^{ii}}$$

$$= \frac{a^i}{\sqrt{g^{ii}}} = a^i\sqrt{g_{ii}} \qquad (2\text{-}7.19)$$

$$A\langle ij\rangle = \frac{A_{ij}}{\sqrt{(g_{ii}g_{jj})}} = A_{ij}\sqrt{(g^{ii}g^{jj})}$$

$$= \frac{A^{ij}}{\sqrt{(g^{ii}g^{jj})}} = A^{ij}\sqrt{(g_{ii}g_{jj})} = A^i{}_j\sqrt{(g^{jj}/g^{ii})}$$

$$= A^i{}_j\sqrt{(g_{ii}/g_{jj})} = \cdots \qquad (2\text{-}7.20)$$

It may be observed that the physical components of a vector or a tensor have the same physical dimensions as the vector or tensor.† This property is in general not shared by the other components because the basis vectors are, in general, not dimensionless.

For coordinate systems that are not orthogonal we may also speak of physical components, provided a vector basis is chosen which is made of dimensionless vectors having unit length. However, in such a case the choice is not unique. We may take vectors of unit length having the same directions

---

† The physical dimensions of a tensor are determined by interpreting the operational definition of the tensor as a multiplication. In other words, the equality $\mathbf{b} = \mathbf{A} \cdot \mathbf{a}$ is dimensionally correct if the product of the dimensions of $\mathbf{a}$ and $\mathbf{A}$ gives the dimensions of $\mathbf{b}$. For example, from the equality $d\mathbf{t} = \mathbf{T} \cdot d\mathbf{s}$, which defines the stress tensor, we deduce that the dimensions of $\mathbf{T}$ are those of a force per unit area.

as those of the natural basis; alternatively, we may take those which have the directions of the vectors of the dual. Accordingly we define 'physical contravariant' components or 'physical covariant' components of vectors. Analogous remarks can be made for tensors. We shall not use any such components.

(d) Elements of matrix algebra

The operation of matrix algebra that is most frequently used in this context is the matrix multiplication 'row by column.' Such expressions as

$$A_{im}B^m_{\ j}$$

correspond to a row by column multiplication of the matrix A_{ij} times the matrix $B^i_{\ j}$ if the index on the left is interpreted as being the rows.

By defining the transpose of a matrix as the matrix obtained from the given one by exchanging the roles of the indices, we observe that such expressions as

$$A_{mi}B^m_{\ j}, \quad A_{im}B_j^{\ m}, \quad A_{mi}B_j^{\ m}$$

also reduce to a 'row by column' multiplication, provided suitable transpositions are performed first.

Another useful operation of matrix algebra is the matrix inversion. From the tensorial identity,

$$\mathbf{A}^{-1} \cdot \mathbf{A} = \mathbf{1} \qquad (2\text{-}7.21)$$

we have

$$(\mathbf{A}^{-1})^{in} A_{nj} = \delta^i_j \qquad (2\text{-}7.22)$$

Recalling the well-known properties of the cofactors of the elements of a square matrix, eq. (2-7.22) shows that

$$(\mathbf{A}^{-1})^{ij} = \frac{\text{cof } A_{ji}}{\det [A_{ij}]} \qquad (2\text{-}7.23)$$

(Notice carefully the position of indices in eq. (2-7.23).) Equation (2-7.23) shows how the contravariant components of $\mathbf{A}^{-1}$ are obtained from the covariant components of $\mathbf{A}$. The operation involved is called matrix inversion.

Similarly,

$$(\mathbf{A}^{-1})_{ij} = \frac{\text{cof } A^{ji}}{\det [A^{ij}]}$$

$$(\mathbf{A}^{-1})^i_{\ j} = \frac{\text{cof } A_j^{\ i}}{\det [A^i_{\ j}]}$$

(2-7.24)

Equation (2-7.23) can be written in particular for the unit tensor:

$$g^{ij} = \frac{\text{cof } g_{ji}}{\det [g_{ij}]} \qquad (2\text{-}7.25)$$

For an orthogonal coordinate system the matrices $[g_{ij}]$ and $[g^{ij}]$ are diagonal, i.e., all the elements whose two indices are different have zero value, as can be seen from eqs (1-3.31)–(1-3.32). Equation (2-7.25) then gives, for the diagonal elements,

$$g^{ii} = \frac{1}{g_{ii}} \qquad (2\text{-}7.26)$$

which is the same as eq. (2-7.18) above.

(e) Nilpotent tensors

A tensor $\mathbf{A}$ is said to be *nilpotent* if, for some integer N:

$$\mathbf{A}^N = \mathbf{0} \qquad (2\text{-}7.27)$$

Using eq. (2-7.27) and the following identity (see eq. (1-3.43)):

$$\det (\mathbf{A}^N) = (\det \mathbf{A})^N \qquad (2\text{-}7.28)$$

it follows that

$$\det \mathbf{A} = \text{III}_\mathbf{A} = 0 \qquad (2\text{-}7.29)$$

From eq. (2-7.27), using iteratively the Cayley–Hamilton theorem, it follows that

$$\mathbf{A}^3 = \mathbf{0} \qquad (2\text{-}7.30)$$

Finally, using again the Cayley–Hamilton theorem together with the definition of $\text{II}_\mathbf{A}$, eq. (1-3.42), it follows from eq. (2-7.30) that

$$\text{I}_\mathbf{A} = \text{II}_\mathbf{A} = 0 \qquad (2\text{-}7.31)$$

In conclusion, for a nilpotent tensor, the three principal invariant are zero and all the powers $\mathbf{A}^N$ for $N > 2$ are the zero tensor. $\mathbf{A}^2$ and $\mathbf{A}$ are, in general, not zero. (It must be remembered that the equation $\mathbf{A} \cdot \mathbf{B} = \mathbf{0}$ does not imply that either $\mathbf{A}$ or $\mathbf{B}$ is the zero tensor.)

It can be shown that a symmetric tensor, with the exception of the zero tensor, cannot be nilpotent.

The geometrical interpretation of nilpotent tensors is now stated. For any nilpotent tensor, a family of parallel planes α and a family of parallel lines β exist, which are characteristics of the tensor. The lines β lie on the planes α. When $\mathbf{A} \neq \mathbf{0}$, $\mathbf{A}^2 = \mathbf{0}$, the tensor $\mathbf{A}$, when operating on any vector not lying on a plane α, transforms it into a vector lying on a line β, and when operating on a vector lying on α, transforms it into the zero vector. Thus, when operating on *any* vector twice in succession, the tensor $\mathbf{A}$ transforms it into the zero vector. When $\mathbf{A}^2 \neq \mathbf{0}$, $\mathbf{A}^3 = \mathbf{0}$, the tensor $\mathbf{A}$ transforms any vector not lying on α into a vector lying on α, any vector lying on α into a vector lying on β, and any vector lying on β into the zero vector. Thus, when operating three times in succession, $\mathbf{A}$ transforms any vector into the zero vector.

2-8   ILLUSTRATIONS

ILLUSTRATION 2A: *Derivation of the stress in a Reiner–Rivlin fluid in lineal Couette flow (simple shear).*

By a proper choice of a Cartesian coordinate system, the flow is described by eqs (2-1.2)–(2-1.3):

$$v^1 = \gamma x^2 \qquad (2\text{-}1.2)$$

$$v^2 = v^3 = 0 \qquad (2\text{-}1.3)$$

Because in a Cartesian system all Christoffel symbols are zero, the components (of any kind) of the velocity gradient tensor $\nabla \mathbf{v}$ are simply given by the derivatives $\partial v^i / \partial x^j$ (see eq. (1-4.9) or (1-4.14)):

$$[\nabla \mathbf{v}] = \begin{Vmatrix} 0 & \gamma & 0 \\ 0 & 0 & 0 \\ 0 & 0 & 0 \end{Vmatrix} \qquad (2\text{-}8.1)$$

Notice that $\operatorname{tr} \nabla \mathbf{v} = \nabla \cdot \mathbf{v} = 0$, so that the continuity equation is satisfied.

We have also

$$[\nabla \mathbf{v}^{\mathsf{T}}] = [\nabla \mathbf{v}]^{\mathsf{T}} = \begin{Vmatrix} 0 & 0 & 0 \\ \gamma & 0 & 0 \\ 0 & 0 & 0 \end{Vmatrix} \qquad (2\text{-}8.2)$$

From eq. (1-9.5) we have

$$[\mathbf{D}] = \tfrac{1}{2}([\nabla \mathbf{v}] + [\nabla \mathbf{v}^{\mathsf{T}}]) = \begin{Vmatrix} 0 & \gamma/2 & 0 \\ \gamma/2 & 0 & 0 \\ 0 & 0 & 0 \end{Vmatrix} \qquad (2\text{-}8.3)$$

which has appeared previously, eq. (2-1.4).

In order to apply the Reiner–Rivlin constitutive equation, (eq. (2-3.4)), we also need $[\mathbf{D}^2]$. This is obtained by row by column multiplication of $[\mathbf{D}]$ times itself. One has:

$$[\mathbf{D}^2] = \begin{Vmatrix} \gamma^2/4 & 0 & 0 \\ 0 & \gamma^2/4 & 0 \\ 0 & 0 & 0 \end{Vmatrix} \qquad (2\text{-}8.4)$$

Applying eq. (2-3.4), we then have

$$\begin{aligned} &\tau_{11} = \phi_2\gamma^2/4, && \tau_{12} = \tau_{21} = \phi_1\gamma/2, \\ &\tau_{13} = \tau_{31} = 0, && \tau_{22} = \phi_2\gamma^2/4, \\ &\tau_{23} = \tau_{32} = 0, && \tau_{33} = 0 \end{aligned} \qquad (2\text{-}8.5)$$

Because τ is determined only within an isotropic tensor, its normal Cartesian components are determined only within an additive constant. The result of eqs (2-8.5) is thus properly expressed as in eqs (2-3.12)–(2-3.15).

We shall now check whether the assumed combination kinematics–constitutive equation is *controllable* (or dynamically possible); that is, whether the dynamical equation is satisfied. If the answer is affirmative, the checking procedure also allows the (so far) indetermined pressure to be calculated. This point will be discussed in general in section 5-1.

From our previous results, we calculate the matrix of the deviatoric stress:

$$[\tau'] = \begin{Vmatrix} \phi_2\gamma^2/12 & \phi_1\gamma/2 & 0 \\ \phi_1\gamma/2 & \phi_2\gamma^2/12 & 0 \\ 0 & 0 & -\phi_2\gamma^2/6 \end{Vmatrix} \qquad (2\text{-}8.6)$$

The above matrix is constant in space, and the vector basis is also constant in a Cartesian system; thus, τ' is a constant tensor and

$$\nabla \cdot \tau' = 0 \qquad (2\text{-}8.7)$$

Also, the path of each particle being a straight line in the x^1 direction, along which the velocity is constant, we have

$$\frac{D\mathbf{v}}{Dt} = 0 \qquad (2\text{-}8.8)$$

The dynamical equation (eq. (1-7.13)), is thus satisfied, provided that

$$-\nabla p + \rho \mathbf{g} = 0 \qquad (2\text{-}8.9)$$

Equation (2-8.9) shows that the pressure has the hydrostatic distribution of a stagnant liquid. The total stress tensor is obtained summing the deviatoric stress to $-p\mathbf{1}$.

ILLUSTRATION 2B: *Velocity distribution for laminar flow of a 'power-law' fluid in a pipe.*

Equation (2-5.6), with the hypothesis that the fluid is of the power-law type, becomes

$$\tau = K(\gamma^2)^{(n-1)/2}\gamma \qquad (2\text{-}8.10)$$

where

$$\gamma = -\frac{du}{dr} \qquad (2\text{-}8.11)$$

Because for the case at hand γ is non-negative, eq. (2-8.10) is simplified to

$$\tau = K\gamma^n \qquad (2\text{-}8.12)$$

By using eq. (2-5.8), eq. (2-8.12) can be written as

$$-\frac{du}{dr} = \gamma = \left(\frac{\tau_w}{RK}\right)^{1/n} r^{1/n} \qquad (2\text{-}8.13)$$

Integration of eq. (2-8.13), with the condition

$$r = R, \qquad u = u_s \qquad (2\text{-}8.14)$$

gives

$$u = u_s + \frac{n}{n+1}\left(\frac{\tau_w}{K}\right)^{1/n} R\left[1 - \left(\frac{r}{R}\right)^{(1+n)/n}\right] \qquad (2\text{-}8.15)$$

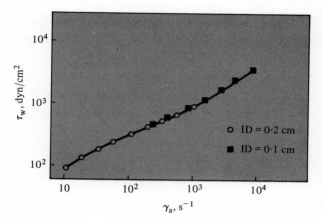

FIGURE 2-2
Stress versus shear rate.

ILLUSTRATION 2C: *Determining the function $\eta(S)$ from capillary viscometer data.*

Using two capillaries (0.2 cm and 0.1 cm i.d.) the flow rate through them has been determined for a number of values of pressure drop. The results are given in the following table.

Internal diameter = 0.2 cm

$\Delta p/L$ (atm/cm)	0.0018	0.0026	0.0036	0.0048	0.0064	0.0083	0.0104	0.0132	0.018
Q (cm^3/s)	0.0086	0.015	0.028	0.046	0.087	0.158	0.27	0.45	0.82

Internal diameter = 0.1 cm

$\Delta p/L$ (atm/cm)	0.012	0.0176	0.022	0.033	0.0455	0.066	0.094	0.152
Q (cm^3/s)	0.010	0.022	0.039	0.085	0.152	0.28	0.45	0.88

We shall first check whether there is any slip at the wall. As shown in eq. (2-5.9), no slip at the wall implies that γ_a is a unique function of τ_w (or vice-versa), independent of tube radius. By using eqs (2-5.8) and (2-5.10), from the raw data of $\Delta p/L$ and Q, the corresponding values of τ_w and γ_a are calculated for the two capillaries. These are plotted in Fig. 2-2. The two sets of data superimpose on each other in the common range of γ_a values, thus showing that no slip at the wall is present.

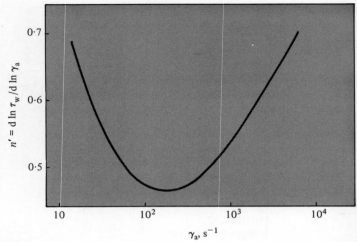

FIGURE 2-3
Flow index versus shear rate.

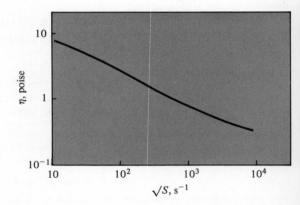

FIGURE 2-4
Viscosity curve.

From the interpolated curve of Fig. 2-2, the local slope

$$n' = \frac{d \ln \tau_w}{d \ln \gamma_a}$$

is determined. This is plotted in Fig. 2-3 as a function of γ_a.

From eqs (2-5.16)–(2-5.17) the curve $\eta(S)$ is then evaluated and plotted in Fig. 2-4. Pseudoplastic behavior of the fluid under test is clearly indicated. Finally, for use in the following illustration, the curve $\eta(\tau)$ is also plotted in Fig. 2-5.

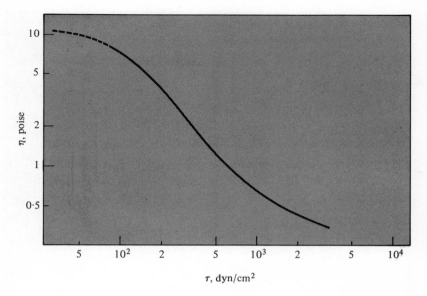

FIGURE 2-5
Viscosity versus shear stress.

ILLUSTRATION 2D: *Pressure drop in a rectangular channel.*

The same fluid as in Illustration 2C is made to flow through a rectangular channel having a cross-section of 10 cm × 0.4 cm. The flow rate is 150 cm³/s and we wish to calculate the value of the corresponding pressure drop.

The aspect ratio of the cross-section of the channel is such that we may assume the flow to be mainly that which would be obtained between two parallel plates of infinite extent.

By a procedure similar to that followed in section 2-5 for a circular tube, one finds (with no slip at the wall) [9]:

$$\frac{Q}{wH^2} = \frac{1}{2\tau_w^2} \int_0^{\tau_w} \tau f(\tau)\, d\tau \qquad (2\text{-}8.16)$$

where Q/w is the flow rate per unit width of the channel, H is the height of the channel, $f(\tau)$ is the function given in eq. (2-5.7), and τ_w is the shear stress at the wall related to the pressure drop by

$$\tau_w = \frac{H}{2} \cdot \frac{\Delta p}{L} \qquad (2\text{-}8.17)$$

Insofar as the results of Illustration 2C are given in terms of the function η

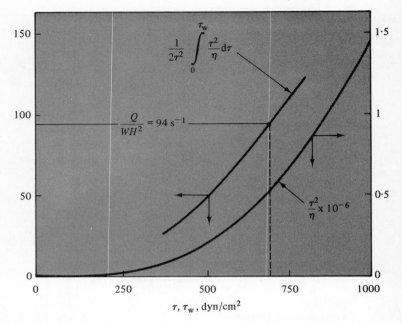

FIGURE 2-6
Graphical solution of eq. (2-8.18).

rather than f, we better rewrite eq. (2-8.16) as

$$\frac{Q}{wH^2} = \frac{1}{2\tau_w^2} \int_0^{\tau_w} \frac{\tau^2}{\eta} \, d\tau \qquad (2\text{-}8.18)$$

For the problem at hand, where Q/w and H are given, we must determine the value of τ_w that satisfies eq. (2-8.18). This is done by means of a graphical procedure. The integral on the right-hand side of eq. (2-8.18), as well as the entire right-hand side, are functions only of τ_w. Figure 2-6 is a plot of the function τ^2/η and of the right-hand side of eq. (2-8.18) thereby obtained by graphical integration. Corresponding to the value of the ordinate equal to Q/wH^2, the value of τ_w is obtained. Finally, from eq. (2-8.17), $\Delta p/L = 0.0035$ atm/cm is calculated.

It must be observed that, in order to evaluate the integral in eq. (2-8.18), the function η must be extrapolated to a zero value of τ_w; this has been done following the dotted line in Fig. 2-5. The arbitrariness that is implicit in such a procedure introduces a possible error in the result, which becomes more negligible the larger the value of τ_w obtained with respect to the minimum value of τ used to determine the function η.

PROBLEMS

2-1   Calculate the components of the vorticity tensor $\mathbf{W}$ for a lineal Couette flow.

2-2   Consider the change of frame identified by the orthogonal tensor $\mathbf{Q}$ whose Cartesian components are:

$$[\mathbf{Q}] = \begin{Vmatrix} 1 & 0 & 0 \\ 0 & 1 & 0 \\ 0 & 0 & -1 \end{Vmatrix}$$

Calculate $\mathbf{D}^*$ and $\mathbf{T}^*$ for the lineal Couette flow. From the principle of material objectivity deduce immediately eq. (2-3.13) for Reiner–Rivlin fluids.

2-3   Which are the dimensions of the quantities ϕ_1 and ϕ_2?

2-4   Is $\mathbf{W}$ indifferent? If not, which is its transformation rule?

2-5   Given an indifferent tensor $\mathbf{A}$, is $\mathbf{A}^n$ indifferent?

2-6   Prove that $\tau':\mathbf{Vv} = \tau':\mathbf{D}$. *Hint*: first prove in general that $\mathbf{A}:\mathbf{B} = 0$ if $\mathbf{A}$ is symmetric and $\mathbf{B}$ is antisymmetric.

2-7   The exponential of a tensor is defined as

$$\exp \mathbf{A} = 1 + \sum_{n=1}^{\infty} \frac{1}{n!}\mathbf{A}^n$$

If $\mathbf{A}$ is an indifferent tensor, is $\exp \mathbf{A}$ indifferent?

BIBLIOGRAPHY

1.   TRUESDELL, C., and NOLL, W.: *The Non-linear Field Theories*, p. 45. Springer-Verlag, Berlin (1965).

2.   *ibid.*, p. 47.

3.   NOLL, W.: *Arch. Ratl Mech. Anal.*, **2**, 197 (1958).

4.   Reference 1, p. 65.

5.   COLEMAN, B. D., and MIZEL, W. J.: *Arch. Ratl Mech. Anal.*, **27**, 255 (1968).

6.   NOLL, W.: *J. Ratl Mech. Anal.*, **4**, 3 (1955).

7.   LEIGH, D. C.: *Phys. Fluids*, **5**, 501 (1962).

8.   TANNER, R. I.: *Ind. Engng Chem. Fund.*, **5**, 55 (1966).

9.   SKELLAND, A. H. P.: *Non-Newtonian Flow and Heat Transfer*. Wiley, New York (1967).

10.   EYRING, H. J.: *J. Chem. Phys.*, **4**, 283 (1936).

11.   CHRISTIANSEN, E. B., RYAN, N. W., and STEVENS, W. E.: *A.I.Ch.E. Jl*, **1**, 544 (1955).

12.   ASTARITA, G., MARRUCCI, G., and PALUMBO, G.: *Ind. Engng Chem. Fund.*, **3**, 333 (1964).

13.   MORRISON, S. R., and HARPER, J. C.: *Ind. Engng Chem. Fund.*, **4**, 176 (1965).

14.   HIROSE, T., and MOO-YOUNG, M.: *Can. J. chem. Engng*, **47**, 265 (1969).

15.   NAKANO, Y., and TIEN, C.: *A.I.Ch.E. Jl*, **14**, 145 (1968).

16.   TRUESDELL, C.: *Phys. Fluids*, **8**, 1936 (1965).

3
KINEMATICS

3-1 DEFORMATION AND STRAIN

Consider a fluid material in flow, and let $X(\tau)$ be the place occupied by a material point at time τ. In order to identify the material point, we may choose some particular instant of time t and use the place $X_t = X(t)$ occupied by the particle at time t as a convenient label identifying the material point considered. The 'motion' is the function

$$X(\tau) = f(X_t, \tau) \qquad (3\text{-}1.1)$$

which describes the trajectory $X(\tau)$ of all material points X_t.

It is sometimes useful to regard t as the present time and τ as any instant in the past. We shall refer to t as the 'instant of observation,' and, whenever no confusion arises, will drop the explicit indication of the τ-dependency; e.g., we will indicate the trajectory simply as X.

The velocity of the particle, v, is

$$v = \dot{X} = \frac{\partial f}{\partial \tau} \qquad (3\text{-}1.2)$$

The deformation gradient

Consider now two infinitesimally close material points, which at time t are located at $\mathbf{X}_t$ and $\mathbf{X}_t + d\mathbf{X}_t$, respectively. At some other time τ the two points are located at $\mathbf{X}$ and $\mathbf{X} + d\mathbf{X}$. We define the 'deformation gradient' $\mathbf{F}$ through the equation

$$d\mathbf{X} = \mathbf{F} \cdot d\mathbf{X}_t \qquad (3\text{-}1.3)$$

where dependency of $d\mathbf{X}$ and $\mathbf{F}$ on τ is understood. It can be proved that $\mathbf{F}$ is indeed a linear operator, and thus a tensor.

One should note that the deformation gradient $\mathbf{F}$, for any given time τ, depends on which time is regarded as the instant of observation. *We shall refer in the following to such tensors as relative tensors.* In dealing with relative tensors, one may wish sometimes to pick a reference which is not the instant of observation (say, e.g., in eq. (3-1.3) one may wish to replace $d\mathbf{X}_t$ on the right-hand side with the distance $d\mathbf{X}_{t'}$ among the same two points in some situation other than the one existing at $\tau = t$). In such cases, we will add as a subscript the reference which has been chosen (say one would write $\mathbf{F}_{t'}$). *Whenever a relative tensor has no subscripts, it is understood that the instant of observation has been chosen as the reference.*

The definition of $\mathbf{F}$ given in eq. (3-1.3) allows it to be identified with the gradient of the 'motion:'†

$$\mathbf{F} = \nabla\mathbf{f} \qquad (3\text{-}1.4)$$

Also, it is evident that

$$\mathbf{F}(t) = \mathbf{1} \qquad (3\text{-}1.5)$$

Let us now express the displacement $d\mathbf{X}$ in terms of the displacement $d\mathbf{X}(t')$ of the same two material points at some other instant of observation t'. We have, using twice eq. (3-1.3),

$$d\mathbf{X} = \mathbf{F} \cdot d\mathbf{X}_t = \mathbf{F} \cdot \mathbf{F}_{t'}(t) \cdot d\mathbf{X}(t') \qquad (3\text{-}1.6)$$

But, again, by definition of $\mathbf{F}$,

$$d\mathbf{X} = \mathbf{F}_{t'} \cdot d\mathbf{X}(t') \qquad (3\text{-}1.7)$$

Comparison of eqs (3-1.6) and (3-1.7) shows that

$$\mathbf{F}_{t'} = \mathbf{F} \cdot \mathbf{F}_{t'}(t) \qquad (3\text{-}1.8)$$

---

† $\mathbf{f}$ is, of course, a point-valued field. For any point-valued field $\mathbf{Y} = \mathbf{g}(\mathbf{X})$, the gradient is defined as

$$d\mathbf{Y} = \nabla\mathbf{g} \cdot d\mathbf{X}$$

Equation (3-1.8) will be frequently used whenever a shift of the instant of observation is required.

From eq. (3-1.8), setting τ (which is understood to be the argument of both $\mathbf{F}$ and $\mathbf{F}_{t'}$) equal to t', we get

$$\mathbf{F}^{-1}(\tau) = \mathbf{F}_{\tau}(t) \qquad (3\text{-}1.9)$$

which shows that $\mathbf{F}$ is invertible. By definition of $\mathbf{F}$ we have

$$d\mathbf{X}_t = \mathbf{F}^{-1} \cdot d\mathbf{X} \qquad (3\text{-}1.10)$$

Stretch and rotation

We now make use of the *polar decomposition theorem*, which states that 'any invertible tensor $\mathbf{F}$ has two unique decompositions:

$$\mathbf{F} = \mathbf{R} \cdot \mathbf{U} \qquad (3\text{-}1.11)$$

$$\mathbf{F} = \mathbf{V} \cdot \mathbf{R} \qquad (3\text{-}1.12)$$

where $\mathbf{R}$ is orthogonal and $\mathbf{U}$ and $\mathbf{V}$ are symmetric and positive-definite.'[†]

When applied to the deformation gradient $\mathbf{F}$ the polar decomposition theorem yields the *rotation* tensor $\mathbf{R}$, the *right stretch* tensor $\mathbf{U}$, and the *left stretch* tensor $\mathbf{V}$. These are all relative tensors, and when written without a subscript, reference to the instant of observation is understood. A geometrical interpretation of $\mathbf{R}$, $\mathbf{U}$, and $\mathbf{V}$ is given in the following.

Consider a material element which at time t has a square shape such as in Fig. 3-1a, and at time τ has the shape shown in Fig. 3-1d. The overall deformation can be considered to be that obtained from the superposition of a rigid rotation and a stretch. One may visualize first a rotation carrying the element to the situation shown in Fig. 3-1b, followed by a stretch, or first a stretch carrying the element to the situation shown in Fig. 3-1c, followed by a rotation. In both cases the situation shown in Fig. 3-1d is the final result, the rotations $(a) \rightarrow (b)$ and $(c) \rightarrow (d)$ being equal. They are identified by $\mathbf{R}$ in the sense that

$$(d\mathbf{X})_b = \mathbf{R} \cdot (d\mathbf{X})_a = \mathbf{R} \cdot d\mathbf{X}_t \qquad (3\text{-}1.13)$$

$$d\mathbf{X} = (d\mathbf{X})_d = \mathbf{R} \cdot (d\mathbf{X})_c \qquad (3\text{-}1.14)$$

The stretches $(b) \rightarrow (d)$ and $(a) \rightarrow (c)$, although representative of the same change of shape, are not describable by the same tensor; we have

$$d\mathbf{X} = (d\mathbf{X})_d = \mathbf{V} \cdot (d\mathbf{X})_b \qquad (3\text{-}1.15)$$

$$(d\mathbf{X})_c = \mathbf{U} \cdot (d\mathbf{X})_a = \mathbf{U} \cdot d\mathbf{X}_t \qquad (3\text{-}1.16)$$

---

† A tensor $\mathbf{A}$ is said to be positive-definite if, for any vector $\mathbf{a}$, the scalar $\mathbf{a} \cdot \mathbf{A} \cdot \mathbf{a}$ is positive.

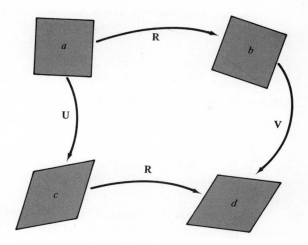

FIGURE 3-1
Decomposition of deformation into stretch and rotation.

Combining eqs (3-1.13) and (3-1.15), and eqs (3-1.14) and (3-1.16), we have

$$dX = V \cdot R \cdot dX_t = R \cdot U \cdot dX_t \qquad (3\text{-}1.17)$$

from which eqs (3-1.11) and (3-1.12) are clearly understood.

Clearly, V and U are different only in the sense that they are 'oriented' differently with respect to the frame of reference; indeed, from eqs (3-1.11) and (3-1.12), the following relation between V and U is immediately established:

$$V = R \cdot U \cdot R^T \qquad (3\text{-}1.18)$$

In general, the components of tensors V and U are irrational functions of easily measured quantities, and it is preferable to introduce the following two new relative tensors:

$$C = U^2 = F^T \cdot F \qquad (3\text{-}1.19)$$

and

$$B = V^2 = F \cdot F^T \qquad (3\text{-}1.20)$$

which will be called the Cauchy and the Green tensors, respectively.† The relationship between B and C is

$$B = R \cdot C \cdot R^T \qquad (3\text{-}1.21)$$

---

† Truesdell and Noll [1] prefer the terminology of right and left Cauchy–Green tensors, respectively.

Inasmuch as the invariants are isotropic functions (see section 1-5), we have

$$I_B = I_C; \quad II_B = II_C; \quad III_B = III_C \quad (3\text{-}1.22)$$

Moreover, whenever the volume is not changed upon deformation, which is always the case for constant density fluids:

$$III_B = III_C = 1 \quad (3\text{-}1.23)$$

The tensors C^{-1} and B^{-1} are often encountered in the literature; we shall call them the Finger and the Piola tensor, respectively. A geometrical direct interpretation of the Cauchy, Green, Finger, and Piola tensors is given below.

Let us calculate the magnitude of the displacement of two neighboring material points. We have, from eq. (3-1.3),

$$dx^2 = d\mathbf{X} \cdot d\mathbf{X} = d\mathbf{X}_t \cdot \mathbf{F}^T \cdot \mathbf{F} \cdot d\mathbf{X}_t$$
$$= d\mathbf{X}_t \cdot \mathbf{C} \cdot d\mathbf{X}_t \quad (3\text{-}1.24)$$

Equation (3-1.24) shows that dx^2 at any time τ can be calculated if the displacement of the two points at the instant of observation, $d\mathbf{X}_t$, and the Cauchy tensor $\mathbf{C}$ are known. Analogously, from eq. (3-1.10),†

$$dx_t^2 = d\mathbf{X}_t \cdot d\mathbf{X}_t = d\mathbf{X} \cdot \mathbf{B}^{-1} \cdot d\mathbf{X} \quad (3\text{-}1.25)$$

which allows the magnitude of the displacement at the instant of observation to be calculated from a knowledge of the displacement at some time τ.

Let us now consider surface elements $d\mathbf{s}$ rather than displacements. A differential volume is measured by the inner product of $d\mathbf{s}$ and $d\mathbf{X}$, see Fig. 3-2. The constant density hypothesis can then be written as

$$d\mathbf{s} \cdot d\mathbf{X} = d\mathbf{s}_t \cdot d\mathbf{X}_t \quad (3\text{-}1.26)$$

where the dependence of $d\mathbf{s}$ on τ is understood. Note that $d\mathbf{s}$ is the material surface which, at time t, is measured by $d\mathbf{s}_t$.

Substituting eq. (3-1.3) into eq. (3-1.26) gives

$$d\mathbf{s} \cdot \mathbf{F} \cdot d\mathbf{X}_t = d\mathbf{s}_t \cdot d\mathbf{X}_t \quad (3\text{-}1.27)$$

which is valid for *any* $d\mathbf{X}_t$. Thus,

$$d\mathbf{s}_t = \mathbf{F}^T \cdot d\mathbf{s}; \quad d\mathbf{s} = (\mathbf{F}^{-1})^T \cdot d\mathbf{s}_t \quad (3\text{-}1.28)$$

---

† In this chapter, use is made of the identities: $(\mathbf{A}^{-1})^T = (\mathbf{A}^T)^{-1}$; $(\mathbf{A} \cdot \mathbf{B})^T = \mathbf{B}^T \cdot \mathbf{A}^T$; $(\mathbf{A} \cdot \mathbf{B})^{-1} = \mathbf{B}^{-1} \cdot \mathbf{A}^{-1}$.

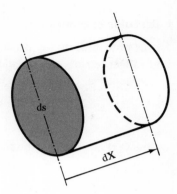

FIGURE 3-2
Elementary volume.

The square of the area of the surface element is

$$ds^2 = ds \cdot ds = ds_t \cdot F^{-1} \cdot (F^{-1})^T \cdot ds_t$$
$$= ds_t \cdot C^{-1} \cdot ds_t \qquad (3\text{-}1.29)$$

which shows that the Finger tensor measures the changes of area just as the Cauchy tensor measures the changes of length.† By an analogous procedure, it is shown that

$$ds_t^2 = ds \cdot B \cdot ds \qquad (3\text{-}1.30)$$

Comparison of eqs (3-1.24) and (3-1.25), and of eqs (3-1.29) and (3-1.30), gives the following relationships between the Cauchy and Piola tensors, and the Finger and Green tensors:

$$C = B_\tau^{-1}(t) \qquad (3\text{-}1.31)$$

$$C^{-1} = B_\tau(t) \qquad (3\text{-}1.32)$$

Finally, from eq. (3-1.5), the following equations are obtained:

$$R(t) = U(t) = V(t) = C(t) = B(t) = 1 \qquad (3\text{-}1.33)$$

When the material has undergone a rigid-body motion between times τ and t, all the tensors considered in this section, with the exception of F and R, coincide with the unit tensor. In the analysis of certain problems, it is convenient to make use of tensors which, for rigid-body motions, reduce to the zero tensor. Therefore, additional tensors (often called strain tensors) have been used in the literature; of these we shall consider only the Cauchy strain tensor G and the Finger strain tensor H:

$$G = C - 1; \qquad H = C^{-1} - 1 \qquad (3\text{-}1.34)$$

† When materials are considered whose density is not constant, the right-hand side of eq. (3-1.29) should be multiplied by III_C. The distinction is irrelevant for constant density materials.

Calculation of components

Let us now indicate the procedure by which the components of the Cauchy and Finger tensors can be calculated from a knowledge of the motion. With reference to a coordinate system x^i let the three scalar components of eq. (3-1.1) be

$$x^i(\tau) = f^i(x_t^j, \tau) \qquad (3\text{-}1.35)$$

where $x^i(\tau)$ are the coordinates of $\mathbf{X}$ and x_t^j the coordinates of $\mathbf{X}_t$. We have

$$d\mathbf{X}_t = \mathbf{e}_j(\mathbf{X}_t)\,dx_t^j \qquad (3\text{-}1.36)$$

$$d\mathbf{X} = \mathbf{e}_i(\mathbf{X})\,dx^i \qquad (3\text{-}1.37)$$

Substitution of eqs (3-1.36) and (3-1.37) into eq. (3-1.3) gives

$$\mathbf{e}_i(\mathbf{X})\,dx^i = \mathbf{F} \cdot [\mathbf{e}_j(\mathbf{X}_t)\,dx_t^j] \qquad (3\text{-}1.38)$$

In the above equations, $\mathbf{e}_j(\mathbf{X}_t)$ is the natural basis at $\mathbf{X}_t$ while $\mathbf{e}_i(\mathbf{X})$ is the natural basis at $\mathbf{X}$. The two bases are, in general, different.

Taking the scalar product of eq. (3-1.38) with $\mathbf{e}^k(\mathbf{X})$, we obtain

$$\mathbf{e}^k(\mathbf{X}) \cdot \mathbf{e}_i(\mathbf{X})\,dx^i = \delta_i^k\,dx^i = dx^k$$
$$= [\mathbf{e}^k(\mathbf{X}) \cdot \mathbf{F} \cdot \mathbf{e}_j(\mathbf{X}_t)]\,dx_t^j \qquad (3\text{-}1.39)$$

But, from eq. (3-1.35),

$$dx^k = \frac{\partial x^k}{\partial x_t^j}\,dx_t^j \qquad (3\text{-}1.40)$$

Comparison of eqs (3-1.39) and (3-1.40) shows that

$$\frac{\partial x^k}{\partial x_t^j} = \mathbf{e}^k(\mathbf{X}) \cdot \mathbf{F} \cdot \mathbf{e}_j(\mathbf{X}_t) \qquad (3\text{-}1.41)$$

or, equivalently,

$$\frac{\partial x^m}{\partial x_t^i} = \mathbf{e}_i(\mathbf{X}_t) \cdot \mathbf{F}^{\mathrm{T}} \cdot \mathbf{e}^m(\mathbf{X}) \qquad (3\text{-}1.42)$$

In general, the right-hand sides of eqs (3-1.41) and (3-1.42) *are not* the mixed components of $\mathbf{F}$ and $\mathbf{F}^{\mathrm{T}}$, because the base vectors are taken at two different places, $\mathbf{X}$ and $\mathbf{X}_t$.

Let us now consider the covariant components at $\mathbf{X}_t$ of the Cauchy tensor. By definition,

$$C_{ij} = \mathbf{e}_i(\mathbf{X}_t) \cdot \mathbf{C} \cdot \mathbf{e}_j(\mathbf{X}_t) \qquad (3\text{-}1.43)$$

Substitution of eq. (3-1.19) gives

$$\mathbf{C}_{ij} = \mathbf{e}_i(\mathbf{X}_t) \cdot \mathbf{F}^{\mathrm{T}} \cdot \mathbf{F} \cdot \mathbf{e}_j(\mathbf{X}_t) \qquad (3\text{-}1.44)$$

and, considering that the sum of dyads $\mathbf{e}^k\mathbf{e}_k$ or $\mathbf{e}^m\mathbf{e}_m$ equals the unit tensor:

$$\mathbf{C}_{ij} = \mathbf{e}_i(\mathbf{X}_t) \cdot \mathbf{F}^{\mathrm{T}} \cdot \mathbf{e}^m(\mathbf{X})\mathbf{e}_m(\mathbf{X}) \cdot \mathbf{e}_k(\mathbf{X})\mathbf{e}^k(\mathbf{X}) \cdot \mathbf{F} \cdot \mathbf{e}_j(\mathbf{X}_t) \qquad (3\text{-}1.45)$$

Substitution of eqs (3-1.41), (3-1.42) and (1-3.32) into eq. (3-1.45) yields

$$\mathbf{C}_{ij} = \frac{\partial x^m}{\partial x_t^i} \cdot \frac{\partial x^k}{\partial x_t^j} g_{km} \qquad (3\text{-}1.46)$$

where g_{km} is the metric at $\mathbf{X}$, while $\mathbf{C}_{ij}$ are components at $\mathbf{X}_t$. Equation (3-1.46) can be used directly to calculate the covariant components of $\mathbf{C}$. The Finger tensor's contravariant components may be obtained either by inverting the matrix $\mathbf{C}_{ij}$ or by the following equation, which is obtained by a similar procedure:

$$(\mathbf{C}^{-1})^{ij} = \frac{\partial x_t^i}{\partial x^m} \cdot \frac{\partial x_t^j}{\partial x^k} g^{km} \qquad (3\text{-}1.47)$$

Of course, in order to use eq. (3-1.47), the motion equations need first to be solved for the x_t^i's.

3-2   HISTORIES. TIME DERIVATIVES. RATES OF STRAIN

Histories

When one considers a quantity which is a function of time, it may be desirable to limit one's attention to the values of the quantity at all times *preceding* the instant of observation t, i.e., to consider only the past. For example, let us focus attention on the temperature of a material point, which is in general a function of time $T(\tau)$. (We shall discuss temperature in more detail in the next chapter.) If one considers the material point at some instant of observation t where the temperature is $T(t)$, there may be an interest in the entire past *history* of temperature, say $T(\tau)$ at $\tau \leq t$. Moreover, it will be shown that what is physically relevant is how long ago had any temperature been assumed, rather than at which particular instant of absolute time it has been assumed. Mathematically, this is obtained by a change of variable: the time lag, $s = t - \tau$, is introduced as the new independent variable.

Given an instant of observation t, and a time-dependent quantity $\psi(\tau)$ (the values of which may be scalars, vectors or tensors), a new function $\psi^t(s)$ can be defined as

$$\psi^t(s) = \psi(\tau) \qquad (3\text{-}2.1)$$

Equation (3-2.1) should be understood as follows: the *value* of the function $\psi(\ )$ at any given τ coincides with the *value* of the function $\psi^t(\ )$ at $s = t - \tau$. The function $\psi^t(s)$ for $s \geq 0$ is called a *history*; obviously, the functional form of a history depends on the choice of the instant of observation. In fact, upon changing the instant of observation, the time lag corresponding to a given instant in the past changes, i.e., one has a shift of the time lag axis. When the values of the function $\psi^t(s)$ are relative tensors, a change in the instant of observation implies not only a time lag shift, but also a change of reference, and thus there is an additional reason for which the form of $\psi^t(s)$ depends on t.

Since quantities with a t superscript will *always* be understood to be functions of the time lag, we will drop the explicit indication of the s dependency and simply write ψ^t when no confusion arises.

In the rest of this section, time differentiation of time-dependent quantities will be considered. In most cases, derivatives $\dot{\psi}(\tau)$ with respect to τ of some function $\psi(\tau)$ are calculated; the derivatives $\dot{\psi}^t(s)$ with respect to s of the history $\psi^t(s)$, for corresponding values of s and τ, have the opposite value:
$\dot{\psi}^t(s) = -\dot{\psi}(\tau)$.

Time differentiation of relative quantities

A trivial but possibly misleading point arises in connection with time differentiation of relative quantities, i.e., quantities defined with a reference taken at the instant of observation. Given a relative quantity $\psi(\tau)$, where the absence of any subscript implies that the instant of observation t has been taken as a reference, the derivative $\dot{\psi}(\tau)$ is still a function of τ. If we now set $\tau = t$ we obtain a particular value of the derivative:

$$\dot{\psi}(t) = \dot{\psi}|_{\tau=t} \qquad (3\text{-}2.2)$$

The quantity $\dot{\psi}(t)$ depends of course on the *choice* of the instant of observation, and may thus be regarded as a function of time $\phi(t)$. Two points must be noted: first, $\phi(t)$ *is not* a relative quantity, although $\dot{\psi}$ is; and $\dot{\phi}(t)$ *is not* the second derivative of ψ at $\tau = t$; in fact,

$$\dot{\phi}(t) = \frac{\mathrm{d}}{\mathrm{d}t}[\dot{\psi}(t)] \qquad (3\text{-}2.3)$$

while

$$\ddot{\psi}(t) = \left[\frac{d^2}{d\tau^2}\psi(\tau)\right]_{\tau=t} \qquad (3\text{-}2.4)$$

and, since the form of the function $\psi(\tau)$ depends on the choice of t, the right-hand side of eq. (3-2.3) has not the same value as the right-hand side of eq. (3-2.4).

Velocity gradient

Let **Y** indicate the point $\mathbf{X} + d\mathbf{X}$, so that eq. (3-1.3) may be written as

$$\mathbf{Y} - \mathbf{X} = \mathbf{F} \cdot (\mathbf{Y}_t - \mathbf{X}_t) \qquad (3\text{-}2.5)$$

Let us now differentiate eq. (3-2.5) with respect to τ; of course, both the instant of observation t and the material points $\mathbf{Y}_t$ and $\mathbf{X}_t$ are kept constant. We have

$$\dot{\mathbf{Y}} - \dot{\mathbf{X}} = \dot{\mathbf{F}} \cdot (\mathbf{Y}_t - \mathbf{X}_t) \qquad (3\text{-}2.6)$$

where $\dot{\mathbf{Y}}$ and $\dot{\mathbf{X}}$ are, because of eq. (3-1.2), the velocities at **Y** and **X**, respectively. Remembering that $\mathbf{Y} = \mathbf{X} + d\mathbf{X}$, we have

$$d\mathbf{v}(\tau) = \dot{\mathbf{F}} \cdot d\mathbf{X}_t = \dot{\mathbf{F}} \cdot \mathbf{F}^{-1} \cdot d\mathbf{X} \qquad (3\text{-}2.7)$$

where $d\mathbf{v}(\tau)$ is the velocity difference at time τ of the two material points $\mathbf{X}_t$ and $\mathbf{X}_t + d\mathbf{X}_t$. Setting $\tau = t$ in eq. (3-2.7), we have

$$d\mathbf{v}(t) = \dot{\mathbf{F}}(t) \cdot d\mathbf{X}_t \qquad (3\text{-}2.8)$$

which, by comparison with the definition of $\nabla\mathbf{v}$, shows that

$$\nabla\mathbf{v} = \dot{\mathbf{F}}(t) \qquad (3\text{-}2.9)$$

The velocity gradient, $\nabla\mathbf{v}$, is in general obtained from a spatial description of flow, i.e., from a knowledge of the velocity field

$$\mathbf{v} = \mathbf{v}(\mathbf{X}, t) \qquad (3\text{-}2.10)$$

Thus, $\nabla\mathbf{v}$ is itself a field, say:

$$\nabla\mathbf{v} = \nabla\mathbf{v}(\mathbf{X}, t) \qquad (3\text{-}2.11)$$

If one wishes to investigate the changes of $\nabla\mathbf{v}$ along the path of a particle, one needs to introduce the time-dependent tensor **L**, defined as

$$\mathbf{L}(\tau) = \nabla\mathbf{v}(\mathbf{f}[\mathbf{X}_t, \tau], \tau) \qquad (3\text{-}2.12)$$

From eqs (3-2.7) and (3-2.12) one obtains

$$L(\tau) = \dot{F} \cdot F^{-1} \qquad (3\text{-}2.13)$$

Note that, although the right-hand side of eq. (3-2.13) is the product of two relative tensors, L is not a relative tensor. In fact, suppose one changes the instant of observation from t to t'. By differentiating eq. (3-1.8), one obtains

$$\dot{F}_{t'} = \dot{F} \cdot F_{t'}(t) \qquad (3\text{-}2.14)$$

and thus, by inverting both sides of eq. (3-1.8),

$$\begin{aligned}
\dot{F}_{t'} \cdot F_{t'}^{-1} &= \dot{F} \cdot F_{t'}(t) \cdot F_{t'}^{-1}(t) \cdot F^{-1} \\
&= \dot{F} \cdot F^{-1}
\end{aligned} \qquad (3\text{-}2.15)$$

which shows that the right-hand side of eq. (3-2.13) is independent of the reference.

Rates of strain

By differentiating eq. (3-1.19) with respect to τ, one obtains

$$\dot{C} = \dot{F}^{T} \cdot F + F^{T} \cdot \dot{F} \qquad (3\text{-}2.16)$$

Setting $\tau = t$ in eq. (3-2.16), and considering eq. (3-2.9), we have

$$\dot{C}(t) = \nabla v + \nabla v^{T} = 2D \qquad (3\text{-}2.17)$$

Furthermore, we have

$$\dot{C} = \overline{\dot{U^2}} = \dot{U} \cdot U + U \cdot \dot{U} \qquad (3\text{-}2.18)$$

Again, setting $\tau = t$, we have

$$\dot{C}(t) = 2\dot{U}(t) \qquad (3\text{-}2.19)$$

Equations (3-2.17) and (3-2.19) show that the stretching tensor D is the rate of stretch at the instant of observation, a concept which was made use of in chapter 2. Of course, considering eq. (3-1.34), D is also identified as the rate of strain.

Let us now differentiate eq. (3-1.11):

$$\dot{F} = \dot{R} \cdot U + R \cdot \dot{U} \qquad (3\text{-}2.20)$$

and again, setting $\tau = t$,

$$\nabla v = \dot{R}(t) + \dot{U}(t) \qquad (3\text{-}2.21)$$

Comparison of eqs (3-2.17), (3-2.19), and (1-9.6) shows that†

$$\mathbf{W} = \dot{\mathbf{R}}(t) \qquad (3\text{-}2.22)$$

which identifies the vorticity $\mathbf{W}$ with the rate of rotation at the instant of observation.

Higher rates of strain

Let us now introduce two types of higher rate-of-strain tensors which are widely used in the literature: the Rivlin–Ericksen tensors and the White–Metzner tensors. We shall use the following notation:

$$\overset{N}{\psi} = \frac{d^N}{d\tau^N}\psi \qquad (3\text{-}2.23)$$

so that

$$\overset{1}{\psi} = \dot{\psi}; \qquad \overset{2}{\psi} = \ddot{\psi} \qquad (3\text{-}2.24)$$

where ψ is a scalar-, vector-, or tensor-valued function of the single scalar variable τ.

Let us calculate the Nth derivative with respect to time of the squared distance between two neighboring points, evaluated at $\tau = t$:

$$\overline{\overset{N}{dx^2}}(t) = \overline{\overset{N}{dx^2}}\Big|_{\tau=t} = d\mathbf{X}_t \cdot \overset{N}{\mathbf{C}}(t) \cdot d\mathbf{X}_t$$
$$= d\mathbf{X}_t \cdot \mathbf{A}_N \cdot d\mathbf{X}_t \qquad (3\text{-}2.25)$$

which defines the Nth Rivlin–Ericksen tensor $\mathbf{A}_N$. Of course, $\mathbf{A}_N$ is a function of t, but is not a relative tensor, although $\overset{N}{\mathbf{C}}$ is. We have

$$\mathbf{A}_N(t) = \overset{N}{\mathbf{C}}(t) \qquad (3\text{-}2.26)$$

From the definition (eq. (3-2.25)), we have

$$\mathbf{A}_0 = \mathbf{1} \qquad (3\text{-}2.27)$$

while comparison of eqs (3-2.17) and (3-2.26) shows that

$$\mathbf{A}_1 = 2\mathbf{D} \qquad (3\text{-}2.28)$$

---

† That $\dot{\mathbf{R}}(t)$ is skew can be seen by differentiating $\mathbf{R} \cdot \mathbf{R}^T = \mathbf{1}$ and considering eq. (3-1.33). At the same time, $\dot{\mathbf{U}}$ is symmetric because $\mathbf{U}$ is also symmetric. The decomposition of a tensor into the sum of a skew and a symmetric tensor is unique; hence eq. (3-2.21) identifies *both* $\dot{\mathbf{U}}(t)$ and $\dot{\mathbf{R}}(t)$ with $\mathbf{D}$ and $\mathbf{W}$.

A recurrence formula for the Rivlin–Ericksen tensors is obtained from their definition:

$$\mathbf{A}_{N+1} = \dot{\mathbf{A}}_N + \nabla \mathbf{v}^T \cdot \mathbf{A}_N + \mathbf{A}_N \cdot \nabla \mathbf{v} \qquad (3\text{-}2.29)$$

In actual calculations, eq. (3-2.26) is easier to use than eq. (3-2.29), but the latter is encountered more frequently in the literature.

The White–Metzner tensors are obtained in a similar way, starting from eq. (3-1.29) rather than (3-1.24). The definition is

$$\overset{N}{\overline{ds^2}}(t) = -d\mathbf{s}_t \cdot \mathbf{B}_N \cdot d\mathbf{s}_t \qquad (3\text{-}2.30)$$

from which one has directly

$$\mathbf{B}_N = -\overset{N}{\overline{\mathbf{C}^{-1}}}(t) \qquad (3\text{-}2.31)$$

The recurrence formula is obtained as

$$\mathbf{B}_{N+1} = \dot{\mathbf{B}}_N - \mathbf{B}_N \cdot \nabla \mathbf{v}^T - \nabla \mathbf{v} \cdot \mathbf{B}_N \qquad (3\text{-}2.32)$$

We have

$$\mathbf{B}_0 = -\mathbf{1} \qquad (3\text{-}2.33)$$

$$\mathbf{B}_1 = 2\mathbf{D} = \mathbf{A}_1 \qquad (3\text{-}2.34)$$

and, in general,

$$\mathbf{B}_N \neq \mathbf{A}_N \qquad (3\text{-}2.35)$$

Of course, analogous equations could be obtained from derivatives of $\mathbf{B}$ and $\mathbf{B}^{-1}$, but these tensors, for reasons to be discussed in detail in chapter 4, will not be used.

The usefulness of the tensors $\mathbf{A}_N$ and $\mathbf{B}_N$ lies in the possibility of expanding the histories of the Cauchy and Finger tensors in a power series about the instant of observation. Under sufficient conditions of smoothness, we have

$$\mathbf{C}^t = \sum_{N=0}^{\infty} (-1)^N \frac{1}{N!} \mathbf{A}_N s^N \qquad (3\text{-}2.36)$$

$$(\mathbf{C}^t)^{-1} = \sum_{N=0}^{\infty} (-1)^{N+1} \frac{1}{N!} \mathbf{B}_N s^N \qquad (3\text{-}2.37)$$

where $s = t - \tau$ is, as already stated, time measured backwards from the instant of observation, and the superscript t implies that s is the independent variable.

A memory fluid, discussed in section 2-6, may be sensitive to past deformations, i.e., in some sense to be defined rigorously in chapter 4, the stress at time t may depend on the entire history of the Cauchy, or Finger, tensor. Equations (3-2.36) and (3-2.37) allow these histories to be expressed in terms of kinematic tensors ($\mathbf{A}_N$ and $\mathbf{B}_N$) measured at the instant of observation.

3-3 TRANSFORMATION OF TENSORS AND THEIR TIME DERIVATIVES UNDER A CHANGE OF FRAME

In this section we wish to investigate the rules of transformation of tensors and their time derivatives under a change of frame. The orthogonal tensor $\mathbf{Q}(t)$ is the one describing the change of frame in the sense of section 1-5.

Transformation of kinematic tensors

Inasmuch as $d\mathbf{X}$ is a geometrical vector, its rule of transformation is

$$d\mathbf{X}^* = \mathbf{Q}(\tau) \cdot d\mathbf{X} \qquad (3\text{-}3.1)$$

In eq. (3-3.1), and later, we consider a change of frame with no shift of the time origin, so that $\tau^* = \tau$. Any extension to the case where $\tau^* = \tau + a$ can be considered trivial, and all the results below also hold in the general case.

From eq. (3-1.3),

$$d\mathbf{X}^* = \mathbf{Q}(\tau) \cdot \mathbf{F} \cdot d\mathbf{X}_t = \mathbf{Q}(\tau) \cdot \mathbf{F} \cdot \mathbf{Q}^T(t) \cdot d\mathbf{X}_t^* \qquad (3\text{-}3.2)$$

so that the rule of transformation of the deformation gradient is obtained:

$$\mathbf{F}^* = \mathbf{Q}(\tau) \cdot \mathbf{F} \cdot \mathbf{Q}^T(t) \qquad (3\text{-}3.3)$$

and, similarly,

$$\mathbf{F}^{*-1} = \mathbf{Q}(t) \cdot \mathbf{F}^{-1} \cdot \mathbf{Q}^T(\tau) \qquad (3\text{-}3.4)$$

The rules of transformation for $\mathbf{R}$ and $\mathbf{U}$ are obtained as follows: eq. (3-3.3) can be written as

$$\mathbf{F}^* = \mathbf{R}^* \cdot \mathbf{U}^* = \mathbf{Q}(\tau) \cdot \mathbf{R} \cdot \mathbf{U} \cdot \mathbf{Q}^T(t) \qquad (3\text{-}3.5)$$

which, $\mathbf{Q}$ being orthogonal, can be written as

$$\mathbf{F}^* = \mathbf{R}^* \cdot \mathbf{U}^* = [\mathbf{Q}(\tau) \cdot \mathbf{R} \cdot \mathbf{Q}^T(t)] \cdot [\mathbf{Q}(t) \cdot \mathbf{U} \cdot \mathbf{Q}^T(t)] \qquad (3\text{-}3.6)$$

The term within the first bracket on the extreme right-hand side of eq. (3-3.6) is an orthogonal tensor; the term within the second bracket is a symmetric positive-definite tensor. But the polar decomposition of $\mathbf{F}$ is unique, and therefore

$$\mathbf{R}^* = \mathbf{Q}(\tau) \cdot \mathbf{R} \cdot \mathbf{Q}^T(t) \qquad (3\text{-}3.7)$$

$$\mathbf{U}^* = \mathbf{Q}(t) \cdot \mathbf{U} \cdot \mathbf{Q}^T(t) \qquad (3\text{-}3.8)$$

The transformation rules for $\mathbf{V}$, $\mathbf{C}$, $\mathbf{C}^{-1}$, $\mathbf{B}$, and $\mathbf{B}^{-1}$ are now obtained immediately from their definitions:

$$\mathbf{V}^* = \mathbf{Q}(\tau) \cdot \mathbf{V} \cdot \mathbf{Q}^T(\tau) \qquad (3\text{-}3.9)$$

$$\mathbf{C}^* = \mathbf{Q}(t) \cdot \mathbf{C} \cdot \mathbf{Q}^T(t) \qquad (3\text{-}3.10)$$

$$\mathbf{C}^{*-1} = \mathbf{Q}(t) \cdot \mathbf{C}^{-1} \cdot \mathbf{Q}^T(t) \qquad (3\text{-}3.11)$$

$$\mathbf{B}^* = \mathbf{Q}(\tau) \cdot \mathbf{B} \cdot \mathbf{Q}^T(\tau) \qquad (3\text{-}3.12)$$

$$\mathbf{B}^{*-1} = \mathbf{Q}(\tau) \cdot \mathbf{B}^{-1} \cdot \mathbf{Q}^T(\tau) \qquad (3\text{-}3.13)$$

Let us consider some relative tensor $\psi(\tau)$ which has its reference at the instant of observation t. $\psi(\tau)$ is, of course, defined with respect to a given frame of reference. To a different 'observer,' the same physical quantity will appear as a different function, $\psi^*(\tau)$. The relationship between ψ and ψ^* may involve either $\mathbf{Q}(\tau)$ or $\mathbf{Q}(t)$ or both. We choose to define as 'indifferent' a relative tensor if the transformation $\psi \to \psi^*$ only involves $\mathbf{Q}(t)$, i.e., the relative orientation of the frames at the instant of observation. In this sense, eqs (3-3.7)–(3-3.13) show that $\mathbf{U}$ and $\mathbf{C}$ are indifferent, while $\mathbf{F}$, $\mathbf{R}$, $\mathbf{V}$, and $\mathbf{B}$ are not. Of course, if any tensor $\mathbf{A}$ is indifferent, so also is $\mathbf{A}^{-1}$, and therefore the Finger tensor is also indifferent, as shown directly by eq. (3-3.11).

Transformation of derivatives and integrals

We now wish to investigate the transformation rules for derivatives and integrals of time-dependent non-relative tensors. Let $\mathbf{J}$ be any time-dependent tensor which is indifferent in the sense that

$$\mathbf{J}^*(t) = \mathbf{Q}(t) \cdot \mathbf{J}(t) \cdot \mathbf{Q}^T(t) \qquad (3\text{-}3.14)$$

When eq. (3-3.14) is differentiated with respect to time, one obtains

$$\dot{\mathbf{J}}^*(t) = \dot{\mathbf{Q}}(t) \cdot \mathbf{J}(t) \cdot \mathbf{Q}(t) + \mathbf{Q}(t) \cdot \dot{\mathbf{J}}(t) \cdot \mathbf{Q}(t) + \mathbf{Q}(t) \cdot \mathbf{J}(t) \cdot \dot{\mathbf{Q}}(t) \qquad (3\text{-}3.15)$$

which, of course, is not the rule of transformation of an indifferent tensor. One thus has in general: *The time derivative of a time-dependent indifferent non-relative tensor is not indifferent.*

Let us now consider the weighed integral of $\mathbf{J}$ over a time interval from some fixed lower limit t_0 (frequently t_0 will be taken as $-\infty$) up to the instant of observation t. If $f(\tau)$ is any scalar-valued weighing function, we have

$$\left[\int_{t_0}^{t} f(\tau)\mathbf{J}(\tau)\,d\tau \right]^* = \int_{t_0}^{t} f(\tau)\mathbf{Q}(\tau) \cdot \mathbf{J}(\tau) \cdot \mathbf{Q}^{\mathrm{T}}(\tau)\,d\tau$$

$$\neq \mathbf{Q}(t) \cdot \int_{t_0}^{t} f(\tau)\mathbf{J}(\tau)\,d\tau \cdot \mathbf{Q}^{\mathrm{T}}(t)$$

(3-3.16)

The integral considered is a function of t (not of τ, because τ is now a dummy variable), which is not indifferent. Indeed, the expression following the sign $\neq$ in eq. (3-3.16) is the one which should hold if the integral were indifferent.

Thus again: *The weighed integral up to time t of an indifferent non-relative time-dependent tensor is not indifferent.*

In contrast, time derivatives at $\tau = t$ of indifferent relative tensors are indifferent. In fact, if $\mathbf{J}$ is an indifferent relative tensor, its transformation rule, according to the discussion following eq. (3-3.13), is

$$\mathbf{J}^* = \mathbf{Q}(t) \cdot \mathbf{J} \cdot \mathbf{Q}^{\mathrm{T}}(t) \qquad (3\text{-}3.17)$$

The derivative of $\mathbf{J}$ with respect to τ obeys

$$\dot{\mathbf{J}}^* = \mathbf{Q}(t) \cdot \dot{\mathbf{J}} \cdot \mathbf{Q}^{\mathrm{T}}(t) \qquad (3\text{-}3.18)$$

and, setting $\tau = t$,

$$\dot{\mathbf{J}}^*(t) = \mathbf{Q}(t) \cdot \dot{\mathbf{J}}(t) \cdot \mathbf{Q}^{\mathrm{T}}(t) \qquad (3\text{-}3.19)$$

which is the rule of transformation of an indifferent time-dependent tensor. The same applies to weighed integrals of indifferent histories.

Associated relative tensors and associated derivatives

In view of the results in the preceding subsection, one may wish to associate to a time-dependent indifferent non-relative tensor $\mathbf{J}$ an indifferent relative tensor whose value at $\tau = t$ coincides with $\mathbf{J}$. We shall call such a relative tensor an 'associated relative tensor.' This may be done in several ways. We begin by introducing the three most important associated relative tensors, viz.:

(a) the 'corotated form of $\mathbf{J}$':

$$\tilde{\mathbf{J}}(\tau) = \mathbf{R}^{\mathrm{T}} \cdot \mathbf{J}(\tau) \cdot \mathbf{R} \qquad (3\text{-}3.20)$$

(b) the 'lower convected form of $\mathbf{J}$':

$$\dot{\mathbf{J}}(\tau) = \mathbf{F}^{\mathrm{T}} \cdot \mathbf{J}(\tau) \cdot \mathbf{F} \qquad (3\text{-}3.21)$$

(c) the 'upper convected form of $\mathbf{J}$':

$$\overset{\circ}{\mathbf{J}}(\tau) = \mathbf{F}^{-1} \cdot \mathbf{J}(\tau) \cdot (\mathbf{F}^{-1})^{\mathrm{T}} \qquad (3\text{-}3.22)$$

The three relative tensors so defined are all indifferent if $\mathbf{J}$ is indifferent, as can be seen by simply substituting the rules of transformation for $\mathbf{R}$, $\mathbf{F}$, and $\mathbf{J}$ in the definitions. Moreover, by setting $\tau = t$ in eqs (3-3.20)–(3-3.22), we have

$$\tilde{\mathbf{J}}(t) = \dot{\mathbf{J}}(t) = \overset{\circ}{\mathbf{J}}(t) = \mathbf{J}(t) \qquad (3\text{-}3.23)$$

The derivatives with respect to τ, evaluated at $\tau = t$, of the associated relative tensors are also indifferent, in view of eq. (3-3.19) which holds for *any* indifferent relative tensor. We shall call these derivatives 'associated derivatives.' We thus define three associated derivatives of $\mathbf{J}$, viz.:

(a) the corotational derivative:

$$\overset{\circ}{\mathbf{J}}_N = \overset{N}{\tilde{\mathbf{J}}}(t) \qquad (3\text{-}3.24)$$

(b) the lower convected derivative:

$$\overset{\wedge}{\mathbf{J}}_N = \overset{N}{\dot{\mathbf{J}}}(t) \qquad (3\text{-}3.25)$$

(c) the upper convected derivative:

$$\overset{\triangledown}{\mathbf{J}}_N = \overset{N}{\overset{\circ}{\mathbf{J}}}(t) \qquad (3\text{-}3.26)$$

The three types of associated derivatives defined above share the same property: they are time-dependent indifferent non-relative tensors if $\mathbf{J}$ is a time-dependent indifferent non-relative tensor.

A question of nomenclature is involved here. The corotational derivative is often also called the Jaumann derivative, and is indicated by the symbol $\mathscr{D}/\mathscr{D}t$. The two convected derivatives are also called Oldroyd derivatives, and are *both* indicated by the symbol $\delta/\delta t$; this notation is only used in connection with indicial notation, and the convention is adopted that the lower convected derivative is intended when covariant components are considered, and the upper convected derivative when contravariant components are considered, so that

$$\frac{\delta^N}{\delta t^N} \mathbf{J}_{ij} = (\overset{\wedge}{\mathbf{J}}_N)_{ij} \qquad (3\text{-}3.27)$$

$$\frac{\delta^N}{\delta t^N} \mathbf{J}^{ij} = (\overset{\triangledown}{\mathbf{J}}_N)^{ij} \qquad (3\text{-}3.28)$$

The term 'convected rate,' with the same notation as here (a superimposed $\triangle$) is used by Truesdell and Noll [2] for the lower convected derivative. These authors do not consider explicitly the upper convected derivative, and use the same notation as here for the corotational derivative.

Needless to say, weighed integrals of indifferent associated relative tensors between a fixed lower limit t_0 and the instant of observation t are also indifferent.

It should be remembered that the associated relative tensors, although their value coincides with $\mathbf{J}$ at $\tau = t$, are not to be confused with the tensor $\mathbf{J}$ itself; while the latter is defined in terms of physically relevant quantities, no direct physical interpretation of the associated relative tensors can often be found, with the exception of the corotated form.

Physical interpretation of the corotated form

Consider a change of frame identified by the orthogonal tensor function $\mathbf{Q}(\tau)$ obeying the equation

$$\mathbf{Q}(\tau) = \mathbf{R}^T \qquad (3\text{-}3.29)$$

The new frame is not rotated with respect to the original one at the instant of observation, i.e., $\mathbf{Q}(t) = \mathbf{1}$. At any other time τ, it is rotated in the same way as the material element surrounding the point considered. In fact, applying eq. (3-3.7), we have:

$$\mathbf{R}^* = \mathbf{R}^T \cdot \mathbf{R} = \mathbf{1} \qquad (3\text{-}3.30)$$

which shows that the material element considered does not rotate with respect to the new frame. Such a frame is called a corotating frame.

Equations (3-3.14) and (3-3.20) show immediately that $\mathbf{\bar{J}}$ is simply, the time-dependent tensor $\mathbf{J}(t)$ as observed in the corotating frame, and the corotational derivatives are the time derivatives as observed in the corotating frame. Of course, no analogous interpretation can be given to the convected forms and the convected derivatives, because $\mathbf{F}$ is not orthogonal.

Associated derivatives of the unit tensor and of stress

Let us now apply the concepts discussed so far to the simplest example of indifferent tensor, i.e., the unit tensor (which may be regarded as a function of time, although, of course, the value is constant). From eqs (3-3.21) and (3-3.22) one obtains

$$\overset{\triangle}{\mathbf{1}} = \mathbf{C} \qquad (3\text{-}3.31)$$

$$\overset{\triangledown}{\mathbf{1}} = \mathbf{C}^{-1} \qquad (3\text{-}3.32)$$

which allows us to identify the convected forms of the unit tensor with the Cauchy and Finger tensors. Also,

$$\hat{\mathbf{I}}_N = \mathbf{A}_N \qquad (3\text{-}3.33)$$

$$\overset{\triangledown}{\mathbf{I}}_N = -\mathbf{B}_N \qquad (3\text{-}3.34)$$

which show that the Rivlin–Ericksen and the White–Metzner tensors are the convected derivatives of the unit tensor.

Let us now apply the same concepts to another indifferent time-dependent tensor, i.e., the total stress tensor **T**. We obtain:

(a) a corotated form of stress:

$$\bar{\mathbf{T}}(\tau) = \mathbf{R}^{\mathrm{T}} \cdot \mathbf{T}(\tau) \cdot \mathbf{R} \qquad (3\text{-}3.35)$$

(b) a lower convected form of stress:

$$\overset{\downarrow}{\mathbf{T}}(\tau) = \mathbf{F}^{\mathrm{T}} \cdot \mathbf{T}(\tau) \cdot \mathbf{F} \qquad (3\text{-}3.36)$$

which is called sometimes the 'convected stress' [2], indicated by the symbol $\bar{\mathbf{T}}$; and

(c) an upper convected form of stress:

$$\overset{\uparrow}{\mathbf{T}}(\tau) = \mathbf{F}^{-1} \cdot \mathbf{T}(\tau) \cdot (\mathbf{F}^{-1})^{\mathrm{T}} \qquad (3\text{-}3.37)$$

which is a generalized version of what is sometimes called the second Piola–Kirchhoff stress tensor [3]. Two convected derivatives and a corotational derivative of stress are also defined.

By a similar argument to that leading to eqs (3-2.29) and (3.2.32), it can be shown that

$$\overset{\circ}{\mathbf{J}}_{N+1} = \overset{\cdot}{\mathbf{J}}_N - \mathbf{W} \cdot \overset{\circ}{\mathbf{J}}_N + \overset{\circ}{\mathbf{J}}_N \cdot \mathbf{W} \qquad (3\text{-}3.38)$$

$$\overset{\wedge}{\mathbf{J}}_{N+1} = \overset{\cdot}{\mathbf{J}}_N + \overset{\wedge}{\mathbf{J}}_N \cdot \nabla\mathbf{v} + \nabla\mathbf{v}^{\mathrm{T}} \cdot \overset{\wedge}{\mathbf{J}}_N \qquad (3\text{-}3.39)$$

$$\overset{\triangledown}{\mathbf{J}}_{N+1} = \overset{\cdot}{\mathbf{J}}_N - \overset{\triangledown}{\mathbf{J}}_N \cdot \nabla\mathbf{v}^{\mathrm{T}} - \nabla\mathbf{v} \cdot \overset{\triangledown}{\mathbf{J}}_N \qquad (3\text{-}3.40)$$

The associated relative tensors and derivatives defined above are not the only possible indifferent ones. Two more examples are easily obtained, and precisely:

(a) the 'left convected form of **J**':

$$\bar{\mathbf{J}} = \mathbf{F}^{-1} \cdot \mathbf{J}(\tau) \cdot \mathbf{F} \qquad (3\text{-}3.41)$$

which gives rise to the left convected derivative

$$\overset{\triangledown}{\mathbf{J}}_N = \overset{N}{\bar{\mathbf{J}}}(t) \qquad (3\text{-}3.42)$$

(b) the 'right convected form of $\mathbf{J}$:'

$$\vec{\mathbf{J}} = \mathbf{F}^{\mathrm{T}} \cdot \mathbf{J}(\tau) \cdot (\mathbf{F}^{-1})^{\mathrm{T}} \qquad (3\text{-}3.43)$$

which gives rise to the right convected derivative

$$\mathring{\mathbf{J}}_N = \overset{N}{\vec{\mathbf{J}}}(t) \qquad (3\text{-}3.44)$$

Also these two additional convected derivatives are sometimes indicated in the literature by the same symbol $\delta^N/\delta t^N$, and the convention is adopted that the left convected derivative is intended when the left mixed components are considered, and the right convected derivative is intended when the right mixed components are considered.† Thus:

$$\frac{\delta^N}{\delta t^N}J^i{}_j = (\mathring{\mathbf{J}}_N)^i{}_j \qquad (3\text{-}3.45)$$

$$\frac{\delta^N}{\delta t^N}J_j{}^i = (\mathring{\mathbf{J}}_N)_j{}^i \qquad (3\text{-}3.46)$$

A further consideration on the associated derivatives is in order. From eqs (3-3.38)–(3-3.40), we obtain

$$\mathring{\mathbf{J}}_1 = \mathring{\mathbf{J}}_1 + \mathbf{J} \cdot \mathbf{D} + \mathbf{D} \cdot \mathbf{J} \qquad (3\text{-}3.47)$$

$$\overset{\triangledown}{\mathbf{J}}_1 = \mathring{\mathbf{J}}_1 - \mathbf{J} \cdot \mathbf{D} - \mathbf{D} \cdot \mathbf{J} \qquad (3\text{-}3.48)$$

The corotational, upper convected, and lower convected derivative of a symmetric tensor are symmetric. In contrast with this, the left and right convected derivatives, as well as the left and right convected histories of a symmetric tensor, are not symmetric. For this reason very little use is found in practice for the last two.

Note that the upper and lower convected derivatives are obtained by summing to the corotational derivative simple combinations of the tensor and the stretching tensor. In effect, *any* linear combination of $\mathbf{J}$ and $\mathbf{D}$, when added to the corotational derivative, yields an expression which can be interpreted as a properly indifferent associated derivative of $\mathbf{J}$. The upper and lower convected derivatives are the two simplest symmetry-preserving derivatives of this type.

---

† Here and elsewhere we call left mixed components of a tensor $\mathbf{A}$ the components $A^i{}_j$, and right mixed components the components $A_j{}^i$.

3-4 THE CONVECTED COORDINATES APPROACH

In this section an entirely different approach to the kinematics of deforming bodies, which has been developed mainly by Oldroyd [4], will be analyzed.

While this entire book is based on the approach, which employs direct methods of vector spaces, the convected coordinates approach which is based on the consideration of a coordinate system embedded in the body and deforming solidally with the latter, has enjoyed widespread use in the literature—and a knowledge of this is necessary if one is to understand much of the published work on non-Newtonian fluid mechanics.

Any constitutive equation written in terms of tensor components with respect to the convected coordinate system automatically satisfies the principle of material objectivity [5]. This point has often been misleadingly extended in the literature to imply that such equations, when written in some algebraically simple form, have some special physical relevance; assumptions of 'linearity,' which are typical of the old non-invariant formulations of linear viscoelasticity, have been made frame-invariant by utilizing the convected coordinates approach, and are assumed therefore to be physically realistic, although there are an infinite number of other possibilities which satisfy the principle of material objectivity and are equally justified (or unjustified) from a phenomenological viewpoint. The confusion between coordinate systems and frames of reference is made even worse by some of the published literature based on the convected coordinates approach, and the distinction between tensors (as linear operators mapping the Euclidean vector space into itself) and matrices of tensor components is often entirely forgotten. Finally, some special physical relevance has often been attributed to the convected derivatives, and sterile discussions about which is the 'true' time derivative have been caused through a misunderstanding of the convected coordinates approach. In this section we will try to place the approach in proper perspective, and to show some of the pitfalls that are readily encountered when the approach is chosen.

Consider a fluid in flow, and let x^i be some coordinate system chosen for the description of flow. Let $x^i(\tau)$ be the coordinates of a material point at time τ. Let ξ^i be defined as

$$\xi^i = x^i(\tau)|_{\tau=t} = x_t^i \qquad (3\text{-}4.1)$$

where t is some fixed time. The x_t^i's may be regarded as triplets of numbers applied to each material point, i.e., the material point is identified by the coordinates x_t^i of the place occupied at the instant of observation t. This is the viewpoint we have adopted in section 3-1.

At the same time, the x_t^i may be regarded from a different viewpoint, and we in fact adopt in this case a different symbol, ξ^i. The ξ^i may be regarded as coordinates embedded in the material, or 'convected coordinates;' one has a coordinate system which moves and *deforms* solidally with the flowing fluid, and at time t coincides with the original fixed coordinate system x^i. Of course, the convected coordinates of the place occupied by a material point do not change with time, because the deformation of the coordinate system exactly matches the deformation of the material.

Let $\gamma_{ij}(\tau)$ be the metric of the convected coordinate system. We have (see eqs (1-3.32) and (1-4.3))

$$dx^2(\tau) = \gamma_{ij}(\tau)\,d\xi^i\,d\xi^j \qquad (3\text{-}4.2)$$

Considering eq. (3-4.1), we have

$$dx^2 = \gamma_{ij}(\tau)\,dx_t^i\,dx_t^j \qquad (3\text{-}4.3)$$

Equation (3-4.3) shows that the components of the metric $\gamma_{ij}(\tau)$ coincide at any instant with the covariant components of the Cauchy strain:

$$\gamma_{ij}(\tau) \Leftrightarrow (\mathbf{C})_{ij} \qquad (3\text{-}4.4)$$

The special symbol $\Leftrightarrow$ has been used in eq. (3-4.4) (and will be used in the following) in order to emphasize the special significance of the equality between the left- and right-hand side of the equation. In fact, $\gamma_{ij}(\tau)$ are the covariant components of the unit tensor in the ξ^i coordinate system; $(\mathbf{C})_{ij}$ are the covariant components of the Cauchy tensor in the x^i system. Although the two matrices coincide at any value of τ, it is clear that two different tensors are involved: equality of components of two tensors *does not* imply equality of the tensors, unless the components are referred to the *same* coordinate system.

The metric $\gamma_{ij}(\tau)$ can be calculated from the rule of transformation of covariant components of tensors, eq. (1-3.24):

$$\gamma_{ij}(\tau) = \frac{\partial x^k}{\partial \xi^i} \cdot \frac{\partial x^m}{\partial \xi^j} g_{km} \qquad (3\text{-}4.5)$$

which, of course, could have been obtained directly from eqs (3-4.4) and (3-1.46).

The area of a surface element at time τ is

$$ds^2(\tau) = \gamma^{ij}(\tau)\,d\sigma_i\,d\sigma_j \qquad (3\text{-}4.6)$$

where $d\sigma_i = ds_i(t)$ are the convected covariant components of the area vector, which do not change with deformation ($d\sigma_i\,d\xi^i$ is the differential volume). Equation (3-4.6) shows that

$$\gamma^{ij}(\tau) \Leftrightarrow (\mathbf{C}^{-1})^{ij} \qquad (3\text{-}4.7)$$

The condition of constant density implies that

$$\det \gamma_{ij} = \det \gamma^{ij} = 1 \qquad (3\text{-}4.8)$$

Let us now turn our attention to time derivatives of tensors. A very important point should be stressed: the time derivatives of the components of a tensor *are not* the components of the time derivative of the tensor. This point is clearly understood by considering that even the components of a *constant* tensor may have non-zero time derivatives. In fact, the bases with respect to which the components are defined may change with time for any one or both of the following reasons:

  (i) When the substantial time derivative is considered, the natural basis may change along the trajectory of a material point because it is not constant in space.
 (ii) When the convected coordinates approach is chosen, the natural basis changes with time because the coordinate system changes with time.

With these in mind, let us differentiate eq. (3-4.3) N times with respect to time:

$$\frac{\mathrm{d}^N}{\mathrm{d}t^N}(\mathrm{d}x^2) = \overset{N}{\gamma_{ij}}\,\mathrm{d}x_t^i\,\mathrm{d}x_t^j \qquad (3\text{-}4.9)$$

It is clear that $\overset{N}{\gamma_{ij}}$ are the Nth derivatives of the covariant components of the unit tensor with respect to the ξ^i system; they *are not* the components of the Nth derivative of the unit tensor!

Comparison of eqs (3-4.9) and (3-2.25) shows that

$$\overset{N}{\gamma_{ij}}(t) \bowtie (\mathbf{A}_N)_{ij} \qquad (3\text{-}4.10)$$

where a new special symbol, $\bowtie$, has been introduced. This symbol warns that, although the matrices on the left- and right-hand sides coincide, and although the coordinates with respect to which components are defined are the same (in fact, at time $\tau = t$, the x^i and ξ^i systems coincide), indices cannot be raised and/or lowered on both sides (see eq. (3-4.12) below). In fact, $\overset{N}{\gamma_{ij}}$ is the Nth derivative of γ_{ij}, and since time differentiation and raising of indices are operations which do not commute, $\overset{N}{\gamma^{ij}}$ cannot be obtained by raising indices on $\overset{N}{\gamma_{ij}}$.

Analogously, differentiation of eq. (3-4.6) gives

$$\frac{d^N}{dt^N}(ds^2) = \overset{N}{\gamma}{}^{ij} d\sigma_i \, d\sigma_j \qquad (3\text{-}4.11)$$

and thus

$$-\overset{N}{\gamma}{}^{ij} \Join (\mathbf{B}_N)^{ij} \qquad (3\text{-}4.12)$$

Comparing eqs (3-4.10) and (3-4.12) clearly shows that indices cannot be raised (or lowered), because, of course, $\mathbf{A}_N \neq -\mathbf{B}_N$.

Let us now consider the time derivatives of the components η_{ij} (or η^{ij}) of a generic tensor $\mathbf{J}$ with respect to the ξ^i coordinate system. From the rule of transformation of tensors, we have

$$\eta_{ij} = \frac{\partial x^k}{\partial \xi^i} \frac{\partial x^m}{\partial \xi^j} J_{km} \qquad (3\text{-}4.13)$$

where J_{km} are the covariant components of $\mathbf{J}$ in the x^i system, evaluated at the place $\mathbf{X}(\tau)$ occupied by the material particle at time τ:

$$J_{km} = \mathbf{e}_k(\mathbf{X}) \cdot \mathbf{J}(\tau) \cdot \mathbf{e}_m(\mathbf{X}) \qquad (3\text{-}4.14)$$

Substituting eqs (3-4.14) and (3-1.41) into eq. (3-4.13), we obtain (after observing that $x_t^i = \xi^i$)

$$\eta_{ij} = \mathbf{e}_i(\mathbf{X}_t) \cdot \mathbf{F}^T \cdot \mathbf{e}^k(\mathbf{X})\mathbf{e}_k(\mathbf{X}) \cdot \mathbf{J} \cdot \mathbf{e}_m(\mathbf{X})\mathbf{e}^m(\mathbf{X}) \cdot \mathbf{F} \cdot \mathbf{e}_j(\mathbf{X}_t) \qquad (3\text{-}4.15)$$

and, considering that the sum of dyads $\mathbf{e}^k\mathbf{e}_k$ is the unit tensor,

$$\eta_{ij} = \mathbf{e}_i(\mathbf{X}_t) \cdot [\mathbf{F}^T \cdot \mathbf{J} \cdot \mathbf{F}] \cdot \mathbf{e}_j(\mathbf{X}_t) \qquad (3\text{-}4.16)$$

Equation (3-4.16) shows that η_{ij} can be identified with the matrix of the covariant components of the lower convected form of J:

$$\eta_{ij} \Leftrightarrow (\overset{\circ}{\mathbf{J}})_{ij} \qquad (3\text{-}4.17)$$

Note that, by definition,

$$\eta_{ij} = \varepsilon_i(\tau) \cdot \mathbf{J} \cdot \varepsilon_j(\tau) \qquad (3\text{-}4.18)$$

where ε_i are the natural basis vectors of the ξ^i system at time τ.

By a strictly analogous procedure it is easy to show that

$$\eta^{ij} \Leftrightarrow (\overset{\circ}{\mathbf{J}})^{ij} \qquad (3\text{-}4.19)$$

Needless to say, indices cannot be raised (or lowered) in eq. (3-4.17) or (3-4.19).

Differentiating eqs (3-4.17) and (3-4.19), and setting $\tau = t$, we have

$$\overset{N}{\eta_{ij}}(t) \rightarrowtail (\mathbf{\mathring{J}}_N)_{ij} \qquad (3\text{-}4.20)$$

$$\overset{N}{\eta^{ij}}(t) \rightarrowtail (\mathbf{\mathring{J}}_N)^{ij} \qquad (3\text{-}4.21)$$

Of course, while the left-hand sides of eq. (3-4.20) and (3-4.21) are Nth derivatives of components of a tensor, the right-hand sides are components of another tensor. Again, needless to say, indices may not be raised or lowered in these equations.

Let us now consider the mixed components $\eta^i{}_j$ and $\eta_j{}^i$. By a reasoning which is analogous to the one leading to eqs (3-4.17) and (3-4.19), we have

$$\eta^i{}_j \Leftrightarrow (\mathbf{\bar{J}})^i{}_j \qquad (3\text{-}4.22)$$

$$\eta_j{}^i \Leftrightarrow (\mathbf{\vec{J}})_j{}^i \qquad (3\text{-}4.23)$$

i.e., 'the left mixed components in the ξ^i system coincide with the left mixed components of the left convected form in the x^i system, while the right mixed components in the ξ^i system coincide with the right mixed components of the right convected form in the x^i system for any tensor $\mathbf{J}$.'

Now consider the special case where $\mathbf{J}$ is symmetric. We have, if $\mathbf{J} = \mathbf{J}^T$,

$$\eta^i{}_j = \eta_j{}^i \qquad (3\text{-}4.24)$$

In spite of this, there exists no symmetric tensor whose components in the x^i system coincide with $\eta^i{}_j$: in fact, the left and the right convected forms of a symmetric tensor are not symmetric, and are not equal to each other. Of course, if $\mathbf{J} = \mathbf{J}^T$, the mixed components may be written as η^i_j; this matrix can then, in view of eqs (3-4.22) and (3-4.23), be interpreted in two different ways.

Differentiating eqs (3-4.22) and (3-4.23), and setting $\tau = t$, we obtain

$$\overset{N}{\eta^i{}_j} \rightarrowtail (\mathbf{\mathring{J}}_N)^i{}_j \qquad (3\text{-}4.25)$$

$$\overset{N}{\eta_j{}^i} \rightarrowtail (\mathbf{\mathring{J}}_N)_j{}^i \qquad (3\text{-}4.26)$$

Again, if $\mathbf{J} = \mathbf{J}^T$, the notation $\overset{N}{\eta^i_j}(t)$ allows for two possible interpretations.

As discussed in section 3-3, the notation $\delta/\delta t$ is often encountered in the literature as indicating time differentiation of the convected components.

In effect, one single symbol is used for *four* different operations; in fact,

$$\frac{\delta^N}{\delta t^N} J^{ij} = (\overset{\triangledown}{\mathbf{J}}_N)^{ij} \qquad (3\text{-}4.27)$$

$$\frac{\delta^N}{\delta t^N} J_{ij} = (\overset{\vartriangle}{\mathbf{J}}_N)_{ij} \qquad (3\text{-}4.28)$$

$$\frac{\delta^N}{\delta t^N} J^i{}_j = (\overset{\circ}{\mathbf{J}}_N)^i{}_j \qquad (3\text{-}4.29)$$

$$\frac{\delta^N}{\delta t^N} J_j{}^i = (\overset{\circ}{\mathbf{J}}_N)_j{}^i \qquad (3\text{-}4.30)$$

This is somewhat misleading, because the left-hand sides of eqs (3-4.27)–(3-4.30) fail to give any indication that the components of four different tensors are being considered. When **J** is symmetric, and the notation $\delta J^i_j/\delta t$ is used, exactly the same notation identifies two (coincident) sets of components of two different tensors.

The convected coordinates approach, discussed in this section, has the great advantage that *any* constitutive equation, when written in terms of convected tensor components, satisfies the principle of material objectivity. The use of this approach is open to some difficulties, which we have tried to illustrate. It should be clear that the choice between the convected coordinates and the vector space approach is a matter of personal preference, and that both approaches, if correctly used, give the same results.

3-5 CONSTANT STRETCH HISTORY FLOWS

In this section we wish to investigate a particular class of flows which is of special physical relevance for memory fluids. These are the 'constant stretch history' flows.

Constant stretch history flows (sometimes also called 'substantially stagnant motions') are, in non-rigorous terms, flows for which the history of stretch does not depend on the instant of observation t, but only on the time lag $s = t - \tau$; that is, the stretch carrying the configuration at time τ into the configuration at the instant of observation t is—apart from irrelevant rotations—independent of the particular value of t, and uniquely determined by the value of s.

The physical relevance of constant stretch history flows is easily understood on the basis of the concepts discussed in section 2-6. In a memory fluid, the stress at the instant of observation is influenced by the entire history of deformation of the neighborhood of the material point considered; in constant stretch history flows, this history is independent of the instant of observation, and thus one may expect the stress, as well as any other dependent variable such as internal energy, to be independent of t. These concepts will be formalized in later chapters, but can be grasped intuitively at this stage.

If flows with a constant stretch history are to have some special physical relevance, their definition must be frame-indifferent, in the sense that a flow with a constant stretch history in one frame of reference must have a constant stretch history in any other frame of reference. Yet, in order to introduce the subject as smoothly as possible, we choose to begin by considering some special frame of reference where equations take a particularly simple form, and to delay the formalized general treatment.

Let asterisks indicate quantities in the special frame of reference mentioned above. Now, consider a flow pattern for which the deformation gradient history fulfils the following condition for any values of t, and t':

$$\mathbf{F}^{t*}(s) = \mathbf{F}^{t'*}(s) \qquad (3\text{-}5.1)$$

Such a flow is clearly a constant stretch history flow, in the sense discussed above: in fact, the tensor transforming $d\mathbf{X}(t)$ into $d\mathbf{X}(\tau)$ only depends on the time lag s, but does not depend on t.

Considering eqs (3-1.19) and (3-1.29), we have

$$\mathbf{C}^{t*} = \mathbf{C}^{t'*} \qquad (3\text{-}5.2)$$

$$(\mathbf{C}^{t*})^{-1} = (\mathbf{C}^{t'*})^{-1} \qquad (3\text{-}5.3)$$

Equation (3-5.1) is satisfied if and only if the deformation gradient history has the following form:

$$\mathbf{F}^{t*} = \exp(-ks\mathbf{N}^*) \qquad (3\text{-}5.4)$$

where $k\mathbf{N}^*$ is a constant tensor. It is useful to choose k so that the magnitude of $\mathbf{N}^*$ is unity, say

$$\sqrt{(\text{tr } \mathbf{N}^* \cdot \mathbf{N}^{*\mathrm{T}})} = 1 \qquad (3\text{-}5.5)$$

In eq. (3-5.4) use has been made of the tensor-valued function of a tensor argument $\exp(\mathbf{A})$. Definition and properties of this function are listed below:

Definition

$$\exp(\mathbf{A}) = 1 + \sum_{N=1}^{\infty} \frac{1}{N!}\mathbf{A}^N \qquad (3\text{-}5.6)$$

If **A** is nilpotent, exp **A** is a finite-order polynomial.

Isotropy

$$\mathbf{Q} \cdot \exp \mathbf{A} \cdot \mathbf{Q}^T = \exp(\mathbf{Q} \cdot \mathbf{A} \cdot \mathbf{Q}^T) \qquad (3\text{-}5.7)$$

Inverse

$$(\exp \mathbf{A})^{-1} = \exp(-\mathbf{A}) \qquad (3\text{-}5.8)$$

Transpose

$$\exp(\mathbf{A}^T) = (\exp \mathbf{A})^T \qquad (3\text{-}5.9)$$

Consideration of eqs (3-5.8) and (3-5.9) shows that if **A** is skew, exp **A** is orthogonal.

Product

$$\exp(\alpha \mathbf{A}) \cdot \exp(\beta \mathbf{A}) = \exp[(\alpha + \beta)\mathbf{A}] \qquad (3\text{-}5.10)$$

but, in general, exp **A** · exp **B** is different from exp (**A** + **B**) unless **A** · **B** = **B** · **A**.

Derivative

$$\frac{d}{d\alpha}[\exp(\alpha \mathbf{A})] = \mathbf{A} \cdot \exp(\alpha \mathbf{A}) \qquad (3\text{-}5.11)$$

$$\left\{\frac{d}{d\alpha}[\exp(\alpha \mathbf{A})]\right\}_{\alpha=0} = \mathbf{A} \qquad (3\text{-}5.12)$$

From eq. (3-2.9) one obtains

$$\nabla \mathbf{v}^* = k\mathbf{N}^* \qquad (3\text{-}5.13)$$

which shows that k is the magnitude of the velocity gradient. A flow has a constant stretch history if the velocity gradient is constant along the path of a material point, i.e., if[†]

$$\dot{\mathbf{L}}^* = \mathbf{0} \qquad (3\text{-}5.14)$$

So far, we have only considered a special frame of reference. Let us now generalize our results by transforming the equations to a non-specified frame of reference, the change of frame from the '*' frame to the generic frame being described by an otherwise arbitrary smooth orthogonal tensor function **Q**(t). In particular, eq. (3-5.4) transforms to:

$$\mathbf{F}^t = \mathbf{Q}(\tau) \cdot [\exp(-ks\mathbf{N}^*)] \cdot \mathbf{Q}^T(t) \qquad (3\text{-}5.15)$$

[†] Huilgol [6] seems to have first noticed explicitly that $\dot{\mathbf{L}} = \mathbf{0}$ is a sufficient condition for a flow to be a constant stretch history flow.

Let us now introduce the orthogonal tensor function $\mathbf{P}(t)$, defined as

$$\mathbf{P}(t) = \mathbf{Q}(t) \cdot \mathbf{Q}^{\mathrm{T}}(0) \qquad (3\text{-}5.16)$$

where $t = 0$ is the (arbitrarily chosen) origin of the time axis. The $\mathbf{P}(t)$ function is such that $\mathbf{P}(0) = 1$.

Substitution of eq. (3-5.16) into (3-5.15) and consideration of the isotropy of the exp ( ) function (see eq. (3-5.7) above) gives

$$\mathbf{F}^t = \mathbf{P}(\tau) \cdot \exp\left[-ks\mathbf{N}\right] \cdot \mathbf{P}^{\mathrm{T}}(t) \qquad (3\text{-}5.17)$$

where

$$\mathbf{N} = \mathbf{Q}^{\mathrm{T}}(0) \cdot \mathbf{N}^* \cdot \mathbf{Q}(0) \qquad (3\text{-}5.18)$$

From eq. (3-5.17), the Cauchy and Finger tensor histories are obtained as

$$\mathbf{C}^t = \mathbf{P}(t) \cdot \mathbf{C}_0^0 \cdot \mathbf{P}^{\mathrm{T}}(t) \qquad (3\text{-}5.19)$$

$$(\mathbf{C}^t)^{-1} = \mathbf{P}^{\mathrm{T}}(t) \cdot (\mathbf{C}_0^0)^{-1} \cdot \mathbf{P}(t) \qquad (3\text{-}5.20)$$

We are now in a position to rigorously formalize the treatment of constant stretch history flows. Following Coleman [7] and Noll [8], we adopt the following definition: 'A flow is said to be a constant stretch history flow when eq. (3-5.19) (or, equivalently, eq. (3-5.20)) holds with $\mathbf{P}(t)$ any smooth orthogonal tensor function with $\mathbf{P}(0) = 1$.'

This definition is properly frame-indifferent. Noll [8] defines constant stretch history flows in terms of eq. (3-5.19); eq. (3-5.20) can equally well be chosen as a starting point. In fact, the choice of the Cauchy tensor in preference to the Finger tensor is justified only on historical grounds, but the two tensors are equally useful in describing completely the deformation history.

It can be proved that a flow has a constant stretch history if and only if[†]

$$\mathbf{F}_0(\tau) = \mathbf{P}(\tau) \cdot \exp(k\tau\mathbf{N}) \qquad (3\text{-}5.21)$$

Notice that $\mathbf{N}$ has unit magnitude, as can be shown immediately from eqs (3-5.5) and (3-5.18).

Equations for all the kinematic tensors defined in the preceding sections are easily obtained from eq. (3-5.17) above, and are collected in Table 3-1 for easy reference.

[†] Proof of sufficiency is immediately obtained by substitution. Of course, eq. (3-5.21) is eq. (3-5.17) for $t = 0$; eq. (3-5.17) can be obtained from eq. (3-5.21). Proof of necessity of eq. (3-5.21) is involved; the reader is referred to Noll [8].

Table 3-1 **EQUATIONS FOR THE KINEMATIC TENSORS IN CONSTANT STRETCH HISTORY FLOWS**

$$\mathbf{F}^t = \mathbf{P}(t - s) \cdot \exp(-ks\mathbf{N}) \cdot \mathbf{P}^T(t) \tag{3-5.17}$$

$$\mathbf{C}^t = \mathbf{P}(t) \cdot \mathbf{C}_0^0 \cdot \mathbf{P}^T(t) \tag{3-5.19}$$

$$(\mathbf{C}^t)^{-1} = \mathbf{P}^T(t) \cdot (\mathbf{C}_0^0)^{-1} \cdot \mathbf{P}(t) \tag{3-5.20}$$

$$\mathbf{D}(t) = \tfrac{1}{2}k\mathbf{P}(t) \cdot (\mathbf{N} + \mathbf{N}^T) \cdot \mathbf{P}^T(t) \tag{3-5.22}$$

$$\mathbf{W}(t) = \tfrac{1}{2}k\mathbf{P}(t) \cdot (\mathbf{N} - \mathbf{N}^T) \cdot \mathbf{P}^T(t) + \dot{\mathbf{P}}(t) \cdot \mathbf{P}^T(t) \tag{3-5.23}$$

$$\mathbf{C}^t = \mathbf{P}(t) \cdot \exp(-ks\mathbf{N}^T) \cdot \exp(-ks\mathbf{N}) \cdot \mathbf{P}^T(t) \tag{3-5.24}$$

$$(\mathbf{C}^t)^{-1} = \mathbf{P}^T(t) \cdot \exp(+ks\mathbf{N}) \cdot \exp(+ks\mathbf{N}^T) \cdot \mathbf{P}(t) \tag{3-5.25}$$

$$\mathbf{A}_N = k^N \mathbf{P}(t) \cdot \left[\sum_{k=0}^{N} \binom{N}{k} (\mathbf{N}^T)^k \cdot \mathbf{N}^{N-k} \right] \cdot \mathbf{P}^T(t) \tag{3-5.26}$$

$$\mathbf{B}_N = (-1)^{N+1} k^N \mathbf{P}(t) \cdot \left[\sum_{k=0}^{N} \binom{N}{k} \mathbf{N}^k \cdot (\mathbf{N}^T)^{N-k} \right] \cdot \mathbf{P}^T(t) \tag{3-5.27}$$

The remainder of this section is dedicated to the analysis of some important subclasses of constant stretch history flows.

Rigid-body motions

A rigid-body motion is characterized by the fact that the deformation gradient is orthogonal. In view of eq. (3-5.17), any constant stretch history flow for which $\mathbf{N}$ is skew is a rigid-body motion (the exponential of a skew tensor is orthogonal). A rigid-body motion does not identify the tensor $\mathbf{N}$; in particular, $\mathbf{N}$ can be taken as the zero tensor.

Viscometric flows

A constant stretch history flow is said to be a viscometric flow if

$$\mathbf{N}^2 = 0; \qquad \mathbf{N} \neq 0 \tag{3-5.28}$$

The matrix of the components of $\mathbf{N}$ with respect to an appropriate orthonormal basis has, for viscometric flows, the form

$$[\mathbf{N}] = \begin{Vmatrix} 0 & 1 & 0 \\ 0 & 0 & 0 \\ 0 & 0 & 0 \end{Vmatrix} \tag{3-5.29}$$

Equations for the kinematic tensors valid for all viscometric flows are easily obtained from the equations in Table 3-1, and are collected in Table 3-2 for easy reference.

It may be worthwhile to point out that, in view of eq. (3-5.35) below, viscometric flows are sometimes referred to as 'second-order' flows.

Table 3-2 EQUATIONS FOR THE KINEMATIC TENSORS IN VISCOMETRIC FLOW

$$\mathbf{F}^t = \mathbf{P}(t - s) \cdot [\mathbf{1} - ks\mathbf{N}] \cdot \mathbf{P}^T(t) \tag{3-5.30}$$

$$\mathbf{C}^t = \mathbf{P}(t) \cdot [\mathbf{1} - ks(\mathbf{N}^T + \mathbf{N}) + k^2 s^2 \mathbf{N}^T \cdot \mathbf{N}] \cdot \mathbf{P}^T(t) \tag{3-5.31}$$

$$(\mathbf{C}^t)^{-1} = \mathbf{P}(t) \cdot [\mathbf{1} + ks(\mathbf{N}^T + \mathbf{N}) + k^2 s^2 \mathbf{N} \cdot \mathbf{N}^T] \cdot \mathbf{P}^T(t) \tag{3-5.32}$$

$$\mathbf{A}_2 = \mathbf{P}(t) \cdot \{2k^2 \mathbf{N}^T \cdot \mathbf{N}\} \cdot \mathbf{P}^T(t) \tag{3-5.33}$$

$$\mathbf{B}_2 = -\mathbf{P}(t) \cdot \{2k^2 \mathbf{N} \cdot \mathbf{N}^T\} \cdot \mathbf{P}^T(t) \tag{3-5.34}$$

$$\mathbf{A}_N = \mathbf{B}_N = 0 \quad \text{if} \quad N > 2 \tag{3-5.35}$$

Fourth-order flows

A constant stretch history flow is said to be a fourth-order flow if:

$$\mathbf{N}^3 = 0; \qquad \mathbf{N}^2 \neq 0 \tag{3-5.36}$$

The matrix of the components of $\mathbf{N}$ with respect to an appropriate orthonormal basis has, for fourth-order flows, the form

$$[\mathbf{N}] = \begin{Vmatrix} 0 & \alpha_1 & \alpha_2 \\ 0 & 0 & \alpha_3 \\ 0 & 0 & 0 \end{Vmatrix} \tag{3-5.37}$$

with $\alpha_1^2 + \alpha_2^2 + \alpha_3^2 = 1$, and α_1 and α_3 are different from zero.

For fourth-order flows, $\mathbf{F}^t$ is a second-order polynomial function of $ks\mathbf{N}$; both the Cauchy and the Finger tensors are fourth-order polynomial functions of ks; and

$$\mathbf{A}_N = \mathbf{B}_N = 0 \quad \text{if} \quad N > 4 \tag{3-5.38}$$

Extensional flows

A constant stretch history flow is said to be an extensional flow if

$$\mathbf{N} = \mathbf{N}^T \tag{3-5.39}$$

Equations for kinematic tensors valid for extensional flow are collected for easy reference in Table 3-3.

Table 3-3 **EQUATIONS FOR KINEMATIC TENSORS IN EXTENSIONAL FLOW**

$$\mathbf{C}^t = \mathbf{P}(t) \cdot \exp(-2ks\mathbf{N}) \cdot \mathbf{P}^{\mathrm{T}}(t) \qquad (3\text{-}5.40)$$

$$\mathbf{A}_N = \mathbf{P}(t) \cdot (2k\mathbf{N})^N \cdot \mathbf{P}^{\mathrm{T}}(t) \qquad (3\text{-}5.41)$$

$$\mathbf{B}_N = (-1)^{N+1} \mathbf{A}_N \qquad (3\text{-}5.42)$$

3-6 ILLUSTRATIONS

ILLUSTRATION 3A: *Kinematic tensors for a lineal Couette flow (simple shear)*.

We have already considered this flow in the previous chapter, where the kinematic tensors $\mathbf{Vv}$ and $\mathbf{D}$ were derived. We now want expressions for the components of the deformation tensors such as $\mathbf{C}$, $\mathbf{C}^{-1}$, etc. In a Cartesian coordinate system, the flow is described by eqs (2-1.2)–(2-1.3):

$$v^1 = \gamma x^2$$

$$v^2 = v^3 = 0$$

These equations are equivalent to the following set of differential equations:

$$\frac{dx^1}{d\tau} = \gamma x^2$$

$$\frac{dx^2}{d\tau} = \frac{dx^3}{d\tau} = 0 \qquad (3\text{-}6.1)$$

The initial conditions are given as

for $\tau = t$,

$$x^1 = x_t^1$$

$$x^2 = x_t^2 \qquad (3\text{-}6.2)$$

$$x^3 = x_t^3$$

Integration gives

$$x^1 = x_t^1 - \gamma x_t^2(t - \tau)$$

$$x^2 = x_t^2 \qquad (3\text{-}6.3)$$

$$x^3 = x_t^3$$

Equations (3-6.3) are the explicit form of the equation of motion in terms of coordinates (eq. (3-1.35)) for the flow we are considering.

We now evaluate the matrix of the derivatives $\partial x^i / \partial x_t^j$:

$$\left[\frac{\partial x^i}{\partial x_t^j}\right] = \left\|\begin{matrix} 1 & -\gamma(t - \tau) & 0 \\ 0 & 1 & 0 \\ 0 & 0 & 1 \end{matrix}\right\| \tag{3-6.4}$$

Because the coordinate system is Cartesian, the basis vectors are the same everywhere, thus the matrix in eq. (3-6.4) coincides with that of the components (of any type) of tensor $\mathbf{F}$ (see eq. (3-1.41)):

$$[\mathbf{F}] = \left[\frac{\partial x^i}{\partial x_t^j}\right] \tag{3-6.5}$$

By differentiating the matrix in eq. (3-6.4) with respect to τ, and setting $\tau = t$, one obtains the matrix of tensor $\nabla \mathbf{v}$, eq. (2-8.1), in accordance with eq. (3-2.9).

Again, the coordinate system being Cartesian, the metric is the unit matrix and thus, from eq. (3-1.46),

$$[\mathbf{C}] = \left[\frac{\partial x^i}{\partial x_t^j}\right]^{\mathrm{T}}\left[\frac{\partial x^i}{\partial x_t^j}\right] = [\mathbf{F}^{\mathrm{T}}][\mathbf{F}] \tag{3-6.6}$$

one obtains:

$$[\mathbf{C}] = \left\|\begin{matrix} 1 & -\gamma(t - \tau) & 0 \\ -\gamma(t - \tau) & 1 + \gamma^2(t - \tau)^2 & 0 \\ 0 & 0 & 1 \end{matrix}\right\| \tag{3-6.7}$$

One may check that $\mathrm{III}_{\mathbf{C}} = 1$. $[\mathbf{C}^{-1}]$ is given by

$$[\mathbf{C}^{-1}] = [\mathbf{C}]^{-1} = \left\|\begin{matrix} 1 + \gamma^2(t - \tau)^2 & \gamma(t - \tau) & 0 \\ \gamma(t - \tau) & 1 & 0 \\ 0 & 0 & 1 \end{matrix}\right\| \tag{3-6.8}$$

Equation (3-1.20) allows us to calculate $[\mathbf{B}]$:

$$[\mathbf{B}] = [\mathbf{F}][\mathbf{F}^{\mathrm{T}}] = \left\|\begin{matrix} 1 + \gamma^2(t - \tau)^2 & -\gamma(t - \tau) & 0 \\ -\gamma(t - \tau) & 1 & 0 \\ 0 & 0 & 1 \end{matrix}\right\| \tag{3-6.9}$$

Hence:

$$[\mathbf{B}^{-1}] = \begin{Vmatrix} 1 & \gamma(t-\tau) & 0 \\ \gamma(t-\tau) & 1+\gamma^2(t-\tau)^2 & 0 \\ 0 & 0 & 1 \end{Vmatrix}. \qquad (3\text{-}6.10)$$

Comparison of eqs (3-6.7) and (3-6.10) (or (3-6.8) and (3-6.9)) allows a check of eq. (3-1.31) (or (3-1.32)). It is also apparent that for $\tau = t$ all the matrices so far considered become the unit matrix.

Finally,

$$[\mathbf{G}] = \begin{Vmatrix} 0 & -\gamma(t-\tau) & 0 \\ -\gamma(t-\tau) & \gamma^2(t-\tau)^2 & 0 \\ 0 & 0 & 0 \end{Vmatrix} \qquad (3\text{-}6.11)$$

$$[\mathbf{H}] = \begin{Vmatrix} \gamma^2(t-\tau)^2 & \gamma(t-\tau) & 0 \\ \gamma(t-\tau) & 0 & 0 \\ 0 & 0 & 0 \end{Vmatrix} \qquad (3\text{-}6.12)$$

which reduce to $[\mathbf{0}]$ for $\tau = t$.

One may notice that the matrices of tensors $\mathbf{R}$, $\mathbf{V}$, and $\mathbf{U}$ have not been derived. They are not used in practical calculations and are generally rather difficult to obtain.

ILLUSTRATION 3B: *Kinematics of a spherical symmetric flow towards a point sink.*

The symmetry of the problem suggests the use of a spherical coordinate system with its center at the sink. The flow is then described by the following equations:

$$v^r = -\frac{a}{r^2}$$
$$v^\theta = v^\phi = 0 \qquad (3\text{-}6.13)$$

In eqs (3-6.13), v^r, v^θ, and v^ϕ are first interpreted as physical components of the velocity. The equation for v^r is deduced from continuity by assuming the fluid to be incompressible; a is a function of time only, but we shall restrict ourselves to the case where a is a positive constant, i.e., steady sink flow.

Because v^θ and v^ϕ are zero and the natural basis vector in the r direction is of unit length, eqs (3-6.13) can also be interpreted as giving the contravariant components of the velocity; thus we can write:

$$\frac{dr}{d\tau} = -\frac{a}{r^2}$$

$$\frac{d\theta}{d\tau} = \frac{d\phi}{d\tau} = 0$$

(3-6.14)

Initial conditions are:

for $\tau = t$,

$$r = r_t$$

$$\theta = \theta_t \qquad (3\text{-}6.15)$$

$$\phi = \phi_t$$

The equations of motion are then obtained:

$$r = [r_t^3 + 3a(t - \tau)]^{1/3}$$

$$\theta = \theta_t \qquad (3\text{-}6.16)$$

$$\phi = \phi_t$$

The matrix of the derivatives $\partial x^i / \partial x_t^j$ is now calculated from eqs (3-6.16). The coordinates are taken in the order r, θ, and ϕ.

$$\left[\frac{\partial x^i}{\partial x_t^j}\right] = \left\|\begin{matrix} [1 + 3a(t - \tau)/r_t^3]^{-2/3} & 0 & 0 \\ 0 & 1 & 0 \\ 0 & 0 & 1 \end{matrix}\right\| \qquad (3\text{-}6.17)$$

In this example, because the basis vectors do change along the path of the particles, i.e., they are different at $\mathbf{X}(\tau)$ and $\mathbf{X}_t$, the above matrix does not coincide with any of the usual component-matrix of tensor $\mathbf{F}$ (see eq. (3-1.41)).

We now use eq. (3-1.46) to calculate the covariant components at $\mathbf{X}_t$ of tensor $\mathbf{C}$. In order to do so, the metric at $\mathbf{X}(\tau)$ is needed:

$$[g_{ij}] = \left\|\begin{matrix} 1 & 0 & 0 \\ 0 & r^2 & 0 \\ 0 & 0 & r^2 \sin^2\theta \end{matrix}\right\| \qquad (3\text{-}6.18)$$

Thus

$$[\mathbf{C}_{ij}] = \begin{Vmatrix} [1 + 3a(t - \tau)/r_t^3]^{-4/3} & 0 & 0 \\ 0 & [r_t^3 + 3a(t - \tau)]^{2/3} & 0 \\ 0 & 0 & [r_t^3 + 3a(t - \tau)]^{2/3} \sin^2 \theta \end{Vmatrix}$$

(3-6.19)

From the covariant components and the metric at $\mathbf{X}_t$ any other types of components of $\mathbf{C}$ can be obtained. Of special relevance are the physical components. Recalling eq. (2-7.20), we have

$$[\mathbf{C}\langle ij\rangle] = \begin{Vmatrix} [1 + 3a(t - \tau)/r_t^3]^{-4/3} & 0 & 0 \\ 0 & [1 + 3a(t - \tau)/r_t^3]^{2/3} & 0 \\ 0 & 0 & [1 + 3a(t - \tau)/r_t^3]^{2/3} \end{Vmatrix}$$

(3-6.20)

It is immediately seen that

$$\det \mathbf{C} = \det [\mathbf{C}\langle ij\rangle] = 1 \qquad (3\text{-}6.21)$$

Tensor $\mathbf{D}$ can be obtained from $\mathbf{C}$ making use of eq. (3-2.17). Because the orthonormal basis of the physical components does not change along the paths of the particles (which are radial), the matrix of the physical components of $\mathbf{D}$ is obtained from

$$[\mathbf{D}] = \tfrac{1}{2}(\overline{[\mathbf{C}]})_{\tau=t} \qquad (3\text{-}6.22)$$

Thus,

$$[\mathbf{D}] = \begin{Vmatrix} 2a/r_t^3 & 0 & 0 \\ 0 & -a/r_t^3 & 0 \\ 0 & 0 & -a/r_t^3 \end{Vmatrix} \qquad (3\text{-}6.23)$$

Of course, tensor $\mathbf{D}$ can also be obtained as the symmetric part of $\nabla\mathbf{v}$. However, the calculation of the components of $\mathbf{D}$ from the equations of the velocity field (eqs (3-6.13)), insofar as the coordinate system is not Cartesian, involves the use of Christoffel symbols. This is done in the next illustration.

ILLUSTRATION 3C: *Evaluation of* [$\nabla\mathbf{v}$] *for the spherical sink flow.*

We make use of eq. (1-4.9) which reads:

$$(\nabla\mathbf{v})^i{}_j = \frac{\partial v^i}{\partial x^j} + \begin{Bmatrix} i \\ j \quad k \end{Bmatrix} v^k \qquad (3\text{-}6.24)$$

Inspection of eqs (3-6.13) shows that eq. (3-6.24) can be reduced to

$$(\mathbf{\nabla v})^i{}_j = \frac{\partial v^i}{\partial x^j} - \left\{ \begin{matrix} i \\ j \ 1 \end{matrix} \right\} \frac{a}{r^2} \qquad (3\text{-}6.25)$$

Thus, the only non-zero Christoffel symbols we are interested in are

$$\left\{ \begin{matrix} 2 \\ 2 \ 1 \end{matrix} \right\} = \left\{ \begin{matrix} 3 \\ 3 \ 1 \end{matrix} \right\} = \frac{1}{r} \qquad (3\text{-}6.26)$$

It is then apparent that all components $(\mathbf{\nabla v})^i{}_j$ for which $i \neq j$ are zero and the diagonal terms are given by

$$(\mathbf{\nabla v})^1{}_1 = \frac{2a}{r^3}$$

$$(3\text{-}6.27)$$

$$(\mathbf{\nabla v})^2{}_2 = (\mathbf{\nabla v})^3{}_3 = -\frac{a}{r^3}$$

Using eq. (2-7.20), the matrix of the physical components is obtained:

$$[\mathbf{\nabla v}] = \left\| \begin{matrix} 2a/r^3 & 0 & 0 \\ 0 & -a/r^3 & 0 \\ 0 & 0 & -a/r^3 \end{matrix} \right\| \qquad (3\text{-}6.28)$$

Comparison of eqs (3-6.28) and (3-6.23) shows that $\mathbf{D} = \mathbf{\nabla v}$ in this case, as one would have expected because the flow is clearly irrotational.

ILLUSTRATION 3D: *The Rivlin–Ericksen tensors in the flows of Illustrations 3A and 3B.*

The Rivlin–Ericksen tensors are obtained from eq. (3-2.26):

$$\mathbf{A}_N(t) = \overset{N}{\mathbf{C}}(t)$$

Because $\mathbf{A}_1 = 2\mathbf{D}$ (eq. (3-2.28)), we start calculating from $\mathbf{A}_2$ onwards.

Simple shear

$$[\overset{1}{\mathbf{C}}(\tau)] = \left\| \begin{matrix} 0 & \gamma & 0 \\ \gamma & -2\gamma^2(t-\tau) & 0 \\ 0 & 0 & 0 \end{matrix} \right\| \qquad (3\text{-}6.29)$$

$$[\overset{2}{\mathbf{C}}(\tau)] = \left\| \begin{matrix} 0 & 0 & 0 \\ 0 & 2\gamma^2 & 0 \\ 0 & 0 & 0 \end{matrix} \right\| \qquad (3\text{-}6.30)$$

Thus,

$$\mathbf{A}_2 = \begin{Vmatrix} 0 & 0 & 0 \\ 0 & 2\gamma^2 & 0 \\ 0 & 0 & 0 \end{Vmatrix}; \qquad \mathbf{A}_3 = \mathbf{A}_4 = \cdots = \mathbf{0} \qquad (3\text{-}6.31)$$

Sink flow

$$[\overset{1}{\mathbf{C}}(\tau)] =$$

$$\begin{Vmatrix} \dfrac{4a}{r_t^3}\left[1 + \dfrac{3a(t-\tau)}{r_t^3}\right]^{-7/3} & 0 & 0 \\[4mm] 0 & -\dfrac{2a}{r_t^3}\left[1 + \dfrac{3a(t-\tau)}{r_t^3}\right]^{-1/3} & 0 \\[4mm] 0 & 0 & -\dfrac{2a}{r_t^3}\left[1 + \dfrac{3a(t-\tau)}{r_t^3}\right]^{-1/3} \end{Vmatrix}$$

$$(3\text{-}6.32)$$

$$[\overset{2}{\mathbf{C}}(\tau)] =$$

$$\begin{Vmatrix} \dfrac{28a^2}{r_t^6}\left[1 + \dfrac{3a(t-\tau)}{r_t^3}\right]^{-10/3} & 0 & 0 \\[4mm] 0 & -\dfrac{2a^2}{r_t^6}\left[1 + \dfrac{3a(t-\tau)}{r_t^3}\right]^{-4/3} & 0 \\[4mm] 0 & 0 & -\dfrac{2a}{r_t^6}\left[1 + \dfrac{3a(t-\tau)}{r_t^3}\right]^{-4/3} \end{Vmatrix}$$

$$(3\text{-}6.33)$$

Thus,

$$[\mathbf{A}_2] = \begin{Vmatrix} 28a^2/r_t^6 & 0 & 0 \\ 0 & -2a^2/r_t^6 & 0 \\ 0 & 0 & -2a^2/r_t^6 \end{Vmatrix} \qquad (3\text{-}6.34)$$

Subsequent Rivlin–Ericksen tensors are obtained by differentiation of eq. (3-6.33) and setting $\tau = t$. They are not zero for this flow.

PROBLEMS

3-1 Determine the matrix of tensor $\mathbf{C}$ in a Cartesian coordinate system for the flow described by: $v_x = \gamma_E x$, $v_y = -\gamma_E y/2$, $v_z = -\gamma_E z/2$.

3-2 Establish whether the flow of problem 3-1 has a constant stretch history and find its category (viscometric, fourth-order, extensional).

3-3 Find the White–Metzner tensors for the flow of Illustrations 3A and 3B (see Illustration 3D).

3-4 Prove eq. (3-2.29). *Hint*: start from the definition of $\dot{\mathbf{A}}_N$, i.e., eq. (3-2.26) plus the definition of derivative.

3-5 For the sink flow of Illustration B, find the expression of $\delta\tau^{ij}/\delta t$ using both possible procedures (eq. (3-4.21)).

BIBLIOGRAPHY

1. TRUESDELL, C., and NOLL, W.: *The Non-linear Field Theories of Mechanics*, p. 53. Springer-Verlag, Berlin (1965).
2. TRUESDELL, C., and NOLL, W.: *ibid.*, pp. 67 and 96.
3. TRUESDELL, C., and NOLL, W.: *ibid.*, p. 124.
4. OLDROYD, J. G.: *Proc. Roy. Soc. London A*, **200**, 523 (1950).
5. TRUESDELL, C., and NOLL, W.: *ibid.*, p. 46.
6. HUILGOL, R. R.: 62nd Annual Meeting A.I.Ch.E., Washington (1969).
7. COLEMAN, B. D.: *Arch. Ratl Mech. Anal.*, **9**, 273 (1962); *Trans. Soc. Rheol.*, **6**, 293 (1962).
8. NOLL, W.: *Arch. Ratl Mech. Anal.*, **11**, 97 (1962).

4

SIMPLE FLUID THEORY

4-1 THE SIMPLE FLUID CONCEPT

In chapter 2 the inadequacy of the Reiner–Rivlin equation to predict the behavior of some real fluids even in such a simple flow as lineal Couette flow has been discussed. The concept of memory in fluid-like materials was also introduced as a necessary consequence of the failure of the Reiner–Rivlin equation, i.e., on the assumption that stress is uniquely determined by the instantaneous rate of deformation.

A much more general assumption is that the stress is determined, in some sense to be made precise, by the entire history of deformation. This assumption is the basis for the theory of simple fluids with fading memory, to be discussed in this chapter. The theory is axiomatic, in the sense that it is developed in logical order from constitutive assumptions which are regarded as definitions of a certain class of materials (i.e., simple fluids with some sort of fading memory), independent of the existence in nature of any materials satisfying the basic assumptions. Nonetheless, the theory is so general in character that almost all constitutive equations that have been proposed in the literature are special cases of it. This generality assures that all the results obtained within the framework of the theory have a very wide validity; on the other hand, few problems of fluid mechanics can be solved within this general framework, and more specific constitutive assumptions are often required in handling practical problems.

This section is dedicated to the physical concepts underlying the simple fluid theory; a mathematical formulation of these concepts will be given in section 4-3. The physical concepts to be discussed take the form of principles, which may be regarded either as postulates (if an axiomatic viewpoint is preferred), or as more or less self-evident statements concerning the behavior of real fluid-like materials (if a naturalistic viewpoint is preferred). These principles are [1]:

    (i) determinism of stress;
    (ii) local action;
    (iii) non-existence of a natural state;
    (iv) fading memory.

Determinism of stress

This principle may be stated as follows: the stress is determined by the history of deformation. It implies the assumption that the stress at a given time is independent of *future* deformations, and only depends on *past* deformations: thus, one is building a theory for materials endowed with memory but not capable of foreseeing the future. Clearly, the idea that the history of deformation determines the stress is much more general than the basic assumption of the Reiner–Rivlin theory, i.e., that the instantaneous rate of deformation determines the stress.

In order to formulate mathematically the principle of determinism of stress, a few mathematical notions on functionals are required. These will be developed in the next section.

Local action

The principle of local action can be stated as follows: the stress at a given point is uniquely determined by the history of deformation of an *arbitrarily small neighborhood* of that material point.

The principle of local action does not imply an assumption of homogeneity. On the contrary, the dependence of the stress on the history of deformation may be different at different material points.

Non-existence of a natural state

This principle is not easily stated in a few words. It implies the formalization of the intuitive but elusive concept of fluidity. Possibly the simplest formulation of the intuitive notion of fluidity is actually just that a fluid material has no preferred configuration or 'natural state.' This implies that all possible configurations are essentially equivalent, so that any difference in stress is only due to a difference in the history of deformation. We shall assume that, for a fluid material, knowledge of the strain carrying every configuration assumed in the past into the present (i.e., knowledge of, for example, the **C** function) suffices in principle to determine the present stress, and it is not necessary to know the strain carrying the present configuration into some preferred 'natural' configuration. Indeed, the differential kinematics discussed in chapter 3 have been developed to such an end.

It is a consequence of the principle of non-existence of a natural state that *every simple fluid is isotropic.* Thus, theories of anisotropic fluids, such as, e.g., those proposed by Ericksen [2], are not included in the framework of simple fluid theory. Anisotropy can only be defined with reference to some preferred directions, and thus in some sense with reference to a natural state of special physical relevance; this is excluded by a notion of fluidity expressed through the principle of non-existence of a natural state. Of course, anisotropic materials capable of flowing are possible, which only shows how the notion of fluidity is elusive.

Fading memory

The principle of fading memory can be stated as follows: the influence of past deformations on the present stress is weaker for the distant past than for the recent past. This principle is required in order to build a theory which can, at least in principle, be subjected to experimental test. In fact, the *entire* history of deformation (up to $s \to \infty$) for any given material can never be known. The principle of fading memory allows us to consider an experiment of finite duration, at the end of which any deformation that has occurred previous to the start of the experiment gives a negligible contribution to the stress. Such an experiment may be used to test predictions of the theory.

It is clear that the principle of fading memory introduces the concept of a 'natural time' for any given material. In some intuitive sense, the natural time is a yardstick of the memory span of the material, say, of the minimum acceptable duration of an experiment like the one discussed above.

The theory of purely viscous fluids (i.e., the Reiner–Rivlin theory) could be obtained as the limiting case where the natural time is zero. Thus, generalized Newtonian fluid mechanics can hopefully be established as being asymptotically valid under suitable conditions. In the following, we will use the symbol Λ for the natural time of a fluid, while the lower-case λ is used for any rheological parameter having units of time, such as, for example, the one appearing in eqs (2-4.6) and (2-4.8).

In the development of the simple fluid theory, we will make use of two further principles, namely the principle of material objectivity (which has already been discussed in chapter 2), and the principle of internal constraints. The latter may be stated as follows: the stress in a material subject to internal kinematic constraints is determined only within an additive stress that does no work in any motion satisfying the constraints.

The simplest example of application of this principle is the case of rigid bodies. No stress of any kind does any work in a rigid-body motion, and therefore the stress in a rigid body is entirely undertermined. We are interested in the special case where the internal kinematic constraint is that of constant density.

The work done by the internal stress $\mathbf{T}$ is:

$$\mathbf{T}:\nabla\mathbf{v} = \text{tr}\,(\mathbf{T}\cdot\mathbf{D}) \qquad (4\text{-}1.1)$$

while any motion satisfying the constant density constraint is characterized by

$$\text{tr}\,(\mathbf{D}) = 0 \qquad (4\text{-}1.2)$$

Any isotropic stress $\alpha\mathbf{1}$ does no work in constant density (or isochoric) motions; in fact,

$$(\alpha\mathbf{1}):\mathbf{D} = \alpha\,\text{tr}\,(\mathbf{1}\cdot\mathbf{D}) = \alpha\,\text{tr}\,\mathbf{D} = 0 \qquad (4\text{-}1.3)$$

One can show that isotropic stresses are the only stresses which do no work in any isochoric motion. Thus, the stress in a constant density material is determined only within an additive isotropic stress.

The mathematical machinery required to handle the principle of fading memory (functionals and their smoothness properties) are discussed in the next section. In section 4-3, the mechanical theory of simple fluids with fading memory is developed in general terms. A purely mechanical theory does not include temperature among the variables, and no energy considerations are taken into account. Although this is satisfactory in handling many fluid mechanics problems, exclusion of energy considerations severely limits even the analysis of isothermal problems; a more complete thermomechanical theory requires thermodynamic considerations, and, in particular, limitations imposed by the second law of thermodynamics need to be analyzed. The thermodynamics of simple fluids is analyzed in section 4-4.

4-2   FUNCTIONALS AND THE PRINCIPLE OF FADING MEMORY

The physical assumption which forms the basis of the Reiner–Rivlin theory is that the stress is uniquely determined by the instantaneous rate of deformation. This is immediately translated into a mathematical formula by writing eq. (2-3.1):

$$\tau = \mathbf{g}(\mathbf{D}) \qquad (4\text{-}2.1)$$

which states that the stress τ equals the *value* of a function $\mathbf{g}$, whose *argument* is the tensor $\mathbf{D}$. Use is made of the mathematical concept of function, which we assume is familiar to the reader: a function is a machine which transforms one or more assigned quantities (called the arguments of the function) into another quantity, which is called the value of the function.

The function is defined for values of the arguments belonging to some set which is called the *domain* of the function; possible values of the function belong to a set called the *range* of the function. The function may be regarded as a *mapping* of the domain into the range.

Of course, the simplest example of function is the one where both the argument (or arguments) and the value are scalar quantities. Nonetheless, extension of the concept to other cases is intuitively simple. In particular, we have regarded tensors as vector-valued functions of vector arguments, enjoying the special property of linearity. Also, we have encountered functions of tensors arguments whose values are scalars, or vectors, or tensors.

In dealing with the principle of determinism of stress, one is faced with the problem that the entire history of deformation determines the present stress. Thus, one requires some rule by means of which (at least in principle), given the history of deformation, the stress can be calculated. The history of deformation is itself a function, and precisely a tensor-valued function of a scalar argument (time). This implies the need for a mapping which transforms a tensor-valued function into a tensor. The scalar counterpart of this is a mapping which transforms ordinary scalar functions into numbers; i.e., some rule by means of which, given a function, a certain number can be calculated. Such a mapping is called a *functional*.

Functionals

The difference between a function and a functional lies in the fact that, while a function's argument is some quantity (be it a scalar, or a vector, or a tensor), the argument of a functional is a function. There is some difficulty here because tensors have themselves been defined as functions; but the special property of linearity implies that tensors are uniquely determined by their nine components, so that indeed a function of a tensor argument can be regarded as a function of nine scalar arguments.

The concept of functional can be better understood by considering a few examples. Consider the following equation:

$$y = \int_a^b f(x)\,dx \qquad (4\text{-}2.2)$$

where a and b are two fixed numbers. The numerical value of y depends on which particular function $f(\ )$ one considers; thus, integration between fixed limits is an example of a functional, which maps the functions $f(\ )$ into the values y. This particular functional has also a very simple geometrical interpretation: y is the area under the curve $f(x)$ in a y–x plane between $x = a$ and $x = b$. Clearly, this area is uniquely determined by the particular function $f(x)$ that one considers.

It is important to avoid the possible confusion between a functional and a composite function, such as

$$y = f[g(x)] \qquad (4\text{-}2.3)$$

Equation (4-2.3) is equivalent to the set of two equations:

$$y = f(z) \qquad (4\text{-}2.4)$$
$$z = g(x) \qquad (4\text{-}2.5)$$

and is again a mapping of the quantity x into the quantity y. The rule to be followed is: Given x, first calculate z by means of the transformation $g(\ )$. The value of the function $g(\ )$ is then taken as the argument of the function $f(\ )$, and the value of y is calculated. The *argument* in eq. (4-2.3) is thus seen to be the quantity x; while in eq. (4-2.2) the argument is the function $f(\ )$.

Another very simple example of a functional is

$$y = f(100) \qquad (4\text{-}2.6)$$

In eq. (4-2.6) the value, y, is uniquely determined by the particular function $f(\ )$. If, for example, $f(x)$ is the function x^2, then $y = 10\,000$; if $f(x)$ is the function $\sqrt{x}$, then $y = 10$; and so on. Clearly, other examples of functionals are easily obtained.

We shall use script symbols (such as $\mathscr{H}$, $\mathscr{P}$, $\mathscr{g}$, etc.) to indicate functionals. Bold-face (lower case and capitals) will be used to identify the tensorial character of the *value* of the functional; so that, for example, $\mathscr{H}$ may be used for a functional whose value is a tensor, $\mathscr{h}$ for a functional whose value is a vector, $\mathscr{h}$ for a scalar-valued functional.

The function which is the argument of a functional will be defined in a given domain of its argument, but the value of the functional may depend on the form of the function only over a set of values of the argument which is 'smaller' than the domain. For example, in eq. (4-2.2) only the interval between a and b needs to be considered, and in eq. (4-2.6) only the value 100, although in both cases the domain of the function $f(\ )$ may be much wider. For this reason, an equation containing a functional will include indication of the interval over which the argument function needs to be considered, such as†

$$y = \mathop{\mathscr{h}}_{x=a}^{x=b} [f(x)] \qquad (4\text{-}2.7)$$

The right-hand side of eq. (4-2.2) is a special example of a functional $\mathscr{h}$ such as in eq. (4-2.7).

Norms and topology

In dealing with functionals, problems of smoothness and continuity will arise. These imply some satisfactory extension of the analogous concepts that are already known in the case of functions. A function is considered continuous basically if, for a sufficiently small variation of its argument, the corresponding variation of its value is also very small (more rigorously, smaller than any arbitrary preassigned value); smoothness implies that the derivative is continuous. In trying to extend this concept to functionals, one is faced with the difficulty of finding a proper definition of what is meant by a small variation of a function. More generally, when concepts of continuity and smoothness are to be considered for a transformation, a satisfactory definition of smallness for the variation of the argument and of the value of the transformation is required.

The reader may notice that, in order to define continuity and smoothness for a transformation, one must assign a precise meaning to the concept that two quantities ψ_1 and ψ_2 are arbitrarily close to each other, where ψ_1 and ψ_2 are either two possible arguments or two possible values of the transforma-

---

† We are here considering the case where the set of interest is an interval; clearly, more general sets can be considered.

tion. Should the ψ be scalars, the meaning is obvious: two scalars are very close to each other when the absolute value of their difference is smaller than some preassigned arbitrarily small positive number ε:

$$|\psi_1 - \psi_2| < \varepsilon \qquad (4\text{-}2.8)$$

When the ψ are not scalars, continuity can be discussed only after a precise mathematical meaning has been given to a statement such as eq. (4-2.8) above.

The quantity $|\psi_1 - \psi_2|$, to be defined when ψ is not a scalar, is called the *distance* between ψ_1 and ψ_2. In order to give a precise meaning to eq. (4-2.8), one only needs to know the conditions under which the distance between ψ_1 and ψ_2 is diminishingly small, while finite distances may remain undefined. In rigorous terms, one only needs to assign the *topology* of the space of the ψ: any transformation of this space which leaves its topology unchanged is irrelevant as far as eq. (4-2.8) is concerned. We may thus conclude by saying that continuity of a transformation is defined in terms of the topology of the domain and of the range.

In order to clarify the point made above, let us consider a case where these concepts have already, albeit intuitively, been used (indeed they are needed when either the arguments or the values of a transformation are not scalars). Consider a scalar-valued field, such as the temperature distribution in a certain region of space. The domain of such a field is the classical Euclidean space of our common experience. The statement that the temperature distribution within a body is continuous implies that the difference of temperature for two points that are infinitesimally close to each other is diminishingly small: if $\mathbf{X}_1$ and $\mathbf{X}_2$ are two such points, i.e., if

$$|\mathbf{X}_1 - \mathbf{X}_2| < \varepsilon \qquad (4\text{-}2.9)$$

the corresponding variation $|T_1 - T_2|$ is diminishingly small.† We here use the intuitive concept of distance between two points. Now consider any deformation of the body we are considering which leaves unaltered both its topology and the temperature of each material point—say a stretching with no cuts or tearings. If the temperature distribution was continuous before such a deformation, it will be continuous afterwards: in fact, although the distance between any two points may be changed, any pair of infinitesimally close points is still divided by an infinitesimally small distance.

---

† The rigorous formulation is as follows: given an arbitrarily small number δ, a positive number ε exists such that, if eq. (4-2.9) is satisfied, then $|T_1 - T_2| < \delta$.

A possible way of defining the topology of the space of some quantity ψ is to assign a *norm* in this space, i.e., to assign a rule which unequivocally transforms any of the entities considered into a non-negative scalar. If the norm for the entities ψ has been defined, the distance $|\psi_1 - \psi_2|$ is then defined as†

$$|\psi_1 - \psi_2| = \text{norm}\,(\psi_1 - \psi_2) \qquad (4\text{-}2.10)$$

Note that, while definition of a norm assigns the topology, different norms may correspond to the same topology : in fact, definition of a norm also assigns definite values to finite distances, while the topology only implies a recognition of infinitesimal distances.

A few examples may clarify the concept of norm. In the space of vectors, the norm is defined as

$$\text{norm}\,(\mathbf{a}) = \sqrt{(\mathbf{a} \cdot \mathbf{a})} = |\mathbf{a}| \qquad (4\text{-}2.11)$$

while in the space of tensors the norm is defined as

$$\text{norm}\,(\mathbf{A}) = \sqrt{(\mathbf{A} : \mathbf{A}^\mathsf{T})} = |\mathbf{A}| \qquad (4\text{-}2.12)$$

When dealing with functionals, one may choose to assign a norm to the argument functions; this norm is itself a functional which transforms functions into scalars. If such a norm is assigned, the topology of the space of functions is also defined, and continuity of the functional is defined in terms of this topology. On the other hand, one should remember that a different choice of the norm may imply the same topology, and that therefore the norm one has chosen is not uniquely determined by the properties of continuity of the functional.

Continuity and differentiability of functionals

Consider a functional $\phi = \mathscr{h}[\psi(x)]$ where the values of both the functional and the argument function may be scalars, or vectors, or tensors. Let ϕ_0 be the value corresponding to the argument function $\psi_0(\ )$, and suppose that the topology of the domain has been assigned, so that eq. (4-2.8) has a precise meaning. The functional $\mathscr{h}$ is said to be continuous at $\psi_0(\ )$ if, given an arbitrarily small positive number δ, a positive number ε exists such that the inequality

$$|\phi_0 - \phi| < \delta \qquad (4\text{-}2.13)$$

is satisfied when

$$|\psi_0 - \psi| < \varepsilon \qquad (4\text{-}2.14)$$

† We here assume that subtraction has been defined for the quantities ψ.

Continuity is, of course, defined in terms of those topologies which assign a precise meaning to eqs (4-2.13) and (4-2.14).

We now wish to introduce the concept of smoothness for a functional.

In order to do so, an extension of the concept of derivative to functionals is required. This subject is perhaps more easily introduced by considering an ordinary function of a scalar argument $h(x)$, which at $x = x_0$ possesses a first derivative $h'(x_0)$. The following equation is obvious:

$$|h(x_0 + y) - h(x_0) - h'(x_0)y| = O(|y|) \qquad (4\text{-}2.15)$$

where the term $O(|y|)$ is such that

$$\lim_{y \to 0} \frac{O(|y|)}{|y|} = 0 \qquad (4\text{-}2.16)$$

Indeed, eq. (4-2.15) can be seen to be equivalent to the usual definition of the derivative $h'(x_0)$.

A functional $\phi = \hbar[\psi(x)]$ is said to possess a *first Fréchet differential* $\delta\hbar[\psi_0(\ )|\psi_1(\ )]$ at $\psi_0(\ )$ if the following equation holds true:

$$|\hbar[\psi_0 + \psi_1] - \hbar[\psi_0] - \delta\hbar[\psi_0(\ )|\psi_1(\ )]| = O(|\psi_1|) \qquad (4\text{-}2.17)$$

where $\delta\hbar$ is a functional which is linear in its second argument $\psi_1(\ )$ and the term $O(|\psi_1|)$ is such that

$$\lim_{|\psi_1| \to 0} \frac{O(|\psi_1|)}{|\psi_1|} = 0 \qquad (4\text{-}2.18)$$

Note that, again, the existence of Fréchet differentials is defined in terms of the topology of the domain of the functional: in fact, eq. (4-2.18) requires only that a precise meaning has been assigned to the concept that the distance between ψ_0 and $\psi_0 + \psi_1$ is diminishingly small.

The Fréchet differential should be regarded as the direct extension of the ordinary differential to functionals: in fact, comparing eqs (4-2.15) and (4-2.17), $\delta\hbar$ is seen to be the analogue of the term $h'y = dh$. Higher order Fréchet differentials, indicated by $\delta^N\hbar$, can be defined in a similar way.

Fading memory

When dealing with the theory of simple fluids, one is faced with the situation where some dependent variable (typically, stress) depends on the history of one or more quantities (typically, the history of deformation). These histories are functions of time, and therefore the constitutive equation takes the form of a functional. The principle of fading memory requires that, given two histories which are almost equal in the recent past but may differ quite widely in the distant past, the corresponding two values of the dependent variable should be rather close. This requirement is satisfied provided that the constitutive functional is assumed to be continuous with respect to an appropriate topology of the space of histories, which assigns a small distance between two such functions. Rigorous formulations of the principle of fading memory are to be given in terms of assumptions of continuity and smoothness for the constitutive functionals.

In the formulation of the principle of fading memory, it is preferable to use the time lag s, rather than time, as the independent variable. Fading memory implies a smaller relevance of phenomena that have taken place in the distant past (i.e., at large values of s) than in the recent past (i.e., at small values of s).

In the mechanical theory of simple fluids with fading memory, to be discussed in the next section, use is made of a formulation of the principle of fading memory due to Coleman and Noll [3], which assigns the topology of the domain of the constitutive functional by introducing an *obliviator*, i.e., a scalar-valued function $h(s)$, which has the following properties:

$$h(0) = 1 \qquad (4\text{-}2.19)$$

$$h(s) \geq 0 \qquad (4\text{-}2.20)$$

$$\lim_{s \to \infty} s^N h(s) = 0 \qquad (4\text{-}2.21)$$

where the integer N is called the order of the obliviator.

The norm of some history $\psi^t(s)$ in the domain of the constitutive functional (indicated by $|\psi^t|_h$) is defined as:

$$|\psi^t|_h = \left[\int_0^\infty h(s)[|\psi^t|]^k \, ds \right]^{1/k} \qquad (4\text{-}2.22)$$

where k is a positive integer and $|\psi^t|$ is the instantaneous value of the norm of the value of ψ^t (e.g., if ψ^t is a tensor-valued function such as $\mathbf{C}^t$, $|\psi^t| = \sqrt{(\mathbf{C}^t:\mathbf{C}^{tT})}$). The right-hand side of eq. (4-2.22) is thus seen to represent a weighed average of the norm of ψ^t over the entire past, where the weight assigned to the distant past is, in view of eq. (4-2.21), much less than the weight assigned to the recent past. It is evident that assumptions of continuity and differentiability of the constitutive functional in terms of the norm defined in eq. (4-2.22) represent a formalization of the principle of fading memory.

It is important to point out that different obliviators may correspond to the same topology of the function space. Therefore, $h(s)$ cannot be regarded as a function characteristic of the material under consideration.

4-3 SIMPLE FLUIDS WITH FADING MEMORY.
MECHANICAL THEORY

We are now in a position to formalize the concepts discussed in section 4-1 and obtain an explicit constitutive equation for constant density simple fluids with fading memory.

We will assume that the history of deformation in an arbitrarily small neighborhood of the material point considered is entirely described by the deformation gradient $\mathbf{F}^t$. This is a restrictive form of the principle of local action, inasmuch as higher gradients of the motion (as defined in eq. (3-3.1)) could also be relevant. The assumption of constant density, the principle of determinism of stress, and the non-existence of a natural state are all satisfied by assuming the following two equations as constitutive definitions of a constant density simple fluid:

$$\det \mathbf{F}^t = 1 \qquad (4\text{-}3.1)$$

$$\tau = \underset{s=0}{\overset{s=\infty}{\mathscr{P}}} [\mathbf{F}^t] \qquad (4\text{-}3.2)$$

where $\mathscr{P}$ is a tensor-valued functional of the tensor-valued function $\mathbf{F}^t$ over the entire past. Homogeneous materials are characterized by the fact that the functional $\mathscr{P}$ is the same for all material points.

Let us now consider the restrictions imposed on the functional $\mathscr{P}$ by the principle of material objectivity. If $\mathbf{Q}(s)$ is any time-dependent orthogonal tensor, the functional $\mathscr{P}$ must, in view of eq. (3-3.3), satisfy the following equation:

$$\mathbf{Q}(0) \cdot \underset{s=0}{\overset{s=\infty}{\mathscr{P}}} [\mathbf{F}^t] \cdot \mathbf{Q}^T(0) = \underset{s=0}{\overset{s=\infty}{\mathscr{P}}} [\mathbf{Q}(s) \cdot \mathbf{F}^t \cdot \mathbf{Q}^T(0)] \qquad (4\text{-}3.3)$$

Let us now consider the special case where the $\mathbf{Q}(\ )$ function is taken as coincident with $(\mathbf{R}^t)^T$:

$$\mathbf{Q}(s) = (\mathbf{R}^t)^T \qquad (4\text{-}3.4)$$

which corresponds (see the discussion following eq. (3-3.29)) to considering a corotating frame. In view of eq. (3-1.33), eq. (4-3.3) reduces to:

$$\underset{s=0}{\overset{s=\infty}{\mathscr{P}}} [\mathbf{F}^t] = \underset{s=0}{\overset{s=\infty}{\mathscr{P}}} [(\mathbf{R}^t)^T \cdot \mathbf{F}^t] \qquad (4\text{-}3.5)$$

Considering that $\mathbf{F}^t$ has the polar decomposition

$$\mathbf{F}^t = \mathbf{R}^t \cdot \mathbf{U}^t \qquad (4\text{-}3.6)$$

and thus

$$(\mathbf{R}^t)^{\mathrm{T}} \cdot \mathbf{F}^t = \mathbf{U}^t \qquad (4\text{-}3.7)$$

eq. (4-3.5) reduces to

$$\underset{s=0}{\overset{s=\infty}{\mathscr{P}}} [\mathbf{F}^t] = \underset{s=0}{\overset{s=\infty}{\mathscr{P}}} [\mathbf{U}^t] \qquad (4\text{-}3.8)$$

Equation (4-3.8) represents the principle of material objectivity as applied to the change of frame from whatever the original frame was to the corotating frame. In the corotating frame, $\mathbf{F}^t$ and $\mathbf{U}^t$ coincide; moreover, the corotating and the original frame coincide at $s = 0$, and therefore the stress at time t must be the same in the two frames. Physically, eq. (4-3.8) implies that the stress at a material point is the same for two histories of deformations which differ from each other only by a superimposed rigid rotation history.

Equations (4-3.2) and (4-3.8) show that the stress τ is uniquely determined by the history $\mathbf{U}^t$. This, in turn, implies that τ is also uniquely determined by the Cauchy strain $\mathbf{G}^t$: in fact, given $\mathbf{G}^t$, the history $\mathbf{C}^t$ is immediately obtained as

$$\mathbf{C}^t = \mathbf{G}^t + \mathbf{1} \qquad (4\text{-}3.9)$$

Moreover, eq. (3-1.19) has a unique solution for $\mathbf{U}$, inasmuch as there is only one symmetric positive-definite tensor whose square equals any given symmetric positive-definite tensor. We may thus define a new functional $\mathscr{H}$ as

$$\underset{s=0}{\overset{s=\infty}{\mathscr{H}}} [\mathbf{G}^t] = \underset{s=0}{\overset{s=\infty}{\mathscr{P}}} [\mathbf{U}^t] \qquad (4\text{-}3.10)$$

and write the constitutive equations for a constant density simple fluid in the more convenient form:

$$\det (\mathbf{G}^t + \mathbf{1}) = 1 \qquad (4\text{-}3.11)$$

$$\tau = \underset{s=0}{\overset{s=\infty}{\mathscr{H}}} [\mathbf{G}^t] \qquad (4\text{-}3.12)$$

In order to fully satisfy the principle of material objectivity, the functional $\mathscr{H}$ must be isotropic, i.e., for any orthogonal tensor,

$$\mathbf{Q} \cdot \underset{s=0}{\overset{s=\infty}{\mathscr{H}}} [\mathbf{G}^t] \cdot \mathbf{Q}^{\mathrm{T}} = \underset{s=0}{\overset{s=\infty}{\mathscr{H}}} [\mathbf{Q} \cdot \mathbf{G}^t \cdot \mathbf{Q}^{\mathrm{T}}] \qquad (4\text{-}3.13)$$

Equations (4-3.11) and (4-3.12), the latter with a functional obeying eq. (4-3.13), constitute the definition of a constant density simple fluid.

Most (if not all) constitutive equations which have been proposed in the literature correspond, if properly frame-invariant, to special choices of the form of the functional $\mathscr{H}$ in eq. (4-3.12). A few problems of non-Newtonian fluid mechanics can be solved without assigning any special form to $\mathscr{H}$; some of these problems will be discussed in the next chapter. In dealing with more complex problems, more special constitutive assumptions, to be discussed in chapter 6, are needed.

Let us now consider one problem which can be solved without assigning any specific form to the functional $\mathscr{H}$, i.e., hydrostatics. Consider a simple fluid which is, and has always been, in a state of rest, so that

$$\mathbf{G}^t = \mathbf{O}(s) \qquad (4\text{-}3.14)$$

where $\mathbf{O}(s)$ is a tensor-valued function whose value is the zero tensor for any value of the argument s. Applying eq. (4-3.13) to this special case, one obtains

$$\mathbf{Q} \cdot \underset{s=0}{\overset{s=\infty}{\mathscr{H}}} [\mathbf{O}(s)] \cdot \mathbf{Q}^{\mathrm{T}} = \underset{s=0}{\overset{s=\infty}{\mathscr{H}}} [\mathbf{O}(s)] \qquad (4\text{-}3.15)$$

A tensor $\mathbf{A}$ satisfying the equation

$$\mathbf{Q} \cdot \mathbf{A} \cdot \mathbf{Q}^{\mathrm{T}} = \mathbf{A} \qquad (4\text{-}3.16)$$

for *any* orthogonal tensor $\mathbf{Q}$ is necessarily an isotropic tensor. Thus, eq. (4-3.16) implies that

$$\tau = \underset{s=0}{\overset{s=\infty}{\mathscr{H}}} [\mathbf{O}(s)] = \alpha\mathbf{1} \qquad (4\text{-}3.17)$$

The conclusion is thus reached that the stress in a simple fluid which has always been at rest is isotropic. Conversely, a simple fluid cannot sustain indefinitely a non-isotropic stress without eventually starting to flow [4]. This conclusion implies that theories of plasticity (i.e., of fluids endowed with a yield stress) are not special cases of the theory of simple fluids.

Note that the value of the functional $\mathscr{H}$ is determined only within an arbitrary isotropic tensor. In other words, a whole family of functionals $\mathscr{H}$ is defined for any given material, whose values differ from each other by isotropic tensors. A particular functional $\mathscr{H}'$ can be identified by the normalization:

$$\mathrm{tr}\,\mathscr{H}' = 0 \qquad (4\text{-}3.18)$$

in which case eq. (4-3.12) reduces to

$$\tau' = \underset{s=0}{\overset{s=\infty}{\mathscr{H}'}} [\mathbf{G}^t] \qquad (4\text{-}3.19)$$

So far, we have not made use of the principle of fading memory. The results to be discussed in the remainder of this section are based on the following simple formulation of the principle of fading memory [3, 5]: 'The functional $\mathcal{H}'$ is continuous and N times Fréchet differentiable at the rest history $\mathbf{G}^t = \mathbf{O}(s)$ with respect to the norm defined in eq. (4-2.22).'

Different formulations of the principle of fading memory will be required in handling the thermodynamic theory to be discussed in the next section. On the basis of the formulation given above, which will be referred to in the following as the 'rest history formulation of the principle of fading memory,' one may obtain approximations to the general constitutive equation in a rigorous way. These are obtainable for the limiting cases of very slow flows [5] and of very small deformations [3].

Let us consider some particular flow pattern, and let $\mathbf{G}^t$ be the deformation history at some instant of observation t; for the development to be given in the following, one needs to assume that $\mathbf{G}^t$ is continuous at $s = 0$. Now, consider another history of deformation ${}_\alpha\mathbf{G}^t$ defined as

$$
{}_\alpha\mathbf{G}^t(s) = \mathbf{G}^t(\alpha s), \quad 0 \le \alpha \le 1 \qquad (4\text{-}3.20)
$$

The flow described by ${}_\alpha\mathbf{G}^t$ is essentially the same as that described by $\mathbf{G}^t$ (which is obtained for $\alpha = 1$), only it takes place at a slower rate: configurations assumed αs time units in the past in the flow described by $\mathbf{G}^t$ have been assumed s time units ago in the flow described by ${}_\alpha\mathbf{G}^t$. By letting $\alpha \to 0$, the behavior of the fluid in any slow flow can be analyzed.

From the assumed continuity of $\mathbf{G}^t$ at $s = 0$, it follows that, as $\alpha \to 0$, any history ${}_\alpha\mathbf{G}^t$ approaches the zero history in the recent past: in fact, $\mathbf{G}^t(0) = \mathbf{0}$. On the basis of the rest history formulation of the principle of fading memory, Nth-order approximations for slow flows to the general constitutive equation of a simple fluid can be obtained. The Nth-order approximation is understood in the sense that the norm of the remainder is of order α^{N+1}. The algebra required to obtain the approximations is cumbersome, and only the results will be given.

The zero-order approximation is simply that, for a sufficiently slow flow, the stress is hydrostatic, and the constitutive equation reduces to

$$
\text{for } N = 0, \qquad\qquad \mathbf{T} = -p\mathbf{1} \qquad (4\text{-}3.21)
$$

Incidentally, eq. (4-3.21) can be regarded as the constitutive equation for incompressible ideal fluids, which is assumed as the basis for classical hydrodynamics.

The first-order approximation is simply the Newtonian constitutive equation:

for $N = 1$, $$\mathbf{T} = -p\mathbf{1} + 2\mu\mathbf{D} \qquad (4\text{-}3.22)$$

This result shows that the classical Newtonian theory has a proper status of asymptotic validity for slow flows of simple fluids with fading memory. Ordinary Newtonian liquids may be regarded as simple fluids whose natural time Λ is so small that *any* flow of pragmatic interest may be considered slow, and thus can be analyzed in terms of eq. (4-3.22).

The second-order approximation to the general constitutive equation takes the following form:

for $N = 2$, $$\mathbf{T} = -p\mathbf{1} + 2\eta_0\mathbf{D} + \beta_1\mathbf{D}^2 + \beta_2\mathbf{A}_2 \qquad (4\text{-}3.23)$$

where η_0, β_1, and β_2 are constants. Equation (4-3.23) can be assumed to be the definition of a 'second-order fluid.' Equation (4-3.23) is particularly useful for obtaining perturbation solutions around purely viscous approximations: the term $\beta_2\mathbf{A}_2$ is the first approximation to the influence of memory on the behavior of real fluids.

A very useful result of the rest history formulation is another form of general approximation to the constitutive equation of simple fluids. Rather than consider slow flows, let us examine the case of small deformations. This is achieved, for example, in oscillatory motions of small amplitude. Such a motion needs only to have taken place in the recent past in order that the norm of $\mathbf{G}^t$ be small. It may then be proved that a first-order approximation to the constitutive equation of a simple fluid with fading memory takes the form

$$\mathbf{T} = -p\mathbf{1} + \int_0^\infty f(s)\mathbf{G}^t \, ds \qquad (4\text{-}3.24)$$

Equation (4-3.24) is the constitutive equation of 'linear viscoelasticity.' The second-order approximation takes the form

$$\mathbf{T} = -p\mathbf{1} + \int_0^\infty f(s)\mathbf{G}^t \, ds + \int_0^\infty \int_0^\infty \{\, \alpha(s_1, s_2)\mathbf{G}^t(s_1) \cdot \mathbf{G}^t(s_2)$$

$$+ \beta(s_1, s_2) \, \mathrm{tr} \; \mathbf{G}^t(s_1)\mathbf{G}^t(s_2)\} \; ds_1 \, ds_2 \qquad (4\text{-}3.25)$$

where the functions $f(\ )$, $\alpha(\ )$, and $\beta(\ )$ are characteristic of the particular material being considered. The functions $\alpha(\ )$ and $\beta(\ )$ are symmetric with respect to their two arguments.

It should be noted that the functions $f(\ )$, $\alpha(\ )$, and $\beta(\ )$ must tend to zero sufficiently rapidly when $s \to \infty$, if the principle of fading memory is to hold. Also, $f(\ )$, $\alpha(\ )$, and $\beta(\ )$ are functions which are characteristic of the material considered, in contrast with the obliviator $h(s)$. In fact, given a functional $\mathscr{H}'$, the approximations in eqs (4-3.24) and (4-3.25) are uniquely determined, while the obliviator $h(s)$ is not.

4-4 THERMODYNAMICS

In this section we introduce the treatment of thermodynamics as applicable to materials with memory. We follow the recent work on the subject by Coleman [6, 7], to which the reader is referred for the rigorous formalism and all mathematical detail.

The subject material is ordered as follows. We first discuss a few introductory points of 'classical' thermodynamics; we assume that the reader is familiar with macroscopic thermodynamics as usually discussed in engineering courses. We then proceed to discuss some *general* thermodynamic results which are applicable to *all* materials (therefore including materials with memory). Next, we consider in a very simple limiting case how the concept of memory influences thermodynamic results, and finally we give the basic results of the thermodynamic theory for simple fluids with fading memory.

Constitutive assumptions of classical thermodynamics

The starting point of the classical thermodynamic theory is the so-called 'equation of state,' i.e., a relationship among pressure p, specific volume V, and absolute temperature T:

$$p = p(V, T) \qquad (4\text{-}4.1)$$

Equation (4-4.1) is really a constitutive assumption: one is considering materials for which the stress is isotropic, and is thus completely identified by the value of one scalar property (pressure), and the constitutive assumption is made that pressure is entirely determined by the instantaneous values of the specific volume and of temperature.

Further, we introduce functions of state, such as internal energy, entropy, etc., each of which is regarded as unequivocally determined by any two of the others, so that, for example, the internal energy can be expressed as

$$U = u_1(V, T) = u_2(p, T) = u_3(S, V) = \cdots \qquad (4\text{-}4.2)$$

In eq. (4-4.2), the functions $u_i(\ )$ are different from each other, although their values coincide. This leads to some confusion if the same symbol is used to indicate a function and its value, a confusion which is avoided by adding indices to partial derivatives, such as

$$\left(\frac{\partial U}{\partial T}\right)_V \equiv \frac{\partial u_1}{\partial T}; \qquad \left(\frac{\partial U}{\partial T}\right)_p \equiv \frac{\partial u_2}{\partial T}; \qquad \cdots \qquad (4\text{-}4.3)$$

In the standard approach, some form of statement of the first and second laws of thermodynamics leads to the so-called Maxwell equations, of which we here only consider the following example (A being Helmholtz' free energy, $A = U - TS$):

$$dA = -S\,dT - p\,dV \qquad (4\text{-}4.4)$$

From eq. (4-4.4) the following two relations are obtained:

$$p = -\left(\frac{\partial A}{\partial V}\right)_T \qquad (4\text{-}4.5)$$

$$S = -\left(\frac{\partial A}{\partial T}\right)_V \qquad (4\text{-}4.6)$$

Equations (4-4.5) and (4-4.6) play a very important role in a logical assessment of the classical thermodynamic theory which is not often fully appreciated. They show that a knowledge of the function $A(T, V)$ completely determines the thermomechanical behavior of the material in the sense that the equation for the pressure (eq. (4-4.1)) is obtained from eq. (4-4.5), while from eq. (4-4.6) the internal energy, heat capacity, etc. can be calculated. For example,

$$C_V \equiv \left(\frac{\partial U}{\partial T}\right)_V = \left(\frac{\partial(A + TS)}{\partial T}\right)_V = \left(\frac{\partial A}{\partial T}\right)_V + T\left(\frac{\partial S}{\partial T}\right)_V + S$$

$$= -T\left(\frac{\partial^2 A}{\partial T^2}\right)_V \qquad (4\text{-}4.7)$$

The function $A(T, V)$, also called the 'entropic equation of state,' must be regarded as a constitutive equation for the material being considered. It assumes that the Helmholtz' free energy only depends on the present values of temperature and specific volume and, if given explicitly, specifies how. It is, however, more general than the 'mechanical' equation of state (eq. (4-4.1)) in that it 'contains' the mechanical equation together with additional information. On the other hand, eq. (4-4.1) does not determine the function $A(T, V)$: a student of classical thermodynamics soon realizes that the thermodynamic behavior of a material is not determined by the p–V–T relationship only, and that an additional relationship such as a heat capacity equation is needed.

Thus the function $A(T, V)$ enjoys a special status which, it must be observed, is not shared by other energetic equations of state. For example, if the internal energy function $u_1(T, V)$ is given, the p–V–T relationship cannot be derived. It is true, however, that a different choice of independent variables such as S and V changes the picture in that the function $u_3(S, V)$ becomes all inclusive. For reasons to be made clear in the following, we let temperature remain an independent variable.

Equations (4-4.4), (4-4.5), and (4-4.6) derive from the first and second laws of thermodynamics as applied to materials whose 'state' (pressure, free energy, etc.) is determined by the present values of T and V only. Equations (4-4.5) and (4-4.6) represent restrictions imposed by the laws of thermodynamics on constitutive assumptions in the sense that it is forbidden to assume constitutive equations, for example, for A and p which do not obey eq. (4-4.5). In the treatment which follows, we shall see how the corresponding equations (or restrictions) for materials with memory are obtained. We shall have the additional complication that the stress cannot, in general, be considered isotropic.

One further observation: The reader familiar with thermodynamics textbooks may remember the sense of uneasiness that was felt when equations such as eq. (4-4.4) were derived, due to some rather vague arguments concerning reversible and irreversible processes which were used somewhere along the line. In the following we shall refer to 'real' processes which are thus irreversible. The relationships obtained belong to the realm of irreversible thermodynamics. Equilibrium relationships (or thermostatics) and linear irreversible thermodynamics (of the type due to Onsager) can be obtained as limiting cases.

The role of the second law of thermodynamics

In section 1-1, the first law of thermodynamics (i.e., the energy balance) was shown to be one of the fundamental equations required in order to be able to solve—at least in principle—any problem of fluid mechanics. It has to be set alongside the mass and momentum balances, while three 'constitutive' equations are needed simultaneously: one for the total stress (which may be decomposed into one for pressure and one for the deviatoric stress), one for the heat flux (which need not be in the simple form of Fourier's law), and one for internal energy (see Table 1-2).

The second law of thermodynamics plays an entirely different role.

In fact, suppose that entropy is added to the list of variables in Table 1-2, as well as an 'entropic' constitutive equation (i.e., an equation giving the entropy S as a function of whatever may be the relevant variables); furthermore, suppose that an appropriate statement of the second law of thermodynamics is considered. One thus has the system of equations and variables listed in Table 4-1. From this list, it is evident that the second law of thermodynamics overdetermines the problem, since one has one more equation than the number of variables. One is thus led to the conclusion that the second law of thermodynamics imposes some restrictions, the nature of which needs to be investigated.†

A subtle point arises here, since one could at first sight take two different viewpoints. One could presume that the second law of thermodynamics places some restrictions on admissible processes, i.e., its fulfilment would require that some transformations of a given material are forbidden. Alternatively, one could assume that restrictions are imposed on constitutive equations; this second viewpoint will hopefully be shown to be the correct one.

Table 4-1 EQUATIONS AND VARIABLES FOR A COMPLETE THERMODYNAMIC THEORY

---

(1) Equations
  (*a*) *Fundamental laws*
  Conservation of mass (scalar)
  Conservation of momentum (vectorial)
  Conservation of energy (scalar)
  Second law
  (*b*) *Constitutive equations*
  Rheological (for the total stress; tensorial)
  Energetic (for internal energy; scalar)
  Thermal (for the heat flux; vectorial)
  Entropic (for entropy, or alternatively for Helmholtz' free energy; scalar)

(2) Variables
  Density (scalar)
  Velocity (vector)
  Internal energy (scalar)
  Total stress (tensor)
  Temperature (scalar)
  Heat flux (vector)
  Entropy, or alternatively Helmholtz' free energy (scalar)

---

† One may notice that the second law of thermodynamics is not an equality, but an inequality; still, its fulfilment does imply some restrictions.

A generalized thermodynamic approach

Let us start with the first law of thermodynamics, which will be written in a general form allowing both compressibility and energy influx by radiation. Let P be the 'stress power,' i.e., the work done by the internal stresses per unit mass:

$$P = \frac{1}{\rho}\mathbf{T}:\mathbf{D} = \frac{1}{\rho}\boldsymbol{\tau}':\mathbf{D} - p\dot{V} \qquad (4\text{-}4.8)$$

where $V = 1/\rho$ is the specific volume, and let Q be the energy influx per unit volume. Then, the first law of thermodynamics takes the following general form:

$$\dot{U} = -\frac{1}{\rho}\mathbf{\nabla}\cdot\mathbf{q} + P + \frac{Q}{\rho} \qquad (4\text{-}4.9)$$

which differs from eq. (1-10.14) because of the term $-p\dot{V}$ accounting for compressibility, and of the term Q/ρ accounting for possible radiation. Note that Q is a quantity which—in principle—can be assigned at will: one may radiate as much energy as desired by controlling the temperature of the external radiation source, hence *independently of the state of the material considered*. The importance of this point will soon become clear.

The second law of thermodynamics will be written in the form of Gibbs–Duhem's inequality [8]:

$$\rho\dot{S} + \left[\mathbf{\nabla}\cdot\left(\frac{\mathbf{q}}{T}\right) - \frac{Q}{T}\right] \geq 0 \qquad (4\text{-}4.10)$$

In eq. (4-4.10) the first term is the rate of increase of entropy in the material element considered. The term in square brackets may be regarded as the rate of increase of entropy of the universe surrounding the material element; in fact, $\mathbf{\nabla}\cdot(\mathbf{q}/T)$ is the entropy flux out of the element due to conduction, and $-Q/T$ is the entropy efflux due to radiation. Equation (4-4.10) can be regarded as a formalization of the statement that for any process the total rate of entropy increase is non-negative.

Combination of eqs (4-4.9) and (4-4.10) (note that $\mathbf{\nabla}\cdot(\mathbf{q}/T) = (1/T)\mathbf{\nabla}\cdot\mathbf{q} - (1/T^2)\mathbf{q}\cdot\mathbf{\nabla}T$) yields

$$T\dot{S} - \dot{U} + P - \frac{1}{\rho T}\mathbf{q}\cdot\mathbf{\nabla}T \geq 0 \qquad (4\text{-}4.11)$$

Let us now introduce the Helmholtz' free energy A, defined as

$$A = U - TS \qquad (4\text{-}4.12)$$

Differentiation of eq. (4-4.12), and substitution into eq. (4-4.11) yields

$$\dot{A} + S\dot{T} - P + \frac{1}{\rho T}\mathbf{q}\cdot\nabla T \leq 0 \qquad (4\text{-}4.13)$$

Let us now consider the entropic constitutive equation in the form of an equation giving the value of A as a function of whatever may be the relevant independent variables. Among the latter, temperature is certainly included, so that one may write, in general,

$$A = a[T, \text{other variables}] \qquad (4\text{-}4.14)$$

Note that, in classical thermodynamics, the 'other variables' would reduce simply to the specific volume V.

Let us assume that the function $a(\ )$ is differentiable with respect to T; this will be seen to be a rather important assumption when materials with memory are considered. One has:

$$\dot{A} = \frac{\partial a}{\partial T}\dot{T} + A_T \qquad (4\text{-}4.15)$$

where A_T is the rate of increase of free energy due to all independent variables except temperature. The physical meaning of A_T will soon become clear; in classical thermodynamics, $A_T = -p\dot{V}$.

Substitution of eq. (4-4.15) into eq. (4-4.13) yields

$$\left(\frac{\partial a}{\partial T} + S\right)\dot{T} + A_T - P + \frac{1}{\rho T}\mathbf{q}\cdot\nabla T \leq 0 \qquad (4\text{-}4.16)$$

A crucial point of the procedure now arises. Consider that, given a material undergoing some process, at some instant of observation, without changing anything of the past history to which the material has been subjected, control of the radiating energy Q allows the value of $\dot{T}$ to be assigned at will.† Hence, in order to fulfil eq. (4-4.16) whatever the value of $\dot{T}$, the coefficient of $\dot{T}$ must be identically zero; i.e.,

$$\frac{\partial a}{\partial T} = -S \qquad (4\text{-}4.17)$$

Equation (4-4.17), which clearly is a generalization of eq. (4-4.6), has three important consequences. First, the internal energy U is obtained as

$$U = A + TS = a(\ ) - T\frac{\partial a(\ )}{\partial T} \qquad (4\text{-}4.18)$$

---

† The reason for choosing T among the independent variables is now clear. It is a variable whose rate of change can be controlled at will by means of the radiating energy Q.

which shows that the energetic constitutive equation (i.e., the equation for U) is not independent of the entropic equation, but is in fact entirely determined by the latter: hence, a restriction imposed by the second law on constitutive equations has been obtained.

Second, eq. (4-4.16) reduces to

$$A_T - P + \frac{1}{\rho T}\mathbf{q} \cdot \nabla T \leq 0 \qquad (4\text{-}4.19)$$

which is an entirely general, and much simpler, statement of the second law of thermodynamics.

Third, the physical meaning of the quantity A_T is now easily obtained; in fact,

$$A_T = \dot{A} - \frac{\partial a}{\partial T}\dot{T} = \dot{A} + S\dot{T} = \dot{U} - T\dot{S} \qquad (4\text{-}4.20)$$

On comparing eq. (4-4.20) with eq. (1-10.17), one obtains

$$A_T = \dot{U}_{el} \qquad (4\text{-}4.21)$$

i.e., A_T is identified with the rate of increase of elastic energy [9].

Energy dissipation

We are now in a position to formalize the somewhat elusive concept of energy dissipation which is somehow embodied in the second law of thermo-dynamics. Let us define a 'rate of energy dissipation,' D, as

$$D = T\dot{S} + \frac{T}{\rho}\nabla \cdot \left(\frac{\mathbf{q}}{T}\right) - \frac{Q}{\rho} \qquad (4\text{-}4.22)$$

Since D equals the left-hand side of eq. (4-4.10) multiplied by the essentially positive quantity T/ρ, D is necessarily non-negative. With this definition, the concept that the energy dissipation is non-negative in any real process is formalized rigorously.

Upon substitution of eqs (4-4.9), (4-4.12), (4-4.15), and (4-4.17), eq. (4-4.22) becomes

$$D = P - A_T - \frac{1}{\rho T}\mathbf{q} \cdot \nabla T \geq 0 \qquad (4\text{-}4.23)$$

Equation (4-4.23) shows that dissipation may take place due to mechanical effects (the first two terms) and to thermal effects (the last term). We may define a mechanical dissipation rate D_M, and a thermal dissipation rate D_T, as

$$D_M = P - A_T \qquad (4\text{-}4.24)$$

$$D_T = -\frac{1}{\rho T}\mathbf{q} \cdot \nabla T \qquad (4\text{-}4.25)$$

$$D = D_M + D_T \geq 0 \qquad (4\text{-}4.26)$$

Equations (4-4.24)–(4-4.26) should be discussed in some detail. First notice that eq. (4-4.26) does not imply that D_T and D_M are separately non-negative, but only that their sum is non-negative. Hence, one has:

(i) for homothermal processes, $\nabla T = \mathbf{0}$, the mechanical dissipation rate is non-negative;

(ii) for materials at rest, $\dot{V} = 0$, $P = 0$; if furthermore $A_T = 0$, the thermal dissipation rate is non-negative, hence $\mathbf{q}$ and ∇T form an obtuse angle: the heat flux vector has a positive component in the direction of decreasing temperatures;

(iii) in general, there is no need for D_T and D_M to be separately non-negative, unless special constitutive assumptions allow the mechanical and the thermal effects to be uncoupled.

Equation (4-4.24) can be verbalized by saying that, in general, the stress power equals the sum of the mechanical dissipation rate and the rate of accumulation of elastic energy, a concept which was already embodied, although on a more intuitive basis, in eq. (1-10.18).

Note that, even in homothermal processes where D_M is necessarily non-negative, there is in general no need for the stress power to be non-negative unless some special constitutive assumptions infer that $A_T = 0$. This is the case of incompressible purely viscous fluids, where $A_T = 0$, $\dot{V} = 0$, and hence $\boldsymbol{\tau}':\mathbf{D}$ is necessarily non-negative: a special conclusion which, in fact, is so often used in fluid mechanics as to be erroneously regarded as general truth.

A final consideration is in order. Upon substitution of eqs (4-4.8), (4-4.17), and (4-4.24) into eq. (4-4.15), one has

$$\dot{A} = -p\dot{V} - S\dot{T} + \frac{1}{\rho}\boldsymbol{\tau}':\mathbf{D} - D_M \qquad (4\text{-}4.27)$$

If one considers ideal fluids, both $\boldsymbol{\tau}'$ and D_M are zero; for purely viscous fluids, $D_M = (1/\rho)\boldsymbol{\tau}':\mathbf{D}$ (see the last part of this section). Hence, in both cases one obtains

$$\dot{A} = -p\dot{V} - S\dot{T} \qquad (4\text{-}4.28)$$

which is, of course, the Maxwell equation for the Helmholtz' free energy of 'classical' thermodynamics, i.e., eq. (4-4.4). It cannot be overemphasized that eq. (4-4.28) does not hold in general, and that simple fluids with fading memory need to be analyzed in terms of eq. (4-4.27).

Thermodynamics and memory

In order to discuss the implications of the concept of memory in thermodynamic theory, let us consider initially a very simple case. Assume that the free energy depends not only on the instantaneous value of temperature, but also on the past history of temperature of the material point considered. Thus, we specify eq. (4-4.14) one step further and write

$$A = \underset{s=0}{\overset{s=\infty}{a}} \; [T^t(s); T, \text{other variables}] \qquad (4\text{-}4.29)$$

In eq. (4-4.29) a is a functional of the argument function $T^t(s)$—i.e., of the temperature history—whose value also depends explicitly on the instantaneous value of temperature as well as on other variables yet to be specified. Notice that, in order to apply the results obtained so far, one needs T to appear explicitly among the independent variables, since we wish eq. (4-4.15) to preserve its meaning. Notice further that the present value of temperature also appears in the history as $T = T^t(0)$. However, this may not constitute a redundancy as clarified by the following discussion.

Assume that the functional a in eq. (4-4.29) is continuous *throughout its domain* with respect to a norm such as that defined in eq. (4-2.22). Now consider two histories, T^t_1 and T^t_2, which differ from each other only on isolated instants of time in the past. Two such histories, according to eq. (4-2.22), are at zero distance from each other, and therefore the value of A is the same for both. The principle of fading memory, as formulated above, implies that isolated peaks of zero duration which may have taken place in the past are irrelevant. Figure 4-1 gives an example of two temperature histories of the type discussed above.

Let us now consider the case where a discontinuity arises at the instant of observation $s = 0$. We cannot refer to such a discontinuity as a peak, because T may or may not, at $\tau > t$, drop back to the previous value; neither can the material at time t foresee which is the case, if the principle of determinism is to be respected. It is perhaps physically intuitive that, when a discontinuity arises at $s = 0$, the response of the material, i.e., the value of A at time t, is in fact influenced by such a discontinuity. Should the reader have

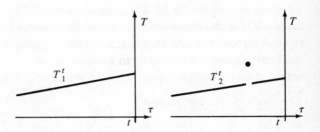

FIGURE 4-1
Two histories differing only over a finite
set of instants of time in the past.

difficulty in accepting this point on the basis of physical intuition, it may be
regarded as a postulate on which the thermodynamic theory is built.†

The norm defined in eq. (4-2.22) assigns a zero distance between two
temperature histories which differ only for the value at $s = 0$ (it should be
stressed that of two such histories, at least one is discontinuous at $s = 0$).
It is therefore necessary to write the constitutive assumption for A in the form
of eq. (4-4.29), which assigns special physical relevance to the instantaneous
value of T.

It is important to point out the significance of the partial derivative of a
with respect to T, which appears in eqs (4-4.15) and (4-4.17) above. Such a
partial derivative implies that a change in temperature is considered at the
instant of observation, albeit differential, *at constant* $T'(s)$, i.e., while the past
history of temperature is kept constant. This means that discontinuities
at the instant of observation are considered.

In the following, we shall refer to partial derivatives of this type as
'instantaneous' derivatives; they measure the change of a dependent variable
in response to an instantaneous change of some independent variable. In
classical thermodynamics, time never appears explicitly because the *rate*
at which phenomena take place is regarded as irrelevant. When fluids
possessing memory are considered, the rate is no more irrelevant, and results
analogous to those of classical thermodynamics (such as eq. (4-4.17)) are
obtained in the limit of transformations taking place at an infinite rate.

---

† This point is discussed in great mathematical detail by Coleman and Mizel [10]; whatever
approach is chosen, some postulate suggested by physical intuition is used at some point.

Another important point needs to be discussed. In order to obtain eq. (4-4.17), it was assumed that the instantaneous value of $\dot{T}$ can be assigned at will by controlling Q, without changing anything of the past history of the material. It may at first sight seem that, upon changing $\dot{T}$, the history T^t would change—an argument which would invalidate the entire treatment. Now, given a history T^t_1, another history T^t_2 may always exist which fulfils the following conditions:

(i) $T^t_2(0) = T^t_1(0)$, i.e., the instantaneous values of T are equal for the two histories;

(ii) $-\dot{T}^t_2(0) = \dot{T}$ has some arbitrarily preassigned value;

(iii) T^t_2 and T^t_1 are at an arbitrarily small distance from each other.

That such a history may exist can generally be proved [6]; we here omit the proof, and only give, in Fig. 4-2, an intuitive example of two histories satisfying (i), (ii), and (iii). Condition (iii) can be satisfied provided the kink at low values of s of T^t_2 is sufficiently small. We can thus conclude that it is indeed possible to assign $\dot{T}$ at will without changing the state of the material; this also applies to the case of fluids with memory. It should be noted that the whole procedure implies that A does not depend explicitly on $\dot{T}$, i.e., that *materials which are sensitive to the instantaneous rate of change of the independent variables are excluded* from the theory. It will be seen that this excludes purely viscous fluids.

The entropic constitutive equation for simple fluids with fading memory

We must now specify the 'other variables' appearing in eq. (4-4.29) for the special case of simple fluids with fading memory. We do this by making use of the so-called 'principle of equipresence,' which was originally formulated by Truesdell ([11], see also [14]). Stated in non-rigorous terms, the principle asserts that when some independent variable is known to appear in one constitutive equation of a complete theory, there is no *a priori* reason to suppose that it does not also enter all other constitutive equations. Since the history $\mathbf{F}^t(s)$ appears in the rheological constitutive equation for simple fluids, we shall assume that it also enters the entropic constitutive equation.

We have already seen that the instantaneous value of temperature has a special relevance in determining the present value of the free energy in the sense that, given two temperature histories which are at zero distance (i.e., the norm of the difference defined by eq. (4-2.22) is zero) but have different present temperature values (with the values of 'other variables' being equal), different values of the free energy are obtained.

Analogously, one has the physical intuition that, apart from the influence of past deformations, there should be a special relevance of deformations occurring abruptly at the instant of observation. Because deformations are defined with respect to a reference configuration, let us make the point clearer by considering the following example where a reference configuration is taken which is not the one assumed by the fluid at the instant of observation. Consider two motions which have equal values of a deformation tensor (such as Cauchy's) for all times up to but excluding the present time where the values are different. (Again, as in the example for temperature, at least one of the two histories is discontinuous at the instant of observation.) Physical intuition suggests that, other variables being equal, the present value of the free energy should be different for the two cases.

In the above example, the same reference configuration was assumed for both motions. Should we have considered (as is usual for fluids) the present configuration as a reference, the same two motions would have had deformation histories with different values at all times with the exception of the instant of observation where, by choice of the reference configuration, the deformation gradient would have been unity for both. Thus, by this choice the physically relevant difference between the two motions at the instant of observation would have been obscured by the mathematical symbolism. More important, with the present configuration as a reference, derivatives with respect to strain impulses occurring at the instant of observation would result in complicated operations.

In this section we will use a reference configuration which is not the one assumed at the instant of observation. Such a fixed configuration will be indicated by R.

The deformation gradient at time τ with respect to R, $\mathbf{F}_R(\tau)$, is defined by

$$d\mathbf{X} = \mathbf{F}_R(\tau) \cdot d\mathbf{X}_R \qquad (4\text{-}4.30)$$

where $d\mathbf{X}_R$ is the vector joining two infinitely close material points in the reference configuration and $d\mathbf{X}$ is, as usual, the vector joining the same two points at time τ. The corresponding history is indicated by $\mathbf{F}_R^t(s) = \mathbf{F}_R(t - s)$.

By $\mathbf{F}_R$ we shall indicate the deformation gradient at the instant of observation t:

$$d\mathbf{X}_t = \mathbf{F}_R \cdot d\mathbf{X}_R; \qquad \mathbf{F}_R = \mathbf{F}_R(t) = \mathbf{F}_R^t(0) \qquad (4\text{-}4.31)$$

Transformation from $\mathbf{F}^t(s)$ to $\mathbf{F}_R^t(s)$ or vice versa is accomplished by means of $\mathbf{F}_R$ (see eq. (3-1.8)):

$$\mathbf{F}_R^t(s) = \mathbf{F}^t(s) \cdot \mathbf{F}_R \qquad (4\text{-}4.32)$$

Other kinematic tensors such as Cauchy's, etc. defined with respect to R, are obtained from $\mathbf{F}_R^t$ in the usual way.

When using a reference configuration which is not the one assumed at the instant of observation, the norm defined by eq. (4-2.22) is not affected (as for the case of temperature) by strain impulses occurring at the instant of observation. These are accounted for separately, by putting $\mathbf{F}_R$ among the variables.† We thus write, temporarily,

$$A = \underset{s=0}{\overset{s=\infty}{a}} \; [T^t(s), \mathbf{F}_R^t(s); T, \mathbf{F}_R, \text{other variables}] \qquad (4\text{-}4.33)$$

Before proceeding any further, we need to introduce two simple mathematical concepts, i.e., the derivatives of a scalar-valued function with respect to vector-valued and tensor-valued arguments.

Let $f(\mathbf{a})$ be a scalar-valued function of a vector argument. The derivative $df/d\mathbf{a}$ is a vector defined by

$$\frac{df}{d\mathbf{a}} \cdot \mathbf{b} = \lim_{\alpha \to 0} \frac{1}{\alpha}[f(\mathbf{a} + \alpha\mathbf{b}) - f(\mathbf{a})] \qquad (4\text{-}4.34)$$

where $\mathbf{b}$ is an arbitrary vector. Analogously, let $f(\mathbf{A})$ be a scalar-valued function of a tensor argument. The derivative $df/d\mathbf{A}$ is a tensor defined by

$$\text{tr}\left(\frac{df}{d\mathbf{A}} \cdot \mathbf{B}^T\right) = \lim_{\alpha \to 0} \frac{1}{\alpha}[f(\mathbf{A} + \alpha\mathbf{B}) - f(\mathbf{A})] \qquad (4\text{-}4.35)$$

where $\mathbf{B}$ is an arbitrary tensor. The derivatives defined above follow simple rules of differential calculus which are analogous to those of ordinary scalar derivatives.

Let us now turn to the identification of the independent variables which must appear in eq. (4-4.33). Since the temperature gradient is known to enter the thermal constitutive equation, one should initially assume that it also enters the entropic constitutive equation. When this variable is included and eq. (4-4.33) is differentiated and substituted into eq. (4-4.13), a term

† Conversely, when the present configuration is used as a reference, the same definition of norm given by eq. (4-2.22) accounts for strain impulses occurring at the instant of observation because, if the past motion is left unchanged and a different present impulse is considered, the entire past history is effectively modified. Because of the influence of present impulses, the approximations obtained for slow flows (eqs (4-3.25), (4-3.26), and (4-3.27)) require that the history be continuous at the instant of observation.

Note finally that one can avoid putting T and $\mathbf{F}_R$ as independent variables in eq. (4-4.33) if a different norm is defined which is sensitive to present values. Such a norm is said to introduce an 'atom' at the present.

of the following form results:

$$\frac{\partial a}{\partial (\nabla T)} \cdot \dot{\overline{\nabla T}}$$

Since this term is the only one containing $\dot{\overline{\nabla T}}$, then reasoning similar to that which gave eq. (4-4.17) yields the conclusion that $\partial a/\partial(\nabla T) = 0$, i.e., the free energy does not depend on the temperature gradient.

It might be thought that the *history* of the temperature gradient should also be considered in analogy to the history of the deformation gradient. This idea has often been discussed [12] and in fact a thermodynamic theory based on the inclusion of the temperature gradient history has been published [13]. However, inclusion of the temperature gradient history violates the principle of local action in the restricted form used here. We are considering simple materials, or materials of 'first degree,' which in broad language may be defined as those materials which are sensitive to what is going on and has gone on in the past with regard to temperature and motion in the neighborhood of the point considered up to *the first approximation*. As for the motion, the first approximation we can consider is the first deformation gradient (position of the material point, $\mathbf{X}$, is meaningless by itself). With regard to the temperature of the neighborhood points, the first approximation is the temperature of the material point considered. Inclusion of the first temperature gradient would be a second-order correction comparable to the inclusion of the *second* deformation gradient of the motion.

In conclusion, we will write the entropic constitutive equation for a simple fluid with fading memory as follows:

$$A = \underset{s=0}{\overset{s=\infty}{a}} \ [T^t, \mathbf{F}_R^t ; T, \mathbf{F}_R] \qquad (4\text{-}4.36)$$

The thermodynamic theory based on the constitutive assumption of eq. (4-4.36) will now be developed; no details are given, and the reader is referred to the original works [6, 7].

Thermodynamics of simple fluids with fading memory

We wish here to investigate the consequences of the constitutive assumption in eq. (4-4.36).

First, the results in eqs (4-4.17) and (4-4.18) also obviously hold for the case considered here. Hence, the constitutive equations for entropy and internal energy take the forms:

$$S = -\frac{\partial a}{\partial T} \qquad (4\text{-}4.37)$$

$$U = a - T\frac{\partial a}{\partial T} \qquad (4\text{-}4.38)$$

where $\partial a/\partial T$ is the instantaneous derivative of the functional a.

We shall not go into the mathematical details of the procedure to be followed, and here state without proof that, upon differentiation, eq. (4-4.36) gives

$$\dot{A} = \frac{\partial a}{\partial T}\dot{T} + \left[\mathbf{F}_R \cdot \left(\frac{\partial a}{\partial \mathbf{F}_R} \right)^{\mathrm{T}} \right] : \mathbf{D} + \delta a \qquad (4\text{-}4.39)$$

where $\partial a/\partial \mathbf{F}_R$ is the instantaneous derivative of the functional a with respect to tensor $\mathbf{F}_R$, while δa is a Fréchet differential of the functional a with respect to both argument histories, having the derivative of the histories as second arguments:

$$\delta a = \underset{s=0}{\overset{s=\infty}{\delta a}} \left[T^t, \mathbf{F}_R^t \, ; \, T, \mathbf{F}_R \, \middle| \, -\frac{\mathrm{d}}{\mathrm{d}s} T^t, -\frac{\mathrm{d}}{\mathrm{d}s} \mathbf{F}_R^t \right] \qquad (4\text{-}4.40)$$

Of course, the assumption has been made that the functional is not only continuous, but Fréchet-differentiable throughout its domain with respect to the topology of the space of histories $T^t, \mathbf{F}_R^t$ determined by the norm (4-2.22).

Upon substitution of eqs (4-4.8), (4-4.17), and (4-4.39) into eq. (4-4.13), and considering that for constant density fluids $\dot{V} = 0$, one has

$$\left[-\frac{1}{\rho}\boldsymbol{\tau} + \mathbf{F}_R \cdot \left(\frac{\partial a}{\partial \mathbf{F}_R} \right)^{\mathrm{T}} \right] : \mathbf{D} + \delta a + \frac{1}{\rho T} \mathbf{q} \cdot \nabla T \leq 0 \qquad (4\text{-}4.41)$$

A number of very important conclusions can be drawn from eq. (4-4.41). First, consider that, since the field acceleration $\mathbf{g}$ is in principle controllable independently of the state of the material, the momentum equation tells us that $\mathbf{D}$ can be controlled, and thus plays in eq. (4-4.41) a role similar to $\dot{T}$ in eq. (4-4.16), *provided that it does not enter explicitly as an independent variable in the constitutive equations.* This is indeed our constitutive assumption (see eq. (4-4.36)), but it should be emphasized that purely viscous fluids are at this point excluded from the analysis.† On this basis, in order that eq. (4-4.41) be fulfilled for all processes, the term containing $\mathbf{D}$ must be identically zero, hence the tensor in brackets in eq. (4-4.41) must be isotropic. One thus obtains

$$\boldsymbol{\tau} = \rho \mathbf{F}_R \cdot \left(\frac{\partial a}{\partial \mathbf{F}_R} \right)^{\mathrm{T}} \qquad (4\text{-}4.42)$$

---

† Thermodynamic results are very sensitive to the assumptions of smoothness that one makes with regard to constitutive equations. If a functional is smooth with respect to any topology of the space of deformation histories, its value cannot depend explicitly on the instantaneous rate of strain: see, by analogy, Fig. 4-2 and the discussion related to it, which shows that the value of a functional that is smooth with respect to T^t cannot depend explicitly on $\dot{T}$.

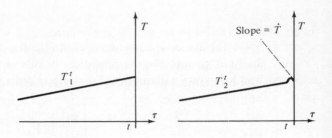

FIGURE 4-2
Two arbitrarily close histories, the second
one having an arbitrarily assigned instantaneous rate of change.

Equation (4-4.42) shows that the rheological constitutive equation
is entirely determined by the entropic equation. It represents the extension
to memory fluids of the 'classical' thermodynamic result of eq. (4-4.5):

$$p = -\left(\frac{\partial A}{\partial V}\right)_T \qquad (4\text{-}4.43)$$

In view of eq. (4-4.42), eq. (4-4.41) reduces to

$$\delta a + \frac{1}{\rho T}\mathbf{q}\cdot\mathbf{\nabla}T \le 0 \qquad (4\text{-}4.44)$$

while the rate of accumulation of elastic energy, A_T, is obtained as

$$A_T = \frac{1}{\rho}\mathbf{\tau}:\mathbf{D} + \delta a = P + \delta a \qquad (4\text{-}4.45)$$

Comparison of eqs (4-4.24) and (4-4.45) allows identification of the
mechanical dissipation rate with $-\delta a$:

$$D_M = -\delta a \qquad (4\text{-}4.46)$$

and hence to have an expression for the mechanical dissipation rate which, in
principle, allows it to be calculated once the entropic constitutive equation
has been assigned.

Furthermore, since the value of a, and hence of δa, is independent of
$\mathbf{\nabla}T$, eq. (4-4.44) implies that δa is non-positive and hence that the mechanical
dissipation rate is non-negative.

In summary, given the constitutive equation for free energy in the form of eq. (4-4.36), the following results are obtained:

  (i) The constitutive equations for entropy, internal energy, and stress are determined unequivocally.

 (ii) The mechanical dissipation rate is determined, and is intrinsically non-negative. This imposes a restriction on the possible forms of the functional a, and thus, because of point (i), also on the possible forms of the constitutive equations for U, S, and τ.

(iii) Through eq. (4-4.26), a restriction on the constitutive equation for the heat flux is also obtained, since D_M is determined by the form of the functional a.

Further thermodynamic results are obtained through routine calculations, involving only the proof that the ordinary chain rules of differential calculus apply also to the instantaneous derivatives and Fréchet differentials that appear in the theory. In particular, one may wish to operate a different choice of independent and dependent variables; in each case, the principle of determinism implies that the deformation history is necessarily to be regarded as an independent variable.

In particular, one may wish to consider the following alternative forms for thermodynamic constitutive equations:

$$U = \underset{s=0}{\overset{s=\infty}{u}} \; [S^t, \mathbf{F}_R^t \, ; S, \mathbf{F}_R] \qquad (4\text{-}4.47)$$

$$S = \underset{s=0}{\overset{s=\infty}{\jmath}} \; [U^t, \mathbf{F}_R^t \, ; U, \mathbf{F}_R] \qquad (4\text{-}4.48)$$

The mechanical dissipation rate D_M is obtained as

$$D_M = -\delta u = T\delta\jmath = -\delta a \qquad (4\text{-}4.49)$$

while the stress τ is given by

$$\tau = -\rho\mathbf{F}_R \cdot \left(\frac{\partial u}{\partial \mathbf{F}_R}\right)^T = \rho T\mathbf{F}_R \cdot \left(\frac{\partial \jmath}{\partial \mathbf{F}_R}\right)^T \qquad (4\text{-}4.50)$$

and the temperature T by

$$T = \frac{\partial u}{\partial S} = \frac{1}{\partial \jmath/\partial U} \qquad (4\text{-}4.51)$$

The reader may have noticed that no mention has been made of two additional thermodynamic quantities, i.e., enthalpy $U + pV$, and the Gibbs free energy $U + pV - TS$. Introduction of both quantities, in classical thermodynamics, is suggested by the appearance of terms such as $p\,dV$ or $\Delta(pV)$, measuring some work done by the system in expansion or in flow. Such work is expressed in the classical way only if the stress is isotropic, so that a thermodynamic theory of materials with non-isotropic stresses (which includes, of course, Newtonian liquids) should really be developed without considering either enthalpy or Gibbs free energy.

As discussed above, the theory developed so far does not include the thermodynamics of purely viscous fluids. This is very briefly reviewed below.

Thermodynamics of purely viscous fluids

By applying the principle of equipresence, the free energy of purely viscous fluids is assumed to depend on temperature, temperature gradient, rate of strain, and specific volume:

$$A = a(T, \nabla T, \mathbf{D}, V) \qquad (4\text{-}4.52)$$

Yet, the argument which allows ∇T to be excluded from the independent variables holds again, and a similar argument can be used to show that $\mathbf{D}$ must also be excluded [14]. Thus, the conclusion is reached that A depends only on V and T,

$$A = a(V, T) \qquad (4\text{-}4.53)$$

From eq. (4-4.53), the value of A_T can be calculated:

$$A_T = \frac{\partial a}{\partial V}\dot{V} \qquad (4\text{-}4.54)$$

and thus from eqs (4-4.8) and (4-4.19):

$$\left(\frac{\partial a}{\partial V} + p\right)\dot{V} - \frac{1}{\rho}\boldsymbol{\tau}:\mathbf{D} + \frac{1}{\rho T}\mathbf{q}\cdot\nabla T \le 0 \qquad (4\text{-}4.55)$$

Equation (4-4.55) implies that

$$\frac{\partial a}{\partial V} = -p \qquad (4\text{-}4.56)$$

so that the classical Maxwell equation for free energy (eq. (4-4.28)) is obtained. The mechanical dissipation rate $\dot{D}_M$ is obtained from eq. (4-4.24):

$$D_M = \frac{1}{\rho}\boldsymbol{\tau}:\mathbf{D} \qquad (4\text{-}4.57)$$

and eq. (4-4.19) reduces to

$$\frac{1}{\rho}\tau:\mathbf{D} - \frac{1}{\rho T}\mathbf{q}\cdot\nabla T \geq 0 \qquad (4\text{-}4.58)$$

Unfortunately, mechanical and thermal effects cannot, in this case, be uncoupled since there is no way to prove that τ does not depend on ∇T, or that $\mathbf{q}$ does not depend on $\mathbf{D}$. Of course, if one wishes to make the additional constitutive *assumption* that τ does not depend on ∇T, then the mechanical dissipation rate must be non-negative. In general, one can only assert that $D_M \geq 0$ in homothermal processes ($\nabla T = \mathbf{0}$); this implies that isothermal (i.e., purely mechanical) constitutive equations for purely viscous fluids must always yield positive values for D_M. In particular, the discussion in section 2-3 is justified.

PROBLEMS

4-1 Consider the following functional:

$$y = \mathop{\hbar}_{\substack{x=b \\ x=a}}^{x=b} [f(x)] = \int_a^b f(x)\,dx$$

where a and b are two given numbers. Calculate its first Fréchet differential, assigning, if necessary, an appropriate norm for the $f(\ )$.

4-2 Solve the same problem as above for an arbitrary linear functional, i.e., a functional satisfying the following equation:

$$\hbar[\alpha f_1(x) + \beta f_2(x)] = \alpha\hbar[f_1(x)] + \beta\hbar[f_2(x)]$$

4-3 Consider a fluid for which the following entropic constitutive equation holds:

$$A = a(T, V, \phi)$$

where ϕ is some (unspecified) parameter. Prove that

$$\left(\frac{\partial U}{\partial \phi}\right)_{V,S} = \frac{\partial a}{\partial \phi}$$

Discuss the implications of this result in (i) the result in eq. (4-4.49), and (ii) the theory of chemical potentials in solutions (let ϕ be the number of moles of the component considered).

BIBLIOGRAPHY

1. NOLL, W.: *Arch. Ratl Mech. Anal.*, **2**, 197 (1958).
2. ERICKSEN, J. L.: *Koll-Zeitschr.*, **173**, 117 (1960); *Arch. Ratl Mech. Anal.*, **4**, 231 (1960), **8**, 1 (1961), **9** (1), 379 (1962), and **10**, 189 (1962); *Int. J. Engng Sci.*, **1**, 157 (1963).
3. COLEMAN, B. D., and NOLL, W.: *Rev. Mod. Phys.*, **33**, 239 (1961).
4. COLEMAN, B. D., MARKOWITZ, H., and NOLL, W.: *Viscometric Flows of Non-Newtonian Fluids*, ch. 2. Springer-Verlag (1966).
5. COLEMAN, B. D., and NOLL, W.: *Arch. Ratl Mech. Anal.*, **6**, 355 (1960).
6. COLEMAN, B. D.: *Arch. Ratl Mech. Anal.*, **17**, 1 (1964).
7. COLEMAN, B. D.: *Arch. Ratl Mech. Anal.*, **17**, 230 (1964).
8. COLEMAN, B. D., and NOLL, W.: *Arch. Ratl Mech. Anal.*, **13**, 167 (1963).
9. ASTARITA, G.: 1st Conv. Italian Society of Rheology, Siena (May 1971).
10. COLEMAN, B. D., and MIZEL, V. J.: *Arch. Ratl Mech. Anal.*, **23**, 87 (1966).
11. TRUESDELL, C.: *J. Math. pures appl.*, **30**, 111 (1951).
12. COLEMAN, B. D.: personal communication (March 1971). Reference is made to B. D. Coleman and M. E. Gurtin, *Z.A.M.P.*, **18**, 199 (1967); B. D. Coleman and V. J. Mizel, *Arch. Ratl Mech. Anal.*, **13**, 245 (1963); and to a discussion at the Bressanone Meeting of the Society of Natural Philosophy (June 1965).
13. ERINGEN, C.: *Int. J. Engng Sci.*, **2**, 179 (1966).
14. COLEMAN, B. D., and MIZEL, V. J.: *J. Chem. Phys.*, **40**, 1116 (1964).

5
RHEOMETRICAL FLOW SYSTEMS

5-1 INTRODUCTION

The experimental determination of the rheological behavior of materials is referred to as 'rheometry.' While rheometry reduces to a very simple set of experiments in the case of Newtonian fluids, it becomes exceedingly difficult for non-Newtonian fluids, and particularly for memory fluids.

The rheological behavior of Newtonian incompressible fluids is entirely determined by the value of one single parameter—viscosity. Viscosity, for a given material, is a function only of temperature. The experimental determination of viscosity consists in the measurement of some easily determined quantity which can be related unequivocally to the viscosity by an equation which is derived theoretically from a solution of the equation of motion. As an example, the pressure gradient $\Delta p/L$ in the axial direction for steady rectilinear flow down a long circular pipe is given by the Hagen–Poiseuille law:

$$\frac{\Delta p}{L} = \frac{16\mu Q}{\pi R^4} \qquad (5\text{-}1.1)$$

where μ is the viscosity, Q the volumetric flow rate, and R the pipe radius.

Such a flow is a rheometrical flow system, because the measurement of Q, R, and $\Delta p/L$ allows us to calculate the viscosity and thus, for Newtonian fluids, to completely characterize their rheological behavior.

When generalized Newtonian fluids are considered, rheometry reduces to the experimental determination of the function $\eta(S)$ in eq. (2-4.1). This is more difficult than the determination of a single value of the viscosity, because a complete apparent viscosity *curve* needs to be determined. The methods of rheometry have in part been discussed in section 2-5, where rheometrical flow systems have been considered which allow the $\eta(S)$ curve to be determined.

In this chapter we focus attention on rheometrical flow systems which are useful for memory fluids. Ideally, rheometry for such fluids should consist of some program of experimental measurements which allows for a complete determination of the functional $\mathscr{H}[\ ]$ in eq. (4-3.12), which is given below:

$$\tau = \mathop{\mathscr{H}}_{s=0}^{s=\infty} [\mathbf{G}^t(s)] \qquad (5\text{-}1.2)$$

Unfortunately, one sees immediately that such a task is hopelessly difficult. In fact, consider first the experimental determination of a function such as $\eta(S)$. What one does is to actually measure the *values* of the function corresponding to a certain finite number of values of the argument. The larger this number, the better the knowledge of the function itself; but clearly the program is possible, and the function can be determined within any desired degree of accuracy, at least over some interval of values of the argument (assumptions of smoothness are made either explicitly or implicitly). Now, consider instead the problem of determining a functional experimentally. The argument in this case being functions, one is faced with the problem of scanning the space of functions, and of making experiments over some interval of that space. Apart from the fact that the topology of the space is not known in advance (in effect, over which topology the functional is smooth is just one of the things one would wish to determine), the space of functions is not countable in the ordinary sense. One just cannot conceive of a program of experiments which exhausts some finite region of the domain of the desired functional, unless such a region is so restricted as to be in a one-to-one correspondence with a finite number of scalar parameters.

Such a restriction is exactly what rheometry of memory fluids amounts to. One restricts attention to some 'classes' of flows, so that the strain histories $\mathbf{G}^t(s)$ are restricted to a class in which each member is entirely determined by the values of some finite number of parameters. The functional

$\mathscr{H}[\ ]$ then reduces to a finite number of *functions*, and rheometry becomes possible. Of course, knowledge of these functions for any given material allows its behavior to be predicted only for those flows that are within the class considered, the behavior in any other flow pattern remaining unpredictable.

Another way of approaching the problem is to *assume* that the fluid considered obeys some preassigned constitutive equation, which is *highly* more specific than eq. (5-1.2). Many such equations have been proposed in the literature, and the next chapter is dedicated to their discussion. Such constitutive equations contain some finite number of parameters and/or functions which need to be determined experimentally; once these are known, the behavior of the material under *any* flow condition can be predicted. One should nevertheless bear in mind that, no matter how well a constitutive equation fits experimental data obtained in some rheometrical flow system, proof has only been obtained that the functional $\mathscr{H}[\ ]$ degenerates into the equation itself for those particular rheometrical flow systems, and nothing can be said *a priori* about other flow conditions, unless the validity of the assumed constitutive equation is believed on grounds of some independent line of reasoning, such as a modeling of the molecular structure of the fluid.

There are basically two classes of flow that have been considered as possible rheometrical flow systems from the viewpoint of fluid mechanics. The first is that of constant stretch history flows, which have been discussed kinematically in section 3-5; the second is that of periodic flows. Rheometrical constant stretch history flow systems are discussed in sections 5-2 and 5-3 below, and periodic flows are discussed in section 5-4.

Constant stretch history flows

For constant stretch history flows, the function $\mathbf{G}^t(s)$ is obtained from eq. (3-5.24) as

$$\mathbf{G}^t(s) = \mathbf{P}(t) \cdot [\exp(-ks\mathbf{N}^T) \cdot \exp(-ks\mathbf{N})] \cdot \mathbf{P}^T(t) - \mathbf{1} \qquad (5\text{-}1.3)$$

Notice that one can choose arbitrarily the origin of the time axis, and $\mathbf{P}(0) = \mathbf{1}$ (see eq. (3-5.16)). One thus has, from eqs (5-1.2) and (5-1.3),

$$\tau(0) = \mathop{\mathscr{H}}_{s=0}^{s=\infty} [\exp(-ks\mathbf{N}^T) \cdot \exp(-ks\mathbf{N}) - \mathbf{1}] = \mathbf{H}(k\mathbf{N}) \qquad (5\text{-}1.4)$$

$$\tau(t) = \mathbf{P}(t) \cdot \tau(0) \cdot \mathbf{P}^T(t) \qquad (5\text{-}1.5)$$

Several rheometrical flow systems are obtained corresponding to different choices of the tensor $\mathbf{N}$, such as viscometric flows, extensional flows, and so on; these flows have already been introduced in section 3-5. For each flow system, the function $\mathbf{H}(\ )$ can be expressed through a certain number of scalar functions, known as 'material functions.'

The thermodynamics of constant stretch history flows pose restrictions upon these material functions. In fact, consider eq. (4-4.36) which we rewrite here in the form

$$A(t) = \mathop{\hat{a}}_{s=0}^{s=\infty} [\mathbf{G}^t(s), T^t(s); T] \qquad (5\text{-}1.6)$$

The functional $\hat{a}$ in eq. (5-1.6) is related to that of eq. (4-4.36) but is not the same for the following reasons: (i) the present configuration is taken as reference (see footnote on p. 145) and (ii) the tensor $\mathbf{G}$ is used instead of $\mathbf{F}$, which is possible because the history of rotation is irrelevant.

The functional $\hat{a}$ is isotropic, so that we may write

$$A(t) = \mathop{\hat{a}}_{s=0}^{s=\infty} [\mathbf{P}^T(t) \cdot \mathbf{G}^t(s) \cdot \mathbf{P}(t), T^t(s); T] \qquad (5\text{-}1.7)$$

In view of eq. (3-5.19) and because the flow is isothermal, we have

$$A(t) = A(0); \qquad \dot{A} = 0 \qquad (5\text{-}1.8)$$

Thus, in an isothermal constant stretch history flow, no free energy is accumulated. From eq. (4-4.27) one then derives that the stress power equals the rate of dissipation:

$$\frac{1}{\rho}\boldsymbol{\tau}:\mathbf{D} = D_M \geq 0 \qquad (5\text{-}1.9)$$

Clearly, eq. (5-1.9) places some definite limitations on the admissible forms of the function $\mathbf{H}(\ )$, and thus on the form of the material functions.

When analyzing constant stretch history flows, it is often useful to consider an orthonormal rotating basis $\mathbf{b}_k(\tau)$ defined as follows:

$$\mathbf{b}_k(\tau) = \mathbf{P}(\tau) \cdot \mathbf{b}_k(0) = \mathbf{P}(\tau) \cdot \mathbf{g}_k \qquad (5\text{-}1.10)$$

where $\mathbf{g}_k$ is some (fixed) orthonormal basis; for instance, $\mathbf{g}_k$ may be a basis with respect to which $\mathbf{N}$ has a matrix of a particularly simple form $[\mathbf{N}]_g$, say,

$$\mathbf{g}_m \cdot \mathbf{N} \cdot \mathbf{g}_k = [\mathbf{N}]_g \qquad (5\text{-}1.11)$$

One may next consider the matrix $[\mathbf{F}^t]_b$ which is defined as follows:

$$[\mathbf{F}^t]_b = \mathbf{b}_m(\tau) \cdot \mathbf{F}^t(s) \cdot \mathbf{b}_k(t) \qquad (5\text{-}1.12)$$

Substitution of eqs (5-1.10) and (3-5.17) shows that†

$$[\mathbf{F}^t]_b = \mathbf{g}_m \cdot \exp(-ks\mathbf{N}) \cdot \mathbf{g}_k = \exp\{-ks[\mathbf{N}]_g\} \qquad (5\text{-}1.13)$$

Equation (5-1.13) shows that the matrix $[\mathbf{F}^t]_b$ is independent of t, and depends on the time lag according to a matrix-exponential law. Conversely, if one can show that, for some given flow pattern, $[\mathbf{F}^t]_b$ has the form given in eq. (5-1.13), with $\mathbf{b}_k$ some rotating orthonormal basis, then the flow considered belongs to that special class of constant stretch history flows defined by the particular form of the matrix $[\mathbf{N}]_g$.

In particular, suppose that the flow is described in some orthogonal coordinate system by the equations:

$$x^i(\tau) = f^i(x_t^j, \tau) \qquad (5\text{-}1.14)$$

According to eq. (3-1.41), one has

$$\frac{\partial x^k}{\partial x_t^m} = \mathbf{e}^k(\mathbf{X}) \cdot \mathbf{F}^t \cdot \mathbf{e}_m(\mathbf{X}_t) \qquad (5\text{-}1.15)$$

Since the coordinate system is orthogonal, one may consider the orthonormal basis:

$$\mathbf{e}\langle k \rangle = \sqrt{g_{kk}}\,\mathbf{e}^k = \frac{\mathbf{e}_k}{\sqrt{g_{kk}}} \qquad \text{(not summed)} \qquad (5\text{-}1.16)$$

and write

$$\frac{\partial x^k}{\partial x_t^m} = \sqrt{\frac{g_{mm}(\mathbf{X}_t)}{g_{kk}(\mathbf{X})}}\,\mathbf{e}\langle k \rangle(\mathbf{X}) \cdot \mathbf{F}^t \cdot \mathbf{e}\langle m \rangle(\mathbf{X}_t) \qquad (5\text{-}1.17)$$

The orthonormal basis $\mathbf{e}\langle k \rangle(\mathbf{X})$ can be identified with a rotating orthonormal basis of the type of eq. (5-1.10) by writing

$$\mathbf{b}_k(\tau) = \mathbf{e}\langle k \rangle(\mathbf{X}) \qquad (5\text{-}1.18)$$

$$\mathbf{g}_k = \mathbf{e}\langle k \rangle(\mathbf{X}_t) \qquad (5\text{-}1.19)$$

In fact, $\mathbf{b}_k$ and $\mathbf{g}_k$ as given by eqs (5-1.18) and (5-1.19) are related to each other by a smooth orthogonal tensor function whose value coincides with the unit tensor at $\tau = t$. Thus one has, with respect to the basis $\mathbf{b}_k$ as defined above:

$$[\mathbf{F}^t]_b = \frac{\partial x^k}{\partial x_t^m}\sqrt{\frac{g_{kk}(\mathbf{X})}{g_{mm}(\mathbf{X}_t)}} \qquad \text{(not summed)} \qquad (5\text{-}1.20)$$

---

† The exponential of a matrix $[A]$ is defined as

$$\exp[A] = \sum_0^\infty \frac{1}{k!}[A]^k$$

Since the right-hand side of eq. (5-1.20) is easily calculated, one can check whether the matrix $[\mathbf{F}^r]_b$ has the form of eq. (5-1.13). Of course, if it *does not* have that form, the flow may still be a constant stretch history flow of the class examined, because some *other* rotating orthonormal basis may exist with respect to which $[\mathbf{F}^r]_b$ has the required form. In practice, however, symmetries of the flow systems of interest suggest directly the coordinate system x^i which leads, through eq. (5-1.18), to the $\mathbf{b}_k$ basis for which $[\mathbf{F}^r]_b$ has the form of eq. (5-1.13).

Note that, if $[\mathbf{F}^r]_b$ has the form of eq. (5-1.13), the tensor function $\mathbf{P}(\tau)$ appearing in eq. (5-1.3) is the one relating the $\mathbf{e}\langle k \rangle$ bases at different points in space:

$$\mathbf{e}\langle k \rangle(\mathbf{X}) = \mathbf{P}(\tau) \cdot \mathbf{e}\langle k \rangle(\mathbf{X}_t) \qquad (5\text{-}1.21)$$

Hence, $\mathbf{P}(\tau)$ may be regarded as a function of space rather than time:

$$\mathbf{P}(\mathbf{X}) = \mathbf{P}[\tau(\mathbf{X})] \qquad (5\text{-}1.22)$$

where $\tau(\mathbf{X})$ is the time at which the material point considered occupies the position $\mathbf{X}$. Equation (5-1.5) can thus be written as

$$\tau[\tau(\mathbf{X})] = \tau(\mathbf{X}) = \mathbf{P}(\mathbf{X}) \cdot \tau_0 \cdot \mathbf{P}^\mathrm{T}(\mathbf{X}) \qquad (5\text{-}1.23)$$

where τ_0 is the stress tensor at $\mathbf{X}(0)$.

Equation (5-1.23) implies that the stress tensor, apart from an inessential rotation, is constant along the trajectory of any material point—a concept which is, of course, intuitively related to the constant stretch history hypothesis. Note that this in no way implies that the stress tensor is constant on points which are not on the same trajectory; indeed, stretch histories of different material points may well be different even if they are constant in time for any given material point.

This fact has important consequences in the actual use of rheometrical flow systems, which can be divided into two basic categories according to whether the stretch history, and hence the stress, is or is not the same for all material points and is thus also on the bounding surfaces. Since, in practice, only the stresses on the bounding surfaces can be measured, rheometrical information can be obtained directly only when the stress is constant in space; when it is not, some form of differentiation of data is required.

Periodic flows

Periodic flows are of particular interest in rheometry in the limit of increasingly small deformations, say, when eq. (4-3.24) can be applied. In fact, with a periodic flow, the total deformation carrying the configuration of the material at any instant of time into the one at any other instant of time can be kept as small as one wishes, although the instantaneous rate of deformation may be high.

Equation (4-3.24) applies when the history $\mathbf{G}^t$ is at a very small distance from the rest history. This is true in practice if, at least in the not very distant past, the magnitude of the value of $\mathbf{G}^t$ is small for any value of s. Indeed, the right-hand side of eq. (4-3.24) is simply the leading term of the expansion in a series of integrals, where the first neglected term is of order two in the magnitude of $\mathbf{G}^t$ (see eq. (4-3.25)). Therefore, evaluation of $\mathbf{G}^t$ for periodic flows used in rheometry needs to be carried out only within terms of order one in its magnitude, since terms of higher order would contribute no more to the stress than terms arising from the neglected integral.

We may define rheometrical periodic flows as those flows for which the function $\mathbf{G}^t$ has, within terms of first order in its magnitude, the following form:†

$$\mathbf{G}^t = \mathrm{Re}\,\{\boldsymbol{\psi}^*(t)[1 - \mathrm{e}^{-i\omega s}]\} \qquad (5\text{-}1.24)$$

where ω is a constant frequency and the time-dependent traceless tensor $\boldsymbol{\psi}^*$ is complex, say,

$$\boldsymbol{\psi}^* = \boldsymbol{\psi}_{\mathrm{R}}(t) + i\boldsymbol{\psi}_{\mathrm{I}}(t) \qquad (5\text{-}1.25)$$

At any given instant of observation t, the strain history $\mathbf{G}^t$ is entirely determined by the values of the frequency ω and of the tensor $\boldsymbol{\psi}^*$; hence,

$$\boldsymbol{\tau}(t) = \mathbf{K}[\omega, \boldsymbol{\psi}^*(t)] \qquad (5\text{-}1.26)$$

where $\mathbf{K}[\ ]$ is an isotropic tensor-valued function.

Since eq. (4-3.24) is assumed to hold, the function $\mathbf{K}[\ ]$ can be obtained explicitly as

$$\boldsymbol{\tau}'(t) = \mathrm{Re}\,\left\{\boldsymbol{\psi}^*(t)\int_0^\infty f(s)[1 - \mathrm{e}^{-i\omega s}]\,\mathrm{d}s\right\} \qquad (5\text{-}1.27)$$

The stress $\boldsymbol{\tau}'$ is therefore determined provided the value of the integral on the right-hand side of eq. (5-1.27) is known. Since this value is uniquely determined by the frequency ω, one can thus define a single complex-valued material function, the complex viscosity η^*, characterizing the behavior of the material in periodic flows. Since η^* is complex, the real and the imaginary

---

† By Re $\{\boldsymbol{\psi}^*\}$ and Im $\{\boldsymbol{\psi}^*\}$ we mean the real part and the coefficient of the imaginary of the complex quantity $\boldsymbol{\psi}^*$; in the literature on periodic flows, the notation Re $\{\ \}$ is often dropped and understood implicitly.

part define two additional material functions, i.e., the dynamic viscosity η' and the dynamic rigidity G':

$$\eta^* = \eta' - \frac{iG'}{\omega} = \frac{i}{\omega} \int_0^\infty f(s)[1 - e^{-i\omega s}] \, ds \qquad (5\text{-}1.28)$$

Substitution of eq. (5-1.28) into eq. (5-1.27) yields

$$\tau'(t) = \text{Re} \{ -i\omega\eta^*\psi^*(t) \} \qquad (5\text{-}1.29)$$

which can be expanded to yield

$$\tau'(t) = -G'\psi_R(t) + \eta'\omega\psi_I(t) \qquad (5\text{-}1.30)$$

An interesting relation is obtained by considering eq. (3-2.17):

$$2\mathbf{D} = \dot{\mathbf{C}}(t) = -\dot{\mathbf{C}}'(0) = -\dot{\mathbf{G}}'(0) \qquad (5\text{-}1.31)$$

From eqs (5-1.24) and (5-1.25) one obtains

$$2\mathbf{D} = -\text{Re} \{i\omega\psi^*(t)\} = \omega\psi_I(t) \qquad (5\text{-}1.32)$$

so that eq. (5-1.30) can be written as

$$\tau'(t) = 2\eta'\mathbf{D} - G'\psi_R(t) \qquad (5\text{-}1.33)$$

In view of eq. (5-1.32), one may wish to *define* a complex rate-of-strain tensor $\mathbf{D}^*$ as

$$2\mathbf{D}^* = -i\omega\psi^*; \qquad \mathbf{D} = \text{Re} \{\mathbf{D}^*\} \qquad (5\text{-}1.34)$$

so that eq. (5-1.29) takes the following pseudo-Newtonian form:

$$\tau' = \text{Re} \{2\eta^*\mathbf{D}^*\} \qquad (5\text{-}1.35)$$

which justifies the name of 'complex viscosity' given to η^*.

Controllable flows

In dealing with rheometrical flow systems, one is faced with the problem of verifying that the flow pattern considered is controllable. By controllable flow one means that a simultaneous solution of the constitutive equation and of the dynamical equation exists, with body forces admitting a potential ψ:

$$\mathbf{g} = -\nabla\psi \qquad (5\text{-}1.36)$$

The limitation to body forces admitting a potential is justified by the fact that in the great majority of cases of interest the only body force acting is gravity, which of course admits a potential. On the other hand, there is no point in considering controllability under arbitrary body forces, since *any* flow is possible under a suitable distribution of body forces. Indeed, the concept that any flow is possible under a suitable body force field was used in the thermodynamic treatment of section 4-4.

Note that if a flow is controllable for some body force field admitting a potential, it is controllable for any such field, and hence, in particular, also in the absence of any body force: this is easily understood from the procedure discussed below for checking controllability.

The check for controllability is obtained by means of a standard procedure. First, a kinematic description of the flow is assumed and the flow is classified; that is, it is identified as, for example, a viscometric flow. The spatial distribution of stresses is then obtained from the constitutive equation. Both the kinematics and the stress distribution are then substituted into the dynamical equation which, under the condition of eq. (5-1.36), reads (see eq. (1-8.5)):

$$\rho\frac{D\mathbf{v}}{Dt} - \mathbf{V}\cdot\tau + \tfrac{1}{3}\mathbf{V}(\text{tr }\tau) = -\mathbf{V}(p + \rho\psi) \qquad (5\text{-}1.37)$$

Equation (5-1.37) shows that the flow is controllable if the left-hand side is expressible as the gradient of a scalar field. Indeed, eq. (5-1.37) determines the pressure field p (within an arbitrary additive constant, see section 1-8). We shall indicate the difference between the actual and the hydrostatic pressure, or the 'extra pressure,' by $\mathscr{P}$:

$$\mathscr{P} = p + \rho\psi \qquad (5\text{-}1.38)$$

The extra pressure $\mathscr{P}$ can also be regarded as the pressure field that would be obtained from the equations of motion in the absence of body forces.

It is important to point out that a flow, even if controllable, may be very difficult to realize in practice because it requires an unrealistic distribution of stresses on the bounding surfaces. A good example is a flow where a non-constant pressure distribution would be required on a free surface.

Stress-relaxation experiments

In addition to the rheometrical flow systems discussed above, an additional type of experiment, namely stress-relaxation, is often performed and yields some well-defined information on the rheological behavior of the material tested.

In an ideal stress-relaxation experiment, a sample of material previously at rest is subjected at time t_0 to a sudden deformation, and is held stationary afterwards. The stress at time $t > t_0$ is measured. The history of deformation G^t (at $t > t_0$) is

$$G^t = 0 \quad \text{for} \quad s < t - t_0 \quad (5\text{-}1.39)$$

$$G^t = G \quad \text{for} \quad s > t - t_0 \quad (5\text{-}1.40)$$

where G is a constant tensor measuring the deformation to which the sample has been subjected. If G^t has a sufficiently small norm (which in practice implies that the magnitude of G is small enough), eq. (4-3.24) applies, and the stress is given by

$$\tau'(t) = \int_0^\infty f(s)G^t \, ds = -G \int_{t-t_0}^\infty -f(s) \, ds \quad (5\text{-}1.41)$$

It is useful to define the function $F(\ )$ as

$$F(t - t_0) = \int_{t-t_0}^\infty -f(s) \, ds \quad (5\text{-}1.42)$$

so that eq. (5-1.41) can be written in the form

$$\tau'(t) = -F(t - t_0)G \quad (5\text{-}1.43)$$

The function $F(\ )$, called the 'stress-relaxation function,' can be obtained from stress-relaxation experiments. It is related to the function $f(\ )$ appearing in eq. (4-3.24) by the following equation, which is immediately obtained from eq. (5-1.42):

$$\frac{dF(s)}{ds} = f(s) \quad (5\text{-}1.44)$$

Equation (4-3.24) can be written in the following form, which is obtained after integration by parts (and observing that $F(s) \to 0$ when $s \to \infty$):

$$\tau' = \int_0^\infty -F(s)\dot{G}^t(s) \, ds \quad (5\text{-}1.45)$$

An experiment which in some sense is the converse of stress-relaxation is also performed; this is known as creep, where the deformation of a sample held under a constant load is determined. In such experiments, the history of deformation is not known in advance, and thus the results do not lead to any useful prediction of the behavior of the material under any flow condition different from the one realized in the experiment.

Of course, if a constitutive equation is assumed (such as those to be discussed in the next chapter), the results of a creep experiment can be predicted from the solution of the appropriate boundary-value problem, in terms of the values of the parameters appearing in the constitutive equation. Such experiments can then be performed in order to assess the reliability of the assumed form of the constitutive equation, and to determine the numerical values of the parameters. Such a procedure can, at least in principle, be applied to *any* flow pattern, and its validity is limited by the considerations made previously.

5-2 VISCOMETRIC FLOWS

The viscometric functions

Viscometric flows (see section 3-5) are constant stretch history flows for which

$$N^2 = 0 \qquad (5\text{-}2.1)$$

The constant k is called the shear rate. Recalling that tensor N is of unit magnitude if it satisfies eq. (5-2.1), an orthonormal basis h_k exists with respect to which the matrix of N has the form

$$[N]_h = \begin{Vmatrix} 0 & 1 & 0 \\ 0 & 0 & 0 \\ 0 & 0 & 0 \end{Vmatrix} \qquad (5\text{-}2.2)$$

Of course, with respect to any orthonormal basis it follows from eq. (5-2.1) that

$$[N]^2 = 0 \qquad (5\text{-}2.3)$$

The stress tensor at $t = 0$ is given by

$$\tau_0 = \tau(0) = H(kN) \qquad (5\text{-}2.4)$$

with $H(\ )$ an isotropic function. In particular, consider an orthogonal tensor Q which, with respect to the same basis for which eq. (5-2.2) holds, has the form

$$[Q]_h = \begin{Vmatrix} 1 & 0 & 0 \\ 0 & 1 & 0 \\ 0 & 0 & -1 \end{Vmatrix} \qquad (5\text{-}2.5)$$

(the tensor Q transforms the orthonormal basis into another orthonormal basis differing from the original only in the orientation of the base vector in the 3-direction; it therefore represents an improper rotation). One sees immediately that

$$\mathbf{Q} \cdot \mathbf{N} \cdot \mathbf{Q}^{\mathrm{T}} = \mathbf{N} \qquad (5\text{-}2.6)$$

and thus, from eq. (5-2.4),

$$\mathbf{Q} \cdot \boldsymbol{\tau}_0 \cdot \mathbf{Q}^{\mathrm{T}} = \boldsymbol{\tau}_0 \qquad (5\text{-}2.7)$$

Equation (5-2.7) holds if and only if

$$\tau_0 \langle 13 \rangle = \tau_0 \langle 23 \rangle = 0 \qquad (5\text{-}2.8)$$

where $\tau_0 \langle ij \rangle$ are the components of $\boldsymbol{\tau}_0$ with respect to $\mathbf{h}_k$. Since the tensor $\boldsymbol{\tau}_0$ is defined only within an arbitrary additive isotropic tensor, eq. (5-2.8) implies that $\boldsymbol{\tau}_0$ is completely determined by three of its components, namely $\tau_0 \langle 12 \rangle$, $\tau_0 \langle 11 \rangle - \tau_0 \langle 22 \rangle$, and $\tau_0 \langle 22 \rangle - \tau_0 \langle 33 \rangle$. These are in turn, as seen from eq. (5-2.4), entirely determined by the value of k; thus one can define the three following viscometric material functions:

$$\tau_0 \langle 12 \rangle = \tau(k) \qquad (5\text{-}2.9)$$

$$\tau_0 \langle 11 \rangle - \tau_0 \langle 22 \rangle = \sigma_1(k) \qquad (5\text{-}2.10)$$

$$\tau_0 \langle 22 \rangle - \tau_0 \langle 33 \rangle = \sigma_2(k) \qquad (5\text{-}2.11)$$

The behavior of a simple fluid in any viscometric flow is entirely determined by the three viscometric functions, which are an intrinsic property of the fluid ([1], pp. 24–25).

By considering an orthogonal tensor $\mathbf{Q}$ whose matrix with respect to $\mathbf{h}_k$ is:

$$[\mathbf{Q}]_h = \left\| \begin{array}{ccc} -1 & 0 & 0 \\ 0 & 1 & 0 \\ 0 & 0 & 1 \end{array} \right\| \qquad (5\text{-}2.12)$$

it is easy to show that $\tau(\ )$ must be an odd function, and the $\sigma_i(\ )$ must be even functions.

For some calculations, it may prove useful to define the viscometric viscosity $\eta_{\mathrm{V}}(\ )$, and the viscometric normal stress coefficients $\psi_i(\ )$ as

$$\eta_{\mathrm{V}}(k^2) = \frac{\tau(k)}{k} \qquad (5\text{-}2.13)$$

$$\psi_1(k^2) = \frac{\sigma_1(k)}{k^2} \qquad (5\text{-}2.14)$$

$$\psi_2(k^2) = \frac{\sigma_2(k)}{k^2} \qquad (5\text{-}2.15)$$

which are even functions of k. Also, in some calculations it is useful to assume that $\tau(\ )$ is invertible, and thus to define a function $\gamma(\ )$ as

$$\gamma(\tau(k)) = k = \gamma(\tau_0\langle 12\rangle) \qquad (5\text{-}2.16)$$

If the function $\tau(\ )$ is invertible, the value of $\tau_0\langle 12\rangle$ determines the value of k, and thus, through eqs (5-2.10) and (5-2.11), also the values of the two normal stress differences. One can therefore define two modified normal stress functions as

$$\hat{\sigma}_1(\tau_0\langle 12\rangle) = \sigma_1[\gamma(\tau_0\langle 12\rangle)] \qquad (5\text{-}2.17)$$

and

$$\hat{\sigma}_2(\tau_0\langle 12\rangle) = \sigma_2[\gamma(\tau_0\langle 12\rangle)] \qquad (5\text{-}2.18)$$

Lineal Couette flow

The simplest example of viscometric flow is lineal Couette flow. This has already been discussed in section 2-1, with reference to Reiner–Rivlin fluids, and the kinematics have been discussed in general in Illustration 3A. In a Cartesian coordinate system x^i, the components of the velocity vector are

$$v^1 = kx^2 \qquad (5\text{-}2.19)$$
$$v^2 = 0 \qquad (5\text{-}2.20)$$
$$v^3 = 0 \qquad (5\text{-}2.21)$$

The trajectories of the material points are

$$x^1 = x_t^1 - ksx_t^2 \qquad (5\text{-}2.22)$$
$$x^2 = x_t^2 \qquad (5\text{-}2.23)$$
$$x^3 = x_t^3 \qquad (5\text{-}2.24)$$

The matrix $\partial x^k/\partial x_t^m$ is calculated as

$$\frac{\partial x^k}{\partial x_t^m} = \begin{Vmatrix} 1 & -ks & 0 \\ 0 & 1 & 0 \\ 0 & 0 & 1 \end{Vmatrix} \qquad (5\text{-}2.25)$$

Since the coordinate system is Cartesian, $\sqrt{(g_{mm}/g_{kk})} = 1$, and eq. (5-1.20) yields

$$[\mathbf{F}^t]_b = [\mathbf{1}] - ks[\mathbf{N}] = \exp\{-ks[\mathbf{N}]\} \qquad (5\text{-}2.26)$$

with $[\mathbf{N}]$ having the form of eq. (5-2.2). Thus, the flow considered is viscometric, and the bases $\mathbf{g}_k$ and $\mathbf{h}_k$ coincide (so that $[\mathbf{N}] = [\mathbf{N}]_g = [\mathbf{N}]_h$). Moreover, the basis $\mathbf{b}_k(\tau)$ defined in eq. (5-1.18) is constant in space, the tensor

$P(\tau)$ is constantly equal to the unit tensor; therefore, from eqs (5-1.4) and (5-1.5),

$$\tau\langle 12\rangle = \tau(k) \qquad (5\text{-}2.27)$$

$$\tau\langle 11\rangle - \tau\langle 22\rangle = \sigma_1(k) \qquad (5\text{-}2.28)$$

$$\tau\langle 22\rangle - \tau\langle 33\rangle = \sigma_2(k) \qquad (5\text{-}2.29)$$

It follows from eqs (5-2.27)–(5-2.29) that $\nabla \cdot \tau = 0$ (since the matrix of τ is constant in space). Hence the reasoning in section 2-8 (see eqs (2-8.7)–(2-8.9)) holds and the flow considered is controllable.

Curvilineal flows

In the following, several examples of viscometric flows which are of interest in rheometry will be discussed. These are more complicated than lineal Couette flow, and can be classified according to the pattern shown in Table 5-1. It appears from the table that all known rheometrical viscometric flow systems of practical interest belong to the class of curvilineal flows, with the exception of cone–torsional flow.

A curvilineal flow is defined as follows. Let x^i be an orthogonal coordinate system, and let the contravariant components of the velocity vector be of the form

$$v^1 = u(x^2) \qquad (5\text{-}2.30)$$

$$v^2 = 0 \qquad (5\text{-}2.31)$$

$$v^3 = v(x^2) \qquad (5\text{-}2.32)$$

Table 5-1   CLASSIFICATION OF VISCOMETRIC FLOWS

A. **Curvilineal flows**
  1. Plane flows
    1.1. Lineal Couette flow
    1.2. Channel flow
  2. Helical flows
    2.1. Poiseuille flow
    2.2. Flows between coaxial cylinders
      2.2.1. Couette flow
      2.2.2. Annular flow
      2.2.3. General
    2.3. Other helical flows
  3. Cone and plate flow
  4. Torsional flow
B. **Non-curvilineal viscometric flows** (the cone–torsional flow is in this category)

---

where $u(\ )$ and $v(\ )$ are smooth functions. The trajectories of a material point are

$$x^1 = x_t^1 - su(x_t^2) \qquad (5\text{-}2.33)$$

$$x^2 = x_t^2 \qquad (5\text{-}2.34)$$

$$x^3 = x_t^3 - sv(x_t^2) \qquad (5\text{-}2.35)$$

If the metric g_{ij} is constant along the trajectories (i.e., along the lines described by eqs (5-2.33)–(5-2.35)), the flow is said to be a *curvilinear flow*.

The matrix $\partial x^k / \partial x_t^m$ is calculated from eqs (5-2.33)–(5-2.35) as

$$\frac{\partial x^k}{\partial x_t^m} = \begin{Vmatrix} 1 & -su' & 0 \\ 0 & 1 & 0 \\ 0 & -sv' & 1 \end{Vmatrix} \qquad (5\text{-}2.36)$$

where u' and v' are the ordinary derivatives of u and v. Note that $\partial x^k / \partial x_t^m$ only depends on x^2, and is thus constant along any path line. Also, by definition, $\sqrt{(g_{mm}/g_{kk})}$ is constant along any path line, and from eq. (5-1.20) one obtains

$$[\mathbf{F}']_b = [\mathbf{1}] - ks[\mathbf{N}]_g = \exp\{-ks[\mathbf{N}]_g\} \qquad (5\text{-}2.37)$$

where

$$k = \sqrt{[(u'^2 g_{11} + v'^2 g_{33})/g_{22}]} \qquad (5\text{-}2.38)$$

$$[\mathbf{N}]_g = \begin{Vmatrix} 0 & \sqrt{\left(\dfrac{g_{11}}{g_{11}u'^2 + g_{33}v'^2}\right)}u' & 0 \\ 0 & 0 & 0 \\ 0 & \sqrt{\left(\dfrac{g_{33}}{g_{11}u'^2 + g_{33}v'^2}\right)}v' & 0 \end{Vmatrix} \qquad (5\text{-}2.39)$$

The matrix $[\mathbf{N}]_g$ obeys eq. (5-2.3), and thus the flow considered is viscometric. $[\mathbf{N}]_g$ *does not* have the form in eq. (5-2.2), which means that the basis $\mathbf{g}_k$ defined in eq. (5-1.19) does not coincide with $\mathbf{h}_k$. Of course, an orthogonal tensor exists which transforms $\mathbf{g}_k$ into $\mathbf{h}_k$:

$$\mathbf{h}_k = \mathbf{Q} \cdot \mathbf{g}_k \qquad (5\text{-}2.40)$$

A simple calculation shows that $\mathbf{Q}$ has the following matrix with respect to $\mathbf{g}_k$:

$$[\mathbf{Q}]_g = \begin{Vmatrix} a & 0 & b \\ 0 & 1 & 0 \\ -b & 0 & a \end{Vmatrix} \qquad (5\text{-}2.41)$$

$$a = \frac{u'}{k}\sqrt{(g_{11}/g_{22})} \qquad (5\text{-}2.42)$$

$$b = \frac{v'}{k}\sqrt{(g_{33}/g_{22})} \qquad (5\text{-}2.43)$$

Since the reasoning leading to eq. (5-1.23) holds for the case considered, the stress field $\tau(\mathbf{X})$ can be calculated, and its physical components with respect to the x^i coordinate system are given by:

$$\tau\langle 12 \rangle = a\tau(k) \qquad (5\text{-}2.44)$$

$$\tau\langle 23 \rangle = b\tau(k) \qquad (5\text{-}2.45)$$

$$\tau\langle 13 \rangle = ab[\sigma_1(k) + \sigma_2(k)] \qquad (5\text{-}2.46)$$

$$\tau\langle 11 \rangle - \tau\langle 22 \rangle = a^2\sigma_1(k) + (a^2 - 1)\sigma_2(k) \qquad (5\text{-}2.47)$$

$$\tau\langle 22 \rangle - \tau\langle 33 \rangle = -b^2\sigma_1(k) - (b^2 - 1)\sigma_2(k) \qquad (5\text{-}2.48)$$

Note that all the components depend on $\mathbf{X}$, since a and b depend on $\mathbf{X}$.

Several examples of curvilineal flows of interest in rheometry are now discussed. For each flow we give a description of the kinematics, of the actual conditions under which the flow can be realized, and of the rheometrical information that can be obtained experimentally (that is, the viscometric functions that can be determined). The relevant equations are given without proof and reference is made to the text of Coleman et al. [1]. We trust that the reader can perform most of the calculations involved.

Channel flow

This flow can be obtained by forcing a fluid to flow between two parallel plates, separated by a distance d. We choose a Cartesian coordinate system with the x^1 axis in the direction of flow and the x^2 axis orthogonal to the plates, which are located at $x^2 = \pm d/2$.

The kinematic description of the flow is

$$v^1 = u(x^2) \qquad (5\text{-}2.49)$$

$$v^2 = v^3 = 0 \qquad (5\text{-}2.50)$$

It is well known that for Newtonian fluids the velocity distribution $u(\ )$ is a parabola; in general, the form of the velocity distribution is determined by the form of the $\tau(\ )$ function.

The flow considered is controllable, and the extra pressure is independent of x^2 and x^3 and varies linearly with x^1:

$$\mathscr{P} = A(t) - fx^1 \qquad (5\text{-}2.51)$$

with $A(t)$ an arbitrary additive constant whose value may depend on time. The quantity f is the extra pressure drop per unit length in the flow direction. This quantity is easily accessible to measurement. In fact, ordinary pressure gauges on the containing plates would read the value of $-T_{22}$; hence a differential pressure measurement over a distance L would read

$$-\Delta T_{22} = -\Delta \tau_{22} + fL - \Delta(\rho\psi) \qquad (5\text{-}2.52)$$

where Δ means the reading at x^1 minus the reading at $x^1 + L$. Since the extra stress τ is independent of x^1, $\Delta \tau_{22} = 0$ and thus f can be obtained if the body force distribution is known. (If the plates are horizontal and the body force is gravity, $\Delta(\rho\psi) = 0$ and the differential pressure reading is directly equal to fL.)

An overall force balance yields the shear stress at the wall, τ_w:

$$\tau_w = \tau_{12}(x^2 = \pm d/2) = \frac{df}{2} \qquad (5\text{-}2.53)$$

It can be shown that the shear rate at the wall, γ_w, is given by (see Illustration 2D)

$$\gamma_w = \gamma(\tau_w) = \frac{6V}{d} \left[\frac{2}{3} + \frac{1}{3} \frac{\partial \ln (6V/d)}{\partial \ln \tau_w} \right] \qquad (5\text{-}2.54)$$

where V is the average velocity, i.e., the flow rate per unit cross-section. Since τ_w and V can both be measured, the $\gamma(\ )$ function, or equivalently the $\tau(\ )$ function, can be obtained. For Newtonian fluids, $6V/d$ is the shear rate at the wall, which is proportional to τ_w, and hence the term in square brackets in eq. (5-2.54) equals unity.

Poiseuille flow

Poiseuille flow (see also section 2-5) is the axial flow in a long cylindrical tube. We choose a cylindrical coordinate system with the z axis in the direction of flow, and the tube wall located at $r = R$. The kinematic description of the flow is

$$v^z = u(r) \qquad (5\text{-}2.55)$$

$$v^r = v^\theta = 0 \qquad (5\text{-}2.56)$$

The flow is easily shown to be controllable. Again, the extra pressure varies linearly with z and is independent of r and θ; let f be the extra pressure drop per unit length of tube. Again f can be easily measured, since τ is independent of z, and the shear stress at the wall is obtained from an overall force balance:

$$\tau_w = \tau\langle rz \rangle (r = R) = \frac{Rf}{2} \qquad (5\text{-}2.57)$$

The shear rate at the wall, γ_w, is given by

$$\gamma_w = \gamma(\tau_w) = \frac{4V}{R}\left[\frac{3}{4} + \frac{1}{4}\frac{\partial \ln(4V/R)}{\partial \ln \tau_w}\right] \quad (5\text{-}2.58)$$

where V is the average velocity, and $4V/R$ is the shear rate at the wall in the case of Newtonian fluids. The $\gamma(\ )$ or the $\tau(\ )$ functions can thus be obtained.

Couette flow

Couette flow is obtained in the annular region between two coaxial cylinders, which rotate with different angular velocities, in the absence of axial extra pressure gradient. We choose a cylindrical coordinate system with the z axis coincident with the axis of the cylinders, which are located at $r = R_1$, and $r = R_2\,(R_1 < R_2)$. The cylinders' angular velocities are Ω_1 and Ω_2. The kinematic description of flow is

$$v^\theta = \omega(r); \qquad v\langle\theta\rangle = r\omega(r) \quad (5\text{-}2.59)$$

$$v^r = v^z = 0 \quad (5\text{-}2.60)$$

The flow is controllable. The torque per unit height, M (which can easily be measured), is constant, and the shear stress distribution is given by

$$\tau\langle r\theta\rangle = \frac{M}{2\pi r^2} \quad (5\text{-}2.61)$$

Definition of the following quantities is useful:

$$\tau_1 = \tau\langle r\theta\rangle(r = R_1) = \frac{M}{2\pi R_1^2} \quad (5\text{-}2.62)$$

$$\tau_2 = \tau\langle r\theta\rangle(r = R_2) = \frac{M}{2\pi R_2^2} \quad (5\text{-}2.63)$$

$$\beta = \frac{R_1^2}{R_2^2} \quad (5\text{-}2.64)$$

It can be shown that the following equation holds:

$$2M\frac{\partial \Delta\Omega}{\partial M} = \gamma(\tau_1) - \gamma(\tau_2) \quad (5\text{-}2.65)$$

where $\Delta\Omega$ is the difference of angular velocities, $\Omega_2 - \Omega_1$. In order to obtain the $\gamma(\ )$ function, one proceeds as follows.

(i) *Small gaps.* If $1 - \beta \ll 1$, the following approximate equation can be used:

$$\Delta\Omega = \frac{R_2 - R_1}{R_1}\gamma(\tau_1) \qquad (5\text{-}2.66)$$

which yields directly the $\gamma(\ )$ function from easily measured quantities.

(ii) *Large gaps.* Given a Couette viscometer with an assigned value of β, the function $\Gamma(\ )$ can be defined as follows:

$$\Gamma(\tau_1) = \gamma(\tau_1) - \gamma(\beta\tau_1) = \gamma(\tau_1) - \gamma(\tau_2) \qquad (5\text{-}2.67)$$

The function $\Gamma(\ )$ is easily obtained from eqs (5-2.65) and (5-2.67). Once the $\Gamma(\ )$ function has been obtained from the data, the $\gamma(\ )$ function is obtained as

$$\gamma(\tau_1) = \sum_{n=0}^{\infty} \Gamma(\beta^n\tau_1) \qquad (5\text{-}2.68)$$

Knowledge of the $\gamma(\ )$ function allows the angular velocity distribution to be calculated from the following equation:

$$\omega(r) - \Omega_1 = \frac{1}{2}\int_{\tau\langle r\theta\rangle}^{\tau_1} \frac{\gamma(\xi)}{\xi}d\xi \qquad (5\text{-}2.69)$$

Another quantity which can be measured is the difference of the normal stress orthogonal to the two cylinders:

$$\Delta T\langle rr\rangle = T\langle rr\rangle(r = R_2) - T\langle rr\rangle(r = R_1) \qquad (5\text{-}2.70)$$

Since part of this normal stress difference is due to centrifugal forces, a 'corrected' value, $\Delta\sigma$, can be calculated as

$$\Delta\sigma = \Delta T\langle rr\rangle + \int_{R_1}^{R_2} \rho r\omega^2\,dr \qquad (5\text{-}2.71)$$

The quantity $\Delta\sigma$ is related to the $\hat{\sigma}_1(\ )$ function through the following equation:

$$\Delta\sigma = \int_{R_1}^{R_2} \frac{\hat{\sigma}_1(\tau\langle r\theta\rangle)}{r}\,dr \qquad (5\text{-}2.72)$$

In order to obtain the $\hat{\sigma}_1(\ )$ function, one proceeds as follows.

(i) *Small gaps.* The $\hat{\sigma}_1(\ )$ function is obtained directly as

$$\hat{\sigma}_1(\tau_1) = \frac{R_1}{R_2 - R_1}\Delta\sigma \qquad (5\text{-}2.73)$$

(ii) *Large gaps.* Define the function $\Psi(\tau_1)$ as

$$\Psi(\tau_1) = 2M \frac{\partial \Delta\sigma}{\partial M} \qquad (5\text{-}2.74)$$

The function $\Psi(\ )$ is directly obtainable from the data by measuring both $\Delta\sigma$ and M. The function $\hat{\sigma}_1(\ )$ is then obtained as

$$\hat{\sigma}_1(\tau_1) = \sum_0^\infty \Psi(\beta^n \tau_1) \qquad (5\text{-}2.75)$$

Hence, a Couette viscometer allows both the viscometric viscosity and the first viscometric normal stress difference to be determined.

Annular flow

Annular flow is the axial flow in the region between two coaxial stationary cylinders. The flow is controllable, and in principle the $\gamma(\ )$ function can be obtained from annular flow experiments, though in practice this is not preferable. The most interesting result that can be obtained from annular flow is the difference of normal stresses orthogonal to the bounding cylinders, which is related to the second normal stress difference through the following equation:

$$\Delta T \langle rr \rangle = -\int_{R_1}^{R_2} \frac{1}{r} \hat{\sigma}_2(\tau) \, dr \qquad (5\text{-}2.76)$$

where R_1 and R_2 are the radii of the inner and outer cylinders, and $\tau(r)$ is the radial shear stress distribution. The $\tau(r)$ distribution is

$$\tau(r) = \frac{b}{r} - \frac{rf}{2} \qquad (5\text{-}2.77)$$

where f is the extra pressure drop per unit axial length, and b is a constant. The value of b is fixed, implicitly, by the following equation:

$$\int_{R_1}^{R_2} \gamma[\tau(r)] \, dr = 0 \qquad (5\text{-}2.78)$$

Cone and plate flow

Cone and plate flow is obtained in the region bounded by a flat plate and a convex cone whose apex touches the plate (see Fig. 5-1). We choose a spherical coordinate system with the origin at the cone apex and the pole on the cone axis (which is orthogonal to the flat plate), and obtain the following

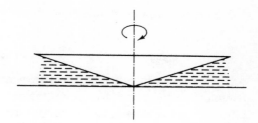

FIGURE 5-1
Cone and plate geometry.

kinematic description of flow:

$$v^\phi = \omega(\theta) \qquad (5\text{-}2.79)$$

$$v^\theta = \overset{\bullet}{v^r} = 0 \qquad (5\text{-}2.80)$$

The flat plate is located at $\theta = \pi/2$, while the cone surface is at $\theta = \pi/2 - \alpha$. The flow is not controllable, unless:

(i) inertia is neglected,

(ii) $\alpha \ll \pi/2$, so that $\cos\theta \approx 1$.

Let $\Delta\Omega$ be the difference of angular velocity of cone and plate. The strain rate tensor has a constant magnitude in space; the shear rate k is, everywhere,

$$k = \frac{\Delta\Omega}{\alpha} \qquad (5\text{-}2.81)$$

If the material fills the region in space at $r \leq R$, the torque (which is easily measured) is given by

$$M = \frac{2\pi}{3} R^3 \tau(k) \qquad (5\text{-}2.82)$$

and therefore the $\tau(k)$ function is obtained directly. Cone and plate viscometers are very popular.

The force required to keep cone and plate in place, F, is directed along the cone axis, and is given by

$$F = \frac{\pi}{2} R^2 \sigma_1(k) \qquad (5\text{-}2.83)$$

(F, here and elsewhere, is positive when squeezing) so that the $\sigma_1(\ )$ function can also be obtained directly (although eq. (5-2.83) is obtained by neglecting surface tension on the free surface of the material).

Finally, it can be shown that

$$\frac{\partial T\langle\theta\theta\rangle}{\partial \ln r} = \sigma_1(k) + 2\sigma_2(k) \qquad (5\text{-}2.84)$$

so that, at least in principle, measurement of the normal stress radial distribution on the plate also allows the $\sigma_2(\ )$ function to be calculated. The latter procedure is open to a series of experimental problems.

Torsional flow

Torsional flow is obtained in the disc-shaped region between two parallel plates, which rotate in their own plane with an angular velocity difference $\Delta\Omega$. If h is the distance between the plates, the kinematic description of flow, in a cylindrical coordinate system with the z axis coincident with the axis of rotation, is:

$$v^\theta = \Delta\Omega\frac{z}{h} + \text{const.} \qquad (5\text{-}2.85)$$

$$v^r = v^z = 0 \qquad (5\text{-}2.86)$$

The shear rate k is given by

$$k = \Delta\Omega\frac{r}{h} \qquad (5\text{-}2.87)$$

If the material fills the region of space at $r \le R$, the $\tau(\ )$ function is obtained from:

$$\tau\left(\frac{\Delta\Omega R}{h}\right) = \frac{2M}{\pi R^3}\left(\frac{3}{4} + \frac{1}{4}\frac{\partial \ln (2M/\pi R^3)}{\partial \ln (\Delta\Omega R/h)}\right) \qquad (5\text{-}2.88)$$

where M is the torque required to keep the flow steady.

From torsional flow experiments, the difference between the σ_1 and σ_2 functions can be obtained from:

$$\sigma_1\left(\frac{R\,\Delta\Omega}{h}\right) - \sigma_2\left(\frac{R\,\Delta\Omega}{h}\right) = \frac{F}{\pi R^2}\left[2 + \frac{\partial \ln (F/\pi R^2)}{\partial \ln (R\,\Delta\Omega/h)}\right] \qquad (5\text{-}2.89)$$

where F is the force orthogonal to the plates required to keep them in place. The force F is given by

$$F = \pi \int_0^R r\{\sigma_1[k(r)] - \sigma_2[k(r)]\}\,dr \qquad (5\text{-}2.90)$$

Equation (5-2.90) is not directly useful (since eq. (5-2.89), which is derived from it, is in practice used to determine σ_2), but is given here for future reference.

Cone–torsional flow

This is the only viscometric rheometrical flow system which is not discussed in reference [1]; the reader is referred to Marsh and Pearson [2] for a detailed analysis.

Cone–torsional flow is obtained in the region between a flat plate and a cone with the axis, which is also the rotation axis, orthogonal to the plate. The cone may be both convex or concave, and, if convex, the tip need not touch the plate (see Fig. 5-2). Let h be the distance of the cone's apex from the flat plate. We choose a cylindrical coordinate system with the z axis along the cone's axis, the plate lying at $z = 0$ and the cone surface lying at $z = h + r \tan \alpha$. The angle α is positive for convex cones and negative for concave cones. Since $\alpha \ll \pi/2$ is the condition for controllability of flow (after neglecting inertia), we will use the approximation $\tan \alpha \approx \alpha$.

The flow field is described by

$$v^\theta = \frac{\Delta \Omega z}{h + r\alpha} + \text{const.} \qquad (5\text{-}2.91)$$

$$v^r = v^z = 0 \qquad (5\text{-}2.92)$$

where $\Delta\Omega$ is the angular velocity difference. The shear rate distribution $k(r)$ is given by

$$k(r) = \frac{\Delta \Omega r}{h + r\alpha} \qquad (5\text{-}2.93)$$

It is interesting to observe that eqs (5-2.91) and (5-2.92) *are not* the equations of a curvilineal flow. We have been unable to identify an orthogonal coordinate system where the flow considered here is described by equations which fulfil the definition of a curvilineal flow, and we suspect that cone–torsional flow is not a curvilineal flow. Nevertheless it is a viscometric flow, and belongs to a very general class of flows discussed in detail by Yin and Pipkin [3].

Cone–torsional flow degenerates, in the limit $\alpha \to 0$, into torsional flow, and in the limit $h \to 0$ into cone-and-plate flow. The shear rate is not constant in space, and since it is not a linear function of the coordinates, the procedure for inverting the integral equations giving the torque M and the normal force F is somewhat lengthy.

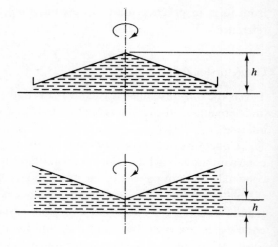

FIGURE 5-2
Cone–torsional flow geometry.

The $\tau(\ )$ function is obtainable from the following equation:

$$\tau\left(\frac{\Delta\Omega R}{h + R\alpha}\right) = \frac{M}{2\pi R^3}\left(3 - \frac{\partial \ln M}{\partial \ln h}\right) \qquad (5\text{-}2.94)$$

where the fluid is supposed to fill the region in space at $r \leq R$, and the derivative with respect to h is taken at constant $\Delta\Omega$ and constant R.

The force required to keep cone and plate in place is given by

$$F = \pi \int_0^R r\left\{\sigma_1[k(r)] - \frac{h}{h + r\alpha}\sigma_2[k(r)]\right\} dr \qquad (5\text{-}2.95)$$

Equation (5-2.95) should be compared with eq. (5-2.90) above, which holds for torsional flow (and is indeed obtainable from eq. (5-2.95) by setting $\alpha = 0$). It is clear that the contribution of the second normal stress difference σ_2 to the value of F depends on the magnitude of the coefficient $h/(h + r\alpha)$. This coefficient equals unity in torsional flow, and can be made larger than unity in a cone–torsional flow by using a concave cone, i.e., by having $\alpha < 0$. Hence, cone–torsional flow can be particularly useful for the experimental determination of the $\sigma_2(\ )$ function.

Equation (5-2.95) can be inverted to yield the $\sigma_2(\ )$ function explicitly:

$$\sigma_2\left(\frac{\Delta\Omega R}{h + R\alpha}\right) = \frac{h + R\alpha}{h}\left[\sigma_1\left(\frac{\Delta\Omega R}{h + R\alpha}\right)\right.$$

$$\left. - \frac{F}{\pi R^2}\left(2 - \frac{\partial \ln (F/\pi R^2)}{\partial \ln h}\right)\right] \quad (5\text{-}2.96)$$

It is easy to show that, on setting $\alpha = 0$, eq. (5-2.96) degenerates into eq. (5-2.89), while on setting $h = 0$ it degenerates into eq. (5-2.83).

General conclusions

The rheometrical flow systems discussed above allow us to determine the viscometric functions for any given material. The function which is most easily obtained is the $\tau(\ )$ function, which can be obtained from all systems with the exception of annular flow. The $\sigma_1(\ )$ function is best obtained from cone-and-plate flow data, but can also be obtained from Couette flow data. The $\sigma_2(\ )$ function is the most difficult to measure, and although annular and torsional flow both lead to its measurement, cone–torsional flow with $\alpha < 0$ seems to offer the best possibility for measuring σ_2.

5-3 EXTENSIONAL FLOWS

The extensiometric functions

Extensional flows (see section 3-5) are constant stretch history flows for which the tensor $\mathbf{N}$ is symmetric:

$$\mathbf{N} = \mathbf{N}^\mathsf{T} \quad (5\text{-}3.1)$$

Since the tensor $\mathbf{N}$ satisfies eq. (5-3.1), an orthonormal basis $\mathbf{h}_k$ exists with respect to which the matrix of $k\mathbf{N}$ has the form

$$[k\mathbf{N}]_h = \left\|\begin{array}{ccc} \gamma_1 & 0 & 0 \\ 0 & \gamma_2 & 0 \\ 0 & 0 & \gamma_3 \end{array}\right\| \quad (5\text{-}3.2)$$

From eq. (3-5.13) one has

$$\nabla \cdot \mathbf{v} = 0 = \text{tr}\,(k\mathbf{N}) \quad (5\text{-}3.3)$$

and therefore the γ_is and k fulfil the following conditions:

$$\gamma_1 + \gamma_2 + \gamma_3 = 0; \quad k = \sqrt{(\gamma_1^2 + \gamma_2^2 + \gamma_3^2)} \quad (5\text{-}3.4)$$

It is important for future reference to point out that, for extensional flows, one has

$$\text{tr}\,(\mathbf{N}^2) = 1 \qquad (5\text{-}3.5)$$

while for viscometric flows one has

$$\text{tr}\,(\mathbf{N}^2) = 0 \qquad (5\text{-}3.6)$$

The stress tensor at $t = 0$ is given by

$$\tau_0 = \tau(0) = \mathbf{H}(k\mathbf{N}) \qquad (5\text{-}3.7)$$

with $\mathbf{H}$ an isotropic function. In particular, consider an orthogonal tensor $\mathbf{Q}$ which, with respect to the same basis for which eq. (5-3.2) holds, has a matrix of the form of eq. (5-2.5). By a reasoning analogous to that leading to eq. (5-2.8), one can prove that the matrix of τ_0 with respect to $\mathbf{h}_k$ has the form

$$[\tau_0]_h = \left\|\begin{matrix} \tau_0\langle 11\rangle & 0 & 0 \\ 0 & \tau_0\langle 22\rangle & 0 \\ 0 & 0 & \tau_0\langle 33\rangle \end{matrix}\right\| \qquad (5\text{-}3.8)$$

i.e., that the tangential stresses are zero.

Since tensor τ_0 is defined only within an arbitrary additive isotropic tensor, eq. (5-3.8) implies that τ_0 is completely determined by two of its components, namely $\tau_0\langle 11\rangle - \tau_0\langle 22\rangle$ and $\tau_0\langle 22\rangle - \tau_0\langle 33\rangle$. These are in turn, as seen from eq. (5-3.7), entirely determined by the value of the γ_is; one can thus define the following two extensiometric material functions:

$$\tau_0\langle 11\rangle - \tau_0\langle 22\rangle = \sigma_{E1}(\gamma_1,\gamma_2,\gamma_3) \qquad (5\text{-}3.9)$$

$$\tau_0\langle 22\rangle - \tau_0\langle 33\rangle = \sigma_{E2}(\gamma_1,\gamma_2,\gamma_3) \qquad (5\text{-}3.10)$$

An additional feature of the functions $\sigma_{E1}(\ )$ and $\sigma_{E2}(\ )$ can be obtained by considering a special case of extensional flow, i.e., 'simple' elongational flow. For elongational flow, two of the γ_i are equal, say, for example,

$$\gamma_2 = \gamma_3 \qquad (5\text{-}3.11)$$

In view of eq. (5-3.4), one thus has only one independent scalar parameter, which is called the rate of elongation:

$$\gamma_E = \gamma_1 = -2\gamma_2 = -2\gamma_3 \qquad (5\text{-}3.12)$$

Consider now an orthogonal tensor $\mathbf{Q}$ whose matrix with respect to the $\mathbf{h}_k$ basis has the following form:

$$[\mathbf{Q}]_h = \left\|\begin{matrix} 1 & 0 & 0 \\ 0 & 1/\sqrt{2} & 1/\sqrt{2} \\ 0 & -1/\sqrt{2} & 1/\sqrt{2} \end{matrix}\right\| \qquad (5\text{-}3.13)$$

Isotropy of the **H**( ) function requires that

$$\tau_0\langle 22\rangle = \tau_0\langle 33\rangle \qquad (5\text{-}3.14)$$

i.e., that

$$\sigma_{E2} = 0 \qquad (5\text{-}3.15)$$

Thus, only one elongational material property needs to be considered; it is generally useful to define an elongational viscosity function, $\eta_E(\ )$, as

$$\frac{\tau_0\langle 11\rangle - \tau_0\langle 22\rangle}{\gamma_E} = \eta_E(\gamma_E) \qquad (5\text{-}3.16)$$

The value of the elongational viscosity for Newtonian fluids was originally determined by Trouton [4], and the elongational viscosity is often referred to as Trouton viscosity. For Newtonian fluids, the elongational viscosity is constant and equal to three times the viscosity. Since the Newtonian constitutive equation holds for all simple fluids with fading memory in the limit of slow flows, one obtains the following general relationship among the elongational and viscometric viscosities:

$$\lim_{\gamma_E \to 0} \eta_E(\gamma_E) = 3 \lim_{k \to 0} \eta_V(k^2) \qquad (5\text{-}3.17)$$

Extensiometric flows

In contrast with viscometric flows, there are no known examples of flows within constraining boundaries which are extensiometric. Flows with free boundaries can be extensiometric; a special class is that which can be described in a Cartesian coordinate system by the following equations for the velocity vector:

$$v^1 = \gamma_1 x^1 \qquad (5\text{-}3.18)$$
$$v^2 = \gamma_2 x^2 \qquad (5\text{-}3.19)$$
$$v^3 = \gamma_3 x^3 \qquad (5\text{-}3.20)$$

with the γ_is related to each other by eq. (5-3.4).

Such flows are irrotational ($\mathbf{W} = \mathbf{0}$), and the natural basis of the Cartesian coordinate system (which is constant in space) is the one with respect to which $[\mathbf{F}^t]_b$ is expressed by eq. (5-1.13), with [**N**] having the form in eq. (5-3.2).

Since $\mathbf{F}^t$ is constant in space, so is the stress tensor τ, and hence eq. (5-1.37) reduces to

$$\rho\frac{D\mathbf{v}}{Dt} = -\nabla\mathscr{P} \qquad (5\text{-}3.21)$$

The acceleration vector $D\mathbf{v}/Dt$ can easily be calculated from eqs (5-3.18)–(5-3.20), and it can be expressed as the gradient of a scalar field. Hence, the flow considered is controllable, with an extra pressure field given by

$$\mathscr{P} = f(t) + \tfrac{1}{2}\rho[(\gamma_1 x^1)^2 + (\gamma_2 x^2)^2 + (\gamma_3 x^3)^2] \qquad (5\text{-}3.22)$$

Unfortunately, the extra pressure field given by eq. (5-3.22) is not realistic for a free-boundary flow, and therefore extensiometric flows cannot easily be realized unless the conditions are such that both gravity and inertia can be neglected. When this is true, a constant pressure makes the flow controllable [5].

5-4 PERIODIC FLOWS

Periodic flows have been defined in section 5-1; it is worth repeating that, since such flows are of interest in rheometry in the limit of very small deformations, the defining equation (eq. (5-1.24)) needs to be fulfilled only within terms of order one in the magnitude of the deformation.

There are basically two types of rheometrical flow systems which are useful for experiments on periodic flows; we shall refer to these two types as Eulerian and Lagrangian. Although both flow types allow the rheometrical determination of the complex viscosity η^*, they are basically different in character: while Lagrangian-periodic flows are constant stretch history flows, Eulerian-periodic flows are not.

Eulerian-periodic flow systems

In Eulerian-periodic flow systems, the sample of material to be tested is subjected to small deformations varying sinusoidally with time by actually imposing on some physical boundary a sinusoidal vibration. To within terms of order one in the magnitude of the deformation, eq. (5-1.24) holds true with tensor $\boldsymbol{\psi}^*(t)$ depending on time sinusoidally, say,

$$\boldsymbol{\psi}^*(t) = -\frac{\boldsymbol{\Gamma}^*}{i\omega} e^{i\omega t} \qquad (5\text{-}4.1)$$

where tensor $\boldsymbol{\Gamma}^*$ is constant:

$$\boldsymbol{\Gamma}^* = \boldsymbol{\Gamma}_R - i\boldsymbol{\Gamma}_I \qquad (5\text{-}4.2)$$

From eq. (5-1.34), the strain rate tensor $\mathbf{D}$ is given by:

$$\mathbf{D} = \text{Re}\left\{-\frac{i\omega \boldsymbol{\psi}^*(t)}{2}\right\} = \tfrac{1}{2}\,\text{Re}\{\boldsymbol{\Gamma}^* e^{i\omega t}\} \qquad (5\text{-}4.3)$$

and is thus seen to vary sinusoidally with time. Analogously, the stress tensor is also sinusoidal in time, as can be seen from eq. (5-1.35):

$$\tau = \text{Re}\{2\eta^* \mathbf{D}^*\} = \text{Re}\{\eta^* \mathbf{\Gamma}^* e^{i\omega t}\} \qquad (5\text{-}4.4)$$

Equations (5-4.3) and (5-4.4) hold of course separately for each component of the tensors involved. If one component of τ (which we will indicate by σ) and one analogous component of $2\mathbf{D}$ (which we will indicate by γ) can be measured, the complex viscosity η^* can be obtained.

In fact, let Γ^* be the component of $\mathbf{\Gamma}^*$ homologous to both σ and γ. Since the latter quantities are periodical in time, with a frequency ω, one may write

$$\gamma = \gamma_m \cos(\omega t + \phi_\gamma) \qquad (5\text{-}4.5)$$

$$\sigma = \sigma_m \cos(\omega t + \phi_\sigma) \qquad (5\text{-}4.6)$$

where γ_m and σ_m are the *amplitudes*, and ϕ_γ and ϕ_σ the *phase angles* of strain rate and of stress, respectively. One has

$$\Gamma^* = \gamma_m \, e^{i\phi_\gamma} \qquad (5\text{-}4.7)$$

$$\eta^* \Gamma^* = \sigma_m \, e^{i\phi_\sigma} \qquad (5\text{-}4.8)$$

The complex viscosity η^* is obtained directly from eqs (5-4.7) and (5-4.8):

$$\eta^* = \frac{\sigma_m}{\gamma_m} e^{i\Phi} \qquad (5\text{-}4.9)$$

where

$$\Phi = \phi_\sigma - \phi_\gamma \qquad (5\text{-}4.10)$$

Hence, the complex viscosity η^* can be calculated if γ_m, σ_m, and Φ are measured on some physical boundary. The difficulty lies in the determination of γ_m from measurable quantities, which requires the solution of the equations of motion, i.e., the calculation of the spatial distribution of the strain rate.

The solution of the equations of motion is generally trivial if the inertia of the fluid is neglected. With such an approximation, one can easily calculate the value of γ_m from the kinematics of the physical boundaries of the system. Indeed, there is a second technique for determining η^*, which is based only on kinematic measurements (while eq. (5-4.9) also implies the measurement of stress). This technique will be discussed in detail only for a geometrically simple situation to be analyzed below; only final results will be given for other geometries.

Periodic steady plane shear

Consider a periodic plane shear, for which the velocity field is described in some Cartesian coordinate system x^i by

$$v^1 = \text{Re}\{a^*(x^2)\,e^{i\omega t}\} \qquad (5\text{-}4.11)$$

$$v^2 = v^3 = 0 \qquad (5\text{-}4.12)$$

$$a^*(x^2) = a_R(x^2) - ia_I(x^2) \qquad (5\text{-}4.13)$$

A standard calculation yields the function $\mathbf{G}^t$; to within terms of order one in the magnitude of the deformation (i.e., of $|a^{*\prime}|/|\omega|$) there is only one non-zero component of $\mathbf{G}^t$, i.e., the 12 component:

$$G^t_{12} = \text{Re}\left\{-\frac{\Gamma^*}{i\omega}\,e^{i\omega t}[1 - e^{-i\omega s}]\right\} \qquad (5\text{-}4.14)$$

where

$$\Gamma^*(x^2) = \frac{\mathrm{d}}{\mathrm{d}x^2}[a^*(x^2)] = a^{*\prime} \qquad (5\text{-}4.15)$$

The flow considered is therefore an Eulerian-periodic flow, and eq. (5-4.9) holds with $\sigma = \tau_{12}$ and $\gamma = 2D_{12}$.

The equations of motion in the 1 and 2 directions are

$$-\frac{\partial \mathscr{P}}{\partial x^1} = \rho\frac{\partial v^1}{\partial t} - \frac{\partial \tau^{12}}{\partial x^2} \qquad (5\text{-}4.16)$$

$$-\frac{\partial \mathscr{P}}{\partial x^2} = 0 \qquad (5\text{-}4.17)$$

The tangential stress $\sigma = \tau_{12}$ is obtained from eq. (5-4.4):

$$\sigma = \text{Re}\{\eta^*\Gamma^*\,e^{i\omega t}\} \qquad (5\text{-}4.18)$$

When eqs (5-4.11), (5-4.15), and (5-4.18) are substituted into eq. (5-4.16), the extra pressure field $\mathscr{P}$ is obtained as

$$\mathscr{P} = f(t) - x^1\,\text{Re}\{\mathscr{P}^*\,e^{i\omega t}\} \qquad (5\text{-}4.19)$$

where

$$\mathscr{P}^* = \rho i\omega a^* - \eta^* a^{*\prime\prime} \qquad (5\text{-}4.20)$$

where $a^{*\prime\prime}$ is the ordinary second derivative of a^*.

Now suppose in particular that the flow is realized by holding the sample between two flat plates, the one at $x^2 = 0$ being stationary, and the one at $x^2 = h$ being vibrated sinusoidally:

$$x^2 = 0, \qquad v^1 = 0 \qquad (5\text{-}4.21)$$

$$x^2 = h, \qquad v^1 = V \cos(\omega t) = \text{Re}\{V e^{i\omega t}\} \qquad (5\text{-}4.22)$$

The small-deformation approximation implies that $V/h\omega$ is small, and eq. (5-1.24) holds to within terms of the first order in $V/h\omega$.

Under these conditions, the extra pressure would be constant in the x^1 direction, $\mathcal{P}^* = 0$, and therefore

$$a^{*\prime\prime} + \alpha^2 a^* = 0 \qquad (5\text{-}4.23)$$

where

$$\alpha^2 = -\frac{i\omega\rho}{\eta^*} \qquad (5\text{-}4.24)$$

The parameter α^2, which has dimensions of an inverse square length, appears in all periodic flow problems, as a yardstick for measuring the relative importance of inertia forces.

In view of eqs (5-4.11), (5-4.21) and (5-4.22), the boundary conditions associated with eq. (5-4.23) are

$$x^2 = 0, \qquad a^* = 0 \qquad (5\text{-}4.25)$$

$$x^2 = h, \qquad a^* = V \qquad (5\text{-}4.26)$$

and therefore the velocity distribution is obtained:

$$a^* = V \frac{\sin(\alpha x^2)}{\sin(\alpha h)} \qquad (5\text{-}4.27)$$

The shear rate distribution is, from eq. (5-4.15),

$$\Gamma^* = \alpha V \frac{\cos(\alpha x^2)}{\sin(\alpha h)} \qquad (5\text{-}4.28)$$

If the stress is measured on the fixed plate $x^2 = 0$, the shear rate γ is

$$\gamma(0) = \text{Re}\{\Gamma^*(0) e^{i\omega t}\} = \text{Re}\left\{\frac{\alpha V e^{i\omega t}}{\sin(\alpha h)}\right\} \qquad (5\text{-}4.29)$$

When αh is small (i.e., when inertia is negligible), a simple expression is

$$\gamma(0) = \text{Re}\left\{\frac{V}{h} e^{i\omega t}\left[1 + \frac{\alpha^2 h^2}{6}\right]\right\} \qquad (5\text{-}4.30)$$

Two points need to be stressed. First, the flow considered here is described by eqs (5-4.11)–(5-4.13) and (5-4.21)–(5-4.22), which are simply the equations that would describe *steady* plane shear between two parallel flat plates, multiplied by a periodic term $e^{i\omega t}$. Equation (5-4.30) shows that, in the limit $\alpha^2 = 0$, the shear rate γ is the one that would occur in steady plane shear multiplied by the same factor. This is generally true for all viscometric flows which give rise to an Eulerian-periodic flow upon multiplication by $e^{i\omega t}$ of the velocity field description. In fact, neglect of inertia makes the differential equation for the velocity distribution in the periodic case (which for plane shear is eq. (5-4.23)) equal to the one that would hold for the analogous steady flow problem.

The second point is that, while the velocity distribution in the limit $\alpha^2 = 0$ does not depend on the properties of the material (say on η^*), even the first-order correction term for inertia does depend on η^* (see eq. (5-4.30)). Therefore, rheometry is best carried out under conditions where inertia at most introduces a correction term, the value of which can be calculated by assuming for η^* the zero-order approximation (i.e., the zero-inertia result).

Let us now consider a second technique for the determination of η^*, which is based only on kinematic measurements. In order to do so, let us define the displacement function $\delta(x^2)$ as

$$\delta = \text{Re} \{\delta^* e^{i\omega t}\}; \qquad \delta^* = \frac{a^*}{i\omega} \qquad (5\text{-}4.31)$$

(All material points in the flow described by eqs (5-4.11)–(5-4.13) move periodically; δ describes the displacement from the mid point of the trajectory.) Again, one may write δ in the form

$$\delta = \delta_m \cos(\omega t + \psi); \qquad \delta^* = \delta_m e^{i\psi} \qquad (5\text{-}4.32)$$

and, therefore, one has

$$\frac{\delta_1^*}{\delta_2^*} = \frac{\delta_{m_1}}{\delta_{m_2}} e^{i\Psi} \qquad (5\text{-}4.33)$$

where

$$\Psi = \psi_1 - \psi_2 \qquad (5\text{-}4.34)$$

Upon measuring the amplitudes of the displacements, δ_{m_1} and δ_{m_2}, and their phase-angle Ψ, the value of η^* can be obtained provided the ratio δ_1^*/δ_2^* can be calculated from the equations of motion. In order to do so, one obtains from eqs (5-4.23) and (5-4.31):

$$\delta^{*''} + \alpha^2 \delta^* = 0 \qquad (5\text{-}4.35)$$

which is the equation of motion expressed in terms of the displacement function. Now, assume that the plate at $x^2 = h$ is vibrated sinusoidally:

$$x^2 = h, \qquad \delta^* = \frac{V}{i\omega} \qquad (5\text{-}4.36)$$

while the plate at $x^2 = 0$ is constrained by an elastic spring of stiffness K and has a mass M (both per unit surface):

$$x^2 = 0, \qquad \sigma = \eta^* i\omega\delta^{*'} \, e^{i\omega t} = S\delta^* \, e^{i\omega t} \qquad (5\text{-}4.37)$$

where

$$S = K - M\omega^2 \qquad (5\text{-}4.38)$$

Equation (5-4.37) is the equation of motion for the plate located at $x^2 = 0$. The differential equation (5-4.35), subject to the boundary conditions (5-4.36) and (5-4.37), has the integral

$$\delta^* = \frac{V}{i\omega} \frac{\cos(\alpha x^2) + (S/\eta^* i\omega\alpha) \sin(\alpha x^2)}{\cos(\alpha h) + (S/\eta^* i\omega\alpha) \sin(\alpha h)} \qquad (5\text{-}4.39)$$

Therefore, one has the following equation for the ratio of the displacements at the two plates (which is related to the measurable quantities δ_{m_0}, δ_{m_h}, and $\Psi = \psi_h - \psi_0$ through eq. (5-4.33)):

$$\frac{\delta_h^*}{\delta_0^*} = \cos(\alpha h) + \frac{S}{\eta^* i\omega\alpha} \sin(\alpha h) \qquad (5\text{-}4.40)$$

Equation (5-4.40) allows us, in principle, to calculate η^* from the measurements of δ_{m_h}, δ_{m_0}, and Ψ. The procedure required to do so is rather involved, unless inertia of the fluid can be neglected; in that case, eq. (5-4.40) reduces to

$$\frac{\delta_{m_h}}{\delta_{m_0}} e^{i\Psi} = 1 + \frac{Sh}{i\omega\eta^*} \qquad (5\text{-}4.41)$$

The relative advantages of the two techniques for determining η^*, i.e., use of eq. (5-4.9) or (5-4.41), are worth discussing. In both cases, the kinematics of the forced plate are measured, but while eq. (5-4.9) implies measurement of the stress on the fixed plate, eq. (5-4.41) implies measurement of the motion of the constrained plate. Since, in practice, the measurement of stress is always obtained by measuring the deflection of some elastic constraint, the two methods basically differ in the following point: eq. (5-4.9) requires the use of a very stiff constraint, so that the constrained plate is almost fixed, while eq. (5-4.41) requires the use of a rather loose constraining mechanism (typically in rotational apparatuses, a torsion wire). When using eq. (5-4.41), care should be taken to avoid the natural frequency of the constrained plate, ω_0. In fact, when $\omega = \omega_0$, $S = 0$, and eq. (5-4.40) or (5-4.41) does not allow us to determine η^*. In the following, only basic results are given for the more complicated geometries; the reader is referred to the technical literature for all details.

Periodic helical flows

A periodic helical flow [6] in a cylindrical coordinate system r, θ, z is described by the following equations for the physical components of the velocity vector:

$$v\langle r \rangle = 0 \tag{5-4.42}$$

$$v\langle \theta \rangle = r \operatorname{Re} \left\{ \Omega^*(r)\, e^{i\omega t} \right\} \tag{5-4.43}$$

$$v\langle z \rangle = \operatorname{Re} \left\{ u^*(r)\, e^{i\omega t} \right\} \tag{5-4.44}$$

To within terms of the first order in the magnitude of the deformation, there are only two non-zero components of stress and of strain rate:

$$\sigma_\theta = \tau\langle r\theta \rangle = r \operatorname{Re} \left\{ \eta^* \Omega^{*\prime}\, e^{i\omega t} \right\} \tag{5-4.45}$$

$$\sigma_z = \tau\langle rz \rangle = \operatorname{Re} \left\{ \eta^* u^{*\prime}\, e^{i\omega t} \right\} \tag{5-4.46}$$

$$\gamma_\theta = 2D\langle r\theta \rangle = r \operatorname{Re} \left\{ \Omega^{*\prime}\, e^{i\omega t} \right\} \tag{5-4.47}$$

$$\gamma_z = 2D\langle rz \rangle = \operatorname{Re} \left\{ u^{*\prime}\, e^{i\omega t} \right\} \tag{5-4.48}$$

The equations of motion yield the following equations for the extra-pressure distribution and the velocity distribution:

$$\mathscr{P} = \operatorname{Re} \left\{ \mathscr{P}^*\, e^{i\omega t} \right\} \tag{5-4.49}$$

$$\mathscr{P}^* = f^*(r) - z\lambda^* \tag{5-4.50}$$

$$\frac{3\Omega^{*\prime}}{r} + \Omega^{*\prime\prime} + \alpha^2 \Omega^* = 0 \tag{5-4.51}$$

$$\frac{u^{*\prime}}{r} + u^{*\prime\prime} + \alpha^2 u^* + \frac{\lambda^*}{\eta^*} = 0 \tag{5-4.52}$$

Note that $\operatorname{Re} \left\{ \lambda^*\, e^{i\omega t} \right\}$ is the extra-pressure gradient in the z direction, which can be non-zero if, for example, a sinusoidally varying pressure differential is applied, as in oscillating Poiseuille flow.

The general integrals of eqs (5-4.51) and (5-4.52) are

$$r\Omega^* = C^*J_1(\alpha r) + D^*Y_1(\alpha r) \qquad (5\text{-}4.53)$$

$$u^* = A^*J_0(\alpha r) + B^*Y_0(\alpha r) + \frac{\lambda^*}{i\omega\rho} \qquad (5\text{-}4.54)$$

where $J_0(\ )$, $J_1(\ )$ and $Y_0(\ )$, $Y_1(\ )$ are Bessel functions of the first and second kind, respectively, and A^*, B^*, C^*, and D^* are complex integration constants to be determined from the boundary conditions. The radial pressure distribution is given by

$$f^* = \rho \int r\Omega^* \, dr \qquad (5\text{-}4.55)$$

First-order corrections to the negligible inertia solutions for the geometries of interest are available in the literature [7].

Torsional and cone-and-plate periodic flows

As indicated in section 5-2, both torsional and cone-and-plate steady flows are not controllable, unless inertia is negligible. The physical interpretation of this is easily obtained by considering that, in order to balance centrifugal forces, a radial pressure distribution is required; since the angular velocity is not constant along the z (torsional flow) or θ (cone-and-plate flow) direction, such a pressure would induce a secondary flow in that direction.

When the periodic form of torsional and cone-and-plate flows are considered, it may at first sight seem useless to look for any but the zero-inertia approximation, since inertial forces make even steady flow uncontrollable. In effect, this is not the case: in fact, inertia in the radial direction (i.e., centrifugal forces) may well be negligible although inertia in the flow direction (due to the oscillatory nature of the periodic flow) is not.

In order to clarify this point, consider the torsional flow between a fixed flat plate and an upper plate subject to oscillatory rotation:

$$z = h, \qquad v^\theta = \mathrm{Re}\,\{\Omega^* \, e^{i\omega t}\} \qquad (5\text{-}4.56)$$

The centrifugal forces and the inertia forces in the flow direction are, at the upper plate:

$$\rho\frac{Dv\langle r\rangle}{Dt} = -\rho r[\mathrm{Re}\,\{\Omega^* \, e^{i\omega t}\}]^2 \qquad (5\text{-}4.57)$$

$$\rho\frac{Dv\langle\theta\rangle}{Dt} = \rho r \, \mathrm{Re}\,\{i\omega\Omega^* \, e^{i\omega t}\} \qquad (5\text{-}4.58)$$

Therefore, inertia forces are much larger than centrifugal forces when the frequency of oscillation is much larger than the amplitude of the rotation rate, i.e., when

$$|\omega| \gg |\Omega^*| \qquad (5\text{-}4.59)$$

Since eq. (5-4.59) is also the condition for the deformation to be small, it is seen that, in the rheometrical determination of η^* with periodic torsional and cone-and-plate flow, inertia forces due to the oscillation are indeed predominant on centrifugal forces, so that consideration of the former and neglect of the latter is plausible.

The cone-and-plate system has been analyzed in detail by Nally [8], and approximate equations have been given by Walters and Kemp [9]. This system is not particularly useful outside the zero-inertia range, where of course the spatial distribution of the strain rate is directly obtained from the steady-flow solution (see the discussion following eq. (5-4.30)). The torsional-periodic flow system has been studied by Walters and Kemp [10]; the relation for η^* based on the measurement of the kinematics of the two plates is again eq. (5-4.40), with

$$S = \frac{2(K - I\omega^2)}{\pi R^4} \qquad (5\text{-}4.60)$$

where K is the restoring constant of the torsion wire, and I the moment of inertia of the constrained plate.

Lagrangian-periodic flow systems

In Lagrangian-periodic flows, the velocity field is steady, in an Eulerian sense, in some frame. In such a frame, each material point travels cyclically along a closed path, and the material elements are subject to periodic deformations. Moreover, Lagrangian-periodic flows are constant stretch history flows, and therefore tensor $\boldsymbol{\psi}^*$ in eq. (5-1.24) does not depend on t:[†]

$$\mathbf{G}^t = \text{Re}\,\{\boldsymbol{\psi}^*[1 - e^{-i\Omega s}]\} \qquad (5\text{-}4.61)$$

From eq. (5-1.30) one obtains

$$\tau = \Omega\eta'\boldsymbol{\psi}_\text{I} - G'\boldsymbol{\psi}_\text{R} \qquad (5\text{-}4.62)$$

Since, in general, non-zero components of $\boldsymbol{\psi}_\text{I}$ correspond to zero components of $\boldsymbol{\psi}_\text{R}$, and vice versa, values of η' and G' can be obtained by measuring two components of τ.

[†] We here indicate the frequency by Ω, because it actually coincides with the angular velocity of some physical boundary rather than with a frequency of imposed oscillation.

As for Eulerian-periodic flow systems, a detailed analysis will be given for only one simple system, while only basic results are given for the more complicated geometries.

Maxwell orthogonal rheometer flow

The Maxwell orthogonal rheometer [11, 12] consists of two flat parallel plates, both rotating in their own plane with the same angular velocity Ω about two parallel but not coincident axes. Let h be the distance between the plates and a the distance between the two axes of rotation; we shall make use of two different coordinate systems. A Cartesian system has the z axis orthogonal to both plates which are located at $z = 0$ and $z = h$, while the axes of rotation are located at $x = 0$ and $y = \pm a/2$. A cylindrical coordinate system has the z axis coincident with the former z axis, and $\theta = 0$ coincides with the z–x plane. The boundary conditions for the velocity distribution are

$$z = 0, \qquad v\langle r \rangle = -\frac{a}{2}\Omega \cos\theta \qquad (5\text{-}4.63)$$

$$z = 0, \qquad v\langle \theta \rangle = \Omega\left[r + \frac{a}{2}\sin\theta\right] \qquad (5\text{-}4.64)$$

$$z = h, \qquad v\langle r \rangle = \frac{a}{2}\Omega \cos\theta \qquad (5\text{-}4.65)$$

$$z = h, \qquad v\langle \theta \rangle = \Omega\left[r - \frac{a}{2}\sin\theta\right] \qquad (5\text{-}4.66)$$

We assume that the velocity distribution has the following form:

$$v\langle r \rangle = a\,\mathrm{Re}\,\{u^*(z)\,e^{i\theta}\} \qquad (5\text{-}4.67)$$

$$v\langle \theta \rangle = r\Omega + a\,\mathrm{Re}\,\{iu^*(z)\,e^{i\theta}\} \qquad (5\text{-}4.68)$$

$$v\langle z \rangle = 0 \qquad (5\text{-}4.69)$$

so that the boundary conditions become

$$z = 0, \qquad u^* = -\tfrac{1}{2}\Omega \qquad (5\text{-}4.70)$$

$$z = h, \qquad u^* = \tfrac{1}{2}\Omega \qquad (5\text{-}4.71)$$

and the equation of continuity is satisfied identically.

We further assume that a/h is small and therefore write all the relevant equations to within terms of order one in a/h; this is tantamount to restricting attention to small deformations. With this simplification, only two non-zero components of $\mathbf{G}^t$ are obtained:

$$G^t\langle rz \rangle = \text{Re} \left\{ \frac{ia}{\Omega}[1 - e^{-i\Omega s}]u^{*\prime} e^{i\theta} \right\} \qquad (5\text{-}4.72)$$

$$G^t\langle z\theta \rangle = \text{Re} \left\{ -\frac{a}{\Omega}[1 - e^{-i\Omega s}]u^{*\prime} e^{i\theta} \right\} \qquad (5\text{-}4.73)$$

Equations (5-4.72) and (5-4.73) show that the assumed flow pattern is indeed that of a Lagrangian-periodic flow system. Of course, there are only two non-zero components of the stress, say from eq. (5-1.29):

$$
\begin{aligned}
\tau\langle rz \rangle \quad &= \text{Re}\,\{\eta^* u^{*\prime} a\, e^{i\theta}\} \\
&= \text{Re}\,\{\eta^* u^{*\prime}\}a\cos\theta - \text{Im}\,\{\eta^* u^{*\prime}\}a\sin\theta
\end{aligned}
\qquad (5\text{-}4.74)
$$

$$
\begin{aligned}
\tau\langle r\theta \rangle \quad &= \text{Re}\,\{ia\eta^* u^{*\prime}\, e^{i\theta}\} \\
&= -\text{Im}\,\{\eta^* u^{*\prime}\}a\cos\theta - \text{Re}\,\{\eta^* u^{*\prime}\}a\sin\theta
\end{aligned}
\qquad (5\text{-}4.75)
$$

At any section $z = $ constant, the total forces exerted in the x and y directions are

$$F_x = \pi R^2[\tau\langle rz \rangle \cos\theta - \tau\langle \theta z \rangle \sin\theta] \qquad (5\text{-}4.76)$$

$$F_y = \pi R^2[\tau\langle rz \rangle \sin\theta + \tau\langle \theta z \rangle \cos\theta] \qquad (5\text{-}4.77)$$

and, upon substitution of eqs (5-4.74) and (5-4.75),

$$F_x - iF_y = a\pi R^2 \eta^* u^{*\prime} \qquad (5\text{-}4.78)$$

Since both F_x and F_y can be measured, the complex viscosity η^* can be determined provided the function $u^*(z)$ can be calculated from the equations of motion.

If the extra-pressure field is assumed to be independent of θ, we may write

$$\mathscr{P} = \mathscr{P}_0 + \frac{\rho\Omega^2 r^2}{2} \qquad (5\text{-}4.79)$$

so that, within terms of order one in a, the centrifugal forces are balanced by pressure,† and the equations of motion degenerate into the single equation,

$$u^{*\prime\prime} = \alpha^2 u^* \qquad (5\text{-}4.80)$$

---

† Note that the pressure field in eq. (5-4.79) implies pressure on the free boundary of a sample to be constant only within term of order zero in a/R. To within the same order F_x and F_y are independent of z.

where

$$\alpha^2 = -\frac{i\Omega\rho}{\eta^*} \qquad (5\text{-}4.81)$$

Equation (5-4.80) with boundary conditions (5-4.70) and (5-4.71) yields

$$u^* = \frac{\Omega}{2} \frac{\sinh(\alpha z) + \sinh[\alpha(z - h)]}{\sinh(\alpha h)} \qquad (5\text{-}4.82)$$

and

$$u^{*\prime} = \frac{\Omega\alpha}{2} \frac{\cosh(\alpha z) + \cosh[\alpha(z - h)]}{\sinh(\alpha h)} \qquad (5\text{-}4.83)$$

If inertia is neglected ($\alpha \to 0$),

$$u^{*\prime} \cong \frac{\Omega}{h} \qquad (5\text{-}4.84)$$

which, together with eq. (5-4.78) allows η' and G'/Ω to be determined immediately from measurable quantities.

Kepes rheometer

In the Kepes rheometer [13] the sample of material is held between two concentric spheres rotating with the same angular velocity Ω about two diametral axes of rotation which form a small angle ε to each other. If x is an axis lying in the plane of both rotation axes, orthogonal to one of them and passing through the center of the sphere, and if y is an axis through the center of the sphere orthogonal to the rotation axes, the couples C_x and C_y are measured. It can be shown that, within terms of the second order in $\alpha^2 d^2$, the following equation holds:

$$C_x - iC_y = 8\pi\varepsilon \frac{r_2^3 r_1^3}{r_2^3 - r_1^3} \eta^*\Omega \left[1 - \frac{\alpha^4 d^4}{1200} \right] \qquad (5\text{-}4.85)$$

where r_2 and r_1 are the radii of the outer and inner sphere, and $d = r_2 - r_1$. It is interesting to observe that the first-order inertia correction on the values of C_x and C_y is zero. The small deformation approximation consists of working to within first order in ε.

Sangamo rheometer

In the Sangamo rheometer [14], the sample is held between two cylinders which rotate with the same angular velocity Ω about their axes which are parallel and at a small distance a from each other. If z is the rotation axis of the inner cylinder, y is orthogonal to z and passing through both rotation axes, and x is orthogonal to both z and y, the forces F_x and F_y are measured. It can be shown that, within terms of the first order in $\alpha^2 d^2$ and in d^2/r_1^2, the following equation holds:

$$F_x - iF_y = \frac{12\pi a r_1^3 \Omega L \eta^*}{d^3}\left[1 - \frac{\alpha^2 d^2}{10}\right] \qquad (5\text{-}4.86)$$

where L is the length of the cylinders, d is the radius difference, and r_1 is the radius of the inner cylinder. The small deformations approximation consists of working to within first order in the quantity a/d.

Non-linear viscoelasticity

In addition to determining the complex viscosity η^*, periodic flow systems lead to the determination of additional properties of the functional $\mathcal{H}$ in the limit of very small deformations. To discuss this point, one needs to consider the second-order approximation to the functional $\mathcal{H}$, given in eq. (4-3.25), which is reported below:

$$\tau = -p\mathbf{1} + \int_0^\infty f(s)\mathbf{G}^t(s)\,ds + \int_0^\infty \int_0^\infty \{\alpha(s_1, s_2)\mathbf{G}^t(s_1) \cdot \mathbf{G}^t(s_2)$$
$$+ \beta(s_1, s_2)\,\mathrm{tr}\,[\mathbf{G}^t(s_1)]\mathbf{G}^t(s_2)\}\,ds_1\,ds_2 \qquad (5\text{-}4.87)$$

Now consider the flow pattern described by eqs (5-4.11)–(5-4.13), with boundary conditions (5-4.25) and (5-4.26), i.e., periodic plane shear between a fixed and a forced flat plate. The matrix of the $\mathbf{F}^t$ function is

$$[\mathbf{F}^t]_b = \begin{Vmatrix} 1 & -\varepsilon F & 0 \\ 0 & 1 & 0 \\ 0 & 0 & 1 \end{Vmatrix} \qquad (5\text{-}4.88)$$

where ε is the magnitude of the deformation:

$$\varepsilon = \frac{V}{h\omega} \qquad (5\text{-}4.89)$$

while F is a function of both t and s:

$$F = \mathrm{Re}\left\{-i\frac{d(a^*/V)}{d(x_t^2/h)}e^{i\omega t}[1 - e^{-i\omega s}]\right\} \qquad (5\text{-}4.90)$$

The value of F is of the order of magnitude of unity; the discussion of periodic plane shear given before was based on evaluating all relevant quantities to within terms of the first order in ε.

Suppose now one wishes to work to within *second* order in ε. This implies that the constitutive equation needs to be written in the form of eq. (5-4.87), since the second integral contributes terms of order ε^2 to the stress. The function $\mathbf{G}^t$ has the matrix

$$[\mathbf{G}^t] = \left\| \begin{matrix} \varepsilon^2 F^2 & -\varepsilon F & 0 \\ -\varepsilon F & 0 & 0 \\ 0 & 0 & 0 \end{matrix} \right\| \qquad (5\text{-}4.91)$$

It is interesting to observe that even the first integral in eq. (5-4.87) gives rise to a first normal stress difference, $\tau_{11} - \tau_{22}$, when the approximation is carried out to the second order in ε. Since the frequency of the G^t_{11} component is 2ω rather than ω, the first normal stress difference has a frequency which is twice that of the tangential stress.

The deformation terms appearing in the second integral in eq. (5-4.87) have the following matrices:

$$\text{tr}\,[\mathbf{G}^t(s_1)]\mathbf{G}^t(s_2) = O(\varepsilon^3) \qquad (5\text{-}4.92)$$

$$[\mathbf{G}^t(s_1) \cdot \mathbf{G}^t(s_2)] = \left\| \begin{matrix} \varepsilon^2 F(s_1)F(s_2) & 0 & 0 \\ 0 & \varepsilon^2 F(s_1)F(s_2) & 0 \\ 0 & 0 & 0 \end{matrix} \right\| + O(\varepsilon^3) \qquad (5\text{-}4.93)$$

Thus, one sees that, to within second order in ε, there is no contribution of the second integral to either τ_{12} or $\tau_{11} - \tau_{22}$; the second normal stress difference, $\tau_{22} - \tau_{33}$, again has a frequency 2ω, and arises from the term containing $\alpha(s_1, s_2)$ in eq. (5-4.87). One can thus draw the following conclusions:

(i) There is no correction to the value of τ_{12} arising from the second-order approximation.

(ii) Since measurement of the $\eta^*(\omega)$ function completely determines the $f(s)$ function (see eq. (5-1.28)), measurement of $\tau_{11} - \tau_{22}$ allows, in principle, a consistency test of the experiment, or of the simple fluid hypothesis, to be made.†

(iii) Measurement of the second normal stress difference allows, in principle, information on the function $\alpha(s_1, s_2)$ to be obtained.

It should be borne in mind that eq. (5-4.87) is based on the hypothesis that the functional $\mathscr{H}$ possesses a second Fréchet differential at the rest history—an assumption which may not hold valid for some materials.

---

† Internal consistency tests of this type are sometimes possible both among and within rheometrical flow systems. Examples are eq. (5-1.44) which related stress relaxation to periodic flow data, or eq. (5-3.17) which relates elongational to viscometric flow data.

PROBLEMS

5-1 Consider a flow system which is described in an orthogonal coordinate system x^i by

$$v^1 = u(x^2, x^3)$$
$$v^2 = v^3 = 0$$

Which conditions should g_{ij} satisfy for such a flow to be viscometric?

5-2 Consider a flow system which is described in a non-orthogonal coordinate system x^i by

$$v^1 = u(x^2)$$
$$v^2 = v^3 = 0$$

Which conditions should g_{ij} satisfy for such a flow to be viscometric?

5-3 Consider a flow which is described in a Cartesian coordinate system by

$$v^1 = \gamma x^2$$
$$v^2 = \gamma x^1$$
$$v^3 = 0$$

Is this a constant stretch history flow? If yes, is it of a known type? If yes, calculate the stress field in terms of well-defined material functions.

5-4 Consider a flow which is described in an orthogonal system x^i where g_{ij} depends only on x^2 by the following equations:

$$v^1 = \text{Re} \{u^*(x^2) \, e^{i\omega t}\}$$
$$v^2 = 0$$
$$v^3 = \text{Re} \{v^*(x^2) \, e^{i\omega t}\}$$

Is this an Eulerian-periodic flow system?

BIBLIOGRAPHY

1. COLEMAN, B. D., MARKOWITZ, H., and NOLL, W.: *Viscometric Flows of Non-Newtonian Fluids.* Springer-Verlag, Berlin (1966).
2. MARSH, B. D., and PEARSON, J. R. A.: *Rheol. Acta*, **7**, 326 (1968).
3. YIN, W. L., and PIPKIN, A. C.: *Arch. Ratl Mech. Anal.*, **37**, 111 (1970).
4. TROUTON, F. T.: *Proc. Roy. Soc. A*, **77**, 426 (1906).
5. COLEMAN, B. D., and NOLL, W.: *Phys. Fluids*, **5**, 840 (1962).
6. MARKOWITZ, H., and COLEMAN, B. D.: *Phys. Fluids*, **7**, 833 (1964).
7. FERRY, J. D.: *Viscoelastic Properties of Polymers.* Wiley, New York (1961).
8. NALLY, M. C.: *Br. J. appl. Phys.*, **16**, 1023 (1965).
9. WALTER, K., and KEMP, R. A.: *Deformation and Flow of High Polymer Systems.* Macmillan (1967).

10.   WALTER, K., and KEMP, R. A.: *Rheol. Acta*, 7, 1 (1967).
11.   BIRD, R. B., and HARRIS, E. K.: *A.I.Ch.E. Jl*, 14, 758 (1968).
12.   ABBOTT, T. N. G., and WALTERS, K.: *J. Fluid Mech.*, 40, 205 (1970).
13.   WALTERS, K.: *J. Fluid Mech.*, 40, 191 (1970).
14.   ABBOTT, T. N. G., and WALTERS, K.: *J. Fluid Mech.*, 43, 257 (1970).

6

CONSTITUTIVE EQUATIONS

6-1 CLASSIFICATION

In this chapter we shall examine some of the numerous constitutive equations for fluids with memory which have been proposed in the literature. They are all special forms of the general constitutive equation of simple fluids, i.e., the functional appearing in eq. (4-3.12) has been assumed to have a somewhat more explicit form; such forms imply smoothness hypotheses for the constitutive functional which may or may not be those discussed in chapter 4. The constitutive equations to be illustrated below are important for the following reasons.

In chapter 5 the predictions of the theory of simple fluids for a number of rheological flows have been discussed. In each of the cases considered, the problem was reduced to the determination of a few material functions which, if no other assumptions are made, must be determined experimentally. In general, no relation can be theoretically derived among material functions relative to different types of rheological flows. Conversely, when a specific constitutive equation has been chosen, the form of the material functions can be derived *a priori* and only a small number of parameters is left to be determined experimentally. Also, relationships are established among the results of different types of rheological flows.

It must also be observed that most flows encountered in engineering applications are kinematically much more complex than those labelled as rheological flows. For such flows, the general theory of simple fluids can make very limited (if any) predictions, while an analysis of these flows with the help of more specific constitutive equations can often be made. In chapter 7 a number of examples of this type will be considered.

Finally, one may mention that many detailed constitutive equations have their foundation, at least in part, in a model of the microscopic structure of the material. Polymeric materials, for instance, can be analyzed with the help of the kinetic theory of rubber elasticity. In this book, however, the emphasis will not be placed on this aspect, which belongs to the realm of so called microrheology. Rather, we shall look at these equations as constitutive assumptions for the continuum under consideration. These assumptions happen to be more or less confirmed by experiments on specific materials.

Actually, the whole process that has gone on during the last few decades— specific constitutive assumptions, derived detailed predictions, and experimental tests—is a trial and error procedure which, it can be said, is still far from being satisfactorily completed. The aim is that of obtaining, for specific materials, a constitutive equation which accounts for the observed properties and is mathematically simple enough for use in engineering applications. It is of course, preferable that the number of parameters to be determined experimentally should be as limited as possible.

Following Truesdell and Noll [1], we shall subdivide the constitutive equations into three categories; i.e., differential, integral, and rate. The first category contains those equations which give the stress tensor as a function of the differential kinematics at the instant of observation only. They are nonetheless equations for fluids with memory because higher order deformation tensors are involved which somehow account for past deformations in the sense already mentioned in section 3-2.

Equations in the second category may be thought of as having been obtained from eq. (4-3.12) for the simple fluid where the functional $\mathscr{H}$ has been specified as one or more integrals. Both the differential and integral types of constitutive equation are explicit in the stress tensor. Conversely, rate equations are not explicit for the stress. In fact, they contain at least one time derivative of the stress tensor. The rate of change of stress that appears in these equations gives the name to the category.

6-2   DIFFERENTIAL EQUATIONS

The most general equation in this category was first considered by Rivlin and Ericksen [2] and is thus generally known as the Rivlin–Ericksen constitutive equation of viscoelastic liquids. The stress is assumed to be a function of the first n Rivlin–Ericksen tensors A_N (see section 3-2):

$$\tau = M(A_1, A_2, \ldots, A_n) \qquad (6\text{-}2.1)$$

For the principle of material objectivity to be fulfilled, the function $M(\ )$ must be isotropic; that is,

$$Q \cdot M(A_1, \ldots, A_n) \cdot Q^T = M(Q \cdot A_1 \cdot Q^T, \ldots, Q \cdot A_n \cdot Q^T) \qquad (6\text{-}2.2)$$

with Q an arbitrary orthogonal tensor. The fact that M is isotropic and tensors $A_1, \ldots, A_n$ are symmetric allows the function M to be represented in more explicit form. For example, if $n = 2$, eq. (6-2.1) becomes

$$\begin{aligned}
\tau = {}&\alpha_0 1 + \alpha_1 A_1 + \alpha_2 A_2 + \alpha_3 A_1^2 + \alpha_4 A_2^2 \\
&+ \alpha_5(A_1 \cdot A_2 + A_2 \cdot A_1) + \alpha_6(A_1^2 \cdot A_2 + A_2 \cdot A_1^2) \\
&+ \alpha_7(A_1 \cdot A_2^2 + A_2^2 \cdot A_1) + \alpha_8(A_1^2 \cdot A_2^2 + A_2^2 \cdot A_1^2)
\end{aligned} \qquad (6\text{-}2.3)$$

In eq. (6-2.3), because of incompressibility, α_0 is an arbitrary scalar. The other eight scalar coefficients are invariant functions of A_1 and A_2.

One might be lead to think that with increasing n—i.e., by including more and more higher order acceleration tensors—the deformation history is in the limit completely represented and, thus, that eq. (6-2.3) becomes, for $n \to \infty$, equivalent to the general constitutive equation of a simple fluid. However, this is not so, unless one restricts the possible histories to those which are sufficiently smooth for expansion (3-2.36) to be valid.

This is indeed a severe restriction. Consider, for example, an arbitrary motion which is suddenly brought to a halt. All tensors become zero after the motion is stopped and so does the deviatoric stress, if eq. (6-2.1) applies. This can be readily seen from eq. (6-2.3) for the case $n = 2$ and from similar representations for $n > 2$. Thus, a fluid obeying eq. (6-2.1), no matter how large is n, does not present the phenomenon of stress relaxation which, conversely, is typical of most polymeric liquids and is displayed by the simple fluid in general. As stated above, this is due to the discontinuity of the history of deformation corresponding to stress relaxation.

As a further example, consider a motion which has a history of the Cauchy tensor, $C(\tau)$, that is continuous with all its derivatives at all times, and a second motion with a history $C'(\tau)$ having the same values of $C(\tau)$ in the interval $\bar{t} \le \tau \le t$ while the values assumed for $\tau < \bar{t}$ are different.

Of course, either $\mathbf{C}'(\tau)$ or one of its derivatives must be discontinuous at $\tau = \bar{t}$. From eq. (6-2.1), whatever the value of n, the same stress is predicted at $\tau = t$ because the $\mathbf{A}_N$ are the same for both motions. Conversely, if the general constitutive equation of simple fluids is used, the two motions give rise in general to different stresses at $\tau = t$. One can further state that for that one of the two motions which has a history that is continuous with all its derivatives, the stress calculated from eq. (4-3.12) and that from eq. (6-2.1) with $n \to \infty$ must coincide.

In the above example, consider now that $\bar{t}$ is sufficiently smaller than t, i.e., that the singular point is sufficiently distant in the past for the deformations previous to $\bar{t}$ to have been 'forgotten' at time t. Clearly, in this case eq. (4-3.12) and eq. (6-2.1) with a sufficiently large n also give approximately the same result for the second motion.

We can now better understand, on an intuitive basis, the meaning of the nth-order approximations of eq. (4-3.12) for 'slow' flows given in section 4-3. Equations (4-3.21), (4-3.22), and (4-3.23) give explicitly the zeroth-, first- and second-order approximations, respectively. One can immediately recognize that such equations are special cases of eq. (6-2.1) (we recall that $\mathbf{A}_1 = 2\mathbf{D}$, see eq. (3-2.28)). The concept of slow flows is made rigorous by the retardation procedure, see eq. (4-3.20). Given any history which is continuous at the instant of observation, the retarded history obtained from it by the factor α is a history which, with decreasing α, becomes continuous with all its derivatives on a wider and wider interval of time previous to the instant of observation. In fact, if a singular point exists in the assigned history, this is displayed further and further in the past with decreasing α. Thus eq. (6-2.1) becomes increasingly more adequate to predict the correct response. Simultaneously, the value of n needed to give an expansion of the history to within a given approximation decreases.

Note that if, say, the second-order approximation is desired—i.e., to within terms of order α^2—the value of n which is needed is 2 but many of the terms in eq. (6-2.3) are also dropped, being of order α^3 or smaller. A comparison between eqs (6-2.3) and (4-3.23) shows that many terms disappear. Furthermore, for the same reason, and because tr $\mathbf{A}_1 = 0$, the coefficients in eq. (4-3.23) must be constant rather than invariant functions of $\mathbf{A}_1$ and $\mathbf{A}_2$.

Equation (4-3.23) is a second-order approximation of the general constitutive equation of simple fluids in the sense explained. From a different viewpoint, it can be taken as the constitutive equation of a restricted class of fluids called 'second-order fluids.' In the rest of this section we shall only consider this constitutive equation which is, among the possible differential equations, the most frequently used. We shall briefly examine the detailed predictions that are obtained from this equation for those rheological flows studied in general in chapter 5.

We shall here rewrite eq. (4-3.23) in the form

$$\tau = \eta_0 \mathbf{A}_1 + \beta_0 \mathbf{A}_1^2 + \gamma_0 \mathbf{A}_2 \qquad (6\text{-}2.4)$$

and all the results will be given in terms of the three material constants $\eta_0, \beta_0, \gamma_0$.

The three viscometric functions defined by (5-2.9), (5-2.10), and (5-2.11) are calculated as [3] (for calculation of $\mathbf{A}_1$ and $\mathbf{A}_2$ in a viscometric flow such as simple shear, see Illustration 3D):

$$\tau(k) = \eta_0 k \qquad (6\text{-}2.5)$$

$$\sigma_1(k) = -2\gamma_0 k^2 \qquad (6\text{-}2.6)$$

$$\sigma_2(k) = (\beta_0 + 2\gamma_0) k^2 \qquad (6\text{-}2.7)$$

Equation (6-2.5) shows that a constant shear viscosity equal to η_0 is predicted from this model. Non-Newtonian effects are displayed by the normal stress differences which are not zero. Experiments on polymeric materials show that σ_1 should be positive and σ_2 probably negative and smaller than σ_1. Hence γ_0 should be negative and β_0 positive with:

$$\beta_0 < -2\gamma_0 \qquad (6\text{-}2.8)$$

The extensiometric material functions (eqs (5-3.9) and (5-3.10)) of a second-order fluid are [4]:

$$\sigma_{E1} = 2\eta_0(\gamma_1 - \gamma_2) + 4(\beta_0 + \gamma_0)(\gamma_1^2 - \gamma_2^2)$$
$$\sigma_{E2} = 2\eta_0(\gamma_2 - \gamma_3) + 4(\beta_0 + \gamma_0)(\gamma_2^2 - \gamma_3^2) \qquad (6\text{-}2.9)$$

The elongational viscosity (eq. (5-3.16)) is given by

$$\eta_E = 3\eta_0 + 3(\beta_0 + \gamma_0)\gamma_E \qquad (6\text{-}2.10)$$

In the limit of diminishingly small values of γ_E, the elongational viscosity reduces to $3\eta_0$, as expected.

In order to calculate the complex viscosity of a second-order fluid let us consider an arbitrary periodic motion of small 'magnitude.' Equation (5-1.24) is rewritten here in the form

$$\mathbf{C} = 1 + \text{Re}\left\{\boldsymbol{\psi}^*(t)[1 - e^{i\omega(t-\tau)}]\right\} \qquad (6\text{-}2.11)$$

The first and second Rivlin–Ericksen tensors are then calculated as

$$\mathbf{A}_1 = \text{Re}\left\{-i\omega\boldsymbol{\psi}^*(t)\right\} \qquad (6\text{-}2.12)$$

$$\mathbf{A}_2 = \text{Re}\left\{\omega^2\boldsymbol{\psi}^*(t)\right\} \qquad (6\text{-}2.13)$$

From eq. (6-2.4), τ' is obtained (within the first order in the magnitude of $\psi^*(t)$):

$$\tau' = \text{Re}\left\{-i\omega(\eta_0 + i\omega\gamma_0)\psi^*(t)\right\} \quad (6\text{-}2.14)$$

Comparison of eq. (6-2.14) with eq. (5-1.29), in view of the arbitrariness of $\psi^*(t)$, shows that

$$\eta^* = \eta_0 + i\omega\gamma_0 \quad (6\text{-}2.15)$$

Equation (6-2.15) can also be written as

$$\eta^* = \sqrt{(\eta_0^2 + \omega^2\gamma_0^2)}\,e^{i\Phi}$$

$$\tan\phi = \frac{\omega\gamma_0}{\eta_0} \quad (6\text{-}2.16)$$

which can best be compared with eqs (5-4.9) and (5-4.10) for experimental determination. Note that η^* does not depend on β_0. Also, if $\gamma_0 \to 0$ or $\omega \to 0$,

$$\eta^* \to \eta_0 \quad (6\text{-}2.17)$$

that is, the Newtonian result is obtained either if the fluid has no elasticity ($\gamma_0 = 0$) or for a sufficiently slow flow ($\omega \to 0$).

Sinusoidal oscillations between parallel plates can be studied for a second-order fluid without using the approximation of small deformations [4]. It is found, in particular, that the normal stress differences oscillate in phase with the square of the velocity gradient.

6-3 INTEGRAL EQUATIONS

Integral equations are those which give the stress in the form of integrals of the deformation history. We have already seen that the general functional of the simple fluid degenerates into an integral equation in the limit of small deformations. The first-order approximation is given by eq. (4-3.24) which is here rewritten as

$$\tau = \int_0^\infty f(s)\mathbf{G}^t(s)\,ds \quad (6\text{-}3.1)$$

If one regards eq. (6-3.1) as valid for an arbitrary history rather than only in the limit of small deformations, it becomes an example of a constitutive integral equation. The physical assumption underlying eq. (6-3.1) is clear: it is assumed that all the deformations which occurred in the past *as measured by the Cauchy tensor* contribute *linearly* to the present value of the stress. The weighing function $f(s)$ is a material function which completely determines a particular material obeying such a linearity rule. The linear relationship expressed by eq. (6-3.1) is also known as Boltzmann's superposition principle.

However, it must be immediately realized that, when dealing with deformations of arbitrary magnitude, the concept of a linear relationship between stress and deformation is not unequivocally defined on physical grounds. This is because a deformation can be measured in infinitely many ways which are equally legitimate and among which there is no means of choosing *a priori* by continuum mechanics arguments. We can use tensors U or C or C^{-1}, or other measures of deformation can be introduced. Now, a linear relationship between stress and, say, C corresponds to a non-linear one between stress and, say, C^{-1}. Thus, a linear relationship is only defined when we know the measurement of deformation with which the relationship is established. Only in the limit of infinitesimal deformation is the concept of linearity unequivocal because a linear relation between τ and one of the deformation measures also implies linearity between τ and any of them.†

In view of the discussion above, it is recognized that eq. (6-3.1) has no special status. An equally possible 'linear' relationship could be:

$$\tau' = \int_0^\infty g(s)\mathbf{H}^t(s)\,\mathrm{d}s; \qquad \mathbf{H}^t = (\mathbf{C}^t)^{-1} - \mathbf{1} \qquad (6\text{-}3.3)$$

Whether eq. (6-3.1) or (6-3.3) or other possible 'linear' relationships hold for a given material under finite deformation conditions is a matter of comparison with experiments. In fact, eq. (6-3.3) gives predictions which are in better agreement with experiments on polymeric materials than those obtained from eq. (6-3.1). Further, eq. (6-3.3) has some foundation in structural theories of polymer solutions and melts [5].

---

† Note that an analogous observation could have been made in the previous section when dealing with equations of the differential type. We could have considered a second-order fluid analogous but different to the one defined by eq. (6-2.4). By using White–Metzner acceleration tensors instead of those by Rivlin–Ericksen, we can postulate the constitutive equation:

$$\tau = \eta_0' \mathbf{B}_1 + \beta_0' \mathbf{B}_1^2 + \gamma_0' \mathbf{B}_2 \qquad (6\text{-}3.2)$$

Equations (6-3.2) and (6-2.4) predict the same stress if a sufficiently slow flow is considered; more precisely, to within terms of order α^2 where α is the retardation coefficient. They, however, give different results if motion of arbitrary 'speed' are considered. One may recall that the Rivlin–Ericksen tensor gives the Taylor expansion of a sufficiently smooth deformation history expressed in terms of the Cauchy tensor C, while White–Metzner tensors are obtained by expanding the history of $\mathbf{C}^{-1}$.

Let us examine the viscometric functions that are predicted from eqs (6-3.1) and (6-3.3). We may refer to the calculations of $\mathbf{G}^t$ and $\mathbf{H}^t$ for lineal Couette flow given in Illustration 3A (eqs (3-6.11) and (3-6.12)). From eq. (6-3.1) one obtains

$$\tau(k) = k \int_0^\infty -sf(s)\,\mathrm{d}s$$

$$\sigma_1(k) = k^2 \int_0^\infty -s^2 f(s)\,\mathrm{d}s \qquad (6\text{-}3.4)$$

$$\sigma_2(k) = -k^2 \int_0^\infty -s^2 f(s)\,\mathrm{d}s$$

while, from eq. (6-3.3),

$$\tau(k) = k \int_0^\infty sg(s)\,\mathrm{d}s$$

$$\sigma_1(k) = k^2 \int_0^\infty s^2 g(s)\,\mathrm{d}s \qquad (6\text{-}3.5)$$

$$\sigma_2(k) = 0$$

In both cases, a shear-rate independent viscosity is predicted. This is a poor result for most if not all polymeric materials; however, if the behavior of very dilute solutions or conditions of not too large shear rate values are considered, it can be accepted. Equation (6-3.1) predicts that the second normal stress difference is equal in value and opposite in sign to the first. This result is largely disproved by all experiments on polymeric materials which display values of σ_2 much smaller than the corresponding ones of σ_1. In fact, the values of σ_2 are so small that they can be measured with great difficulty and a dispute is still open on whether they have the same or the opposite sign of those of σ_1. In this respect, the predictions of eq. (6-3.3) of a zero value of σ_2 can be accepted as a first approximation for many practical purposes.

Disregarding eq. (6-3.1) henceforth, let us examine eq. (6-3.3) a little further. If sufficiently small deformations are considered it is true that†

$$\mathbf{H}^t = -\mathbf{G}^t \qquad (6\text{-}3.7)$$

Thus, eq. (4-3.24) can also be written as

$$\tau' = -\int_0^\infty f(s)\mathbf{H}^t(s)\,\mathrm{d}s \qquad (6\text{-}3.8)$$

† In general, the following relationship exists between $\mathbf{H}^t$ and $\mathbf{G}^t$:

$$\mathbf{H}^t = -\mathbf{G}^t \cdot (\mathbf{1} + \mathbf{G}^t)^{-1} \qquad (6\text{-}3.6)$$

More precisely, eq. (6-3.8) represents an alternative form (with respect to eq. (4-3.24)) of the approximation of the simple fluid functional valid in the limit of a deformation history sufficiently 'close' to the rest history. Comparison of eqs (6-3.3), (6-3.8), and (5-1.44) shows that

$$g(s) = -f(s) = -\frac{d}{ds}F(s) \qquad (6\text{-}3.9)$$

that is, $g(s)$ is recognized as the opposite of the derivative of the stress-relaxation function.

Isothermal stress-relaxation experiments on polymeric materials as well as microrheological theories indicate that $g(s)$ can often be taken of the form [6]:

$$g(s) = G \sum_{j=1}^{z} \frac{1}{\lambda_j} \exp\left(-\frac{s}{\lambda_i}\right) \qquad (6\text{-}3.10)$$

where G is an 'elastic modulus' and the λ_j are 'relaxation times.' λ_1 is the largest relaxation time and the others are related to it. In some cases the following relation can be assumed:

$$\lambda_j = \frac{\lambda_1}{j^2} \qquad (6\text{-}3.11)$$

In any case, the influence of the jth term becomes progressively negligible with increasing values of j, so that a definite value of Z is not needed provided it is assumed to be sufficiently large.

Substitution of eq. (6-3.10) into eq. (6-3.3) gives the explicit constitutive equation:

$$\tau = G \sum_{j=1}^{z} \frac{1}{\lambda_j} \int_0^\infty \exp\left(-\frac{s}{\lambda_j}\right) \mathbf{H}'(s)\,ds \qquad (6\text{-}3.12)$$

In order to find the predictions of eq. (6-3.12) in one of the extensional flows discussed in section 5-3, we consider eqs (3-5.40) and (5-3.2). One obtains for the extensiometric material function σ_{E1} of eq. (5-3.9):

$$\sigma_{E1} = G \sum_{j=1}^{z} \frac{1}{\lambda_j} \int_0^\infty \left\{ \exp\left[s\left(2\gamma_1 - \frac{1}{\lambda_j} \right) \right] \right.$$
$$\left. - \exp\left[s\left(2\gamma_2 - \frac{1}{\lambda_j} \right) \right] \right\} ds \qquad (6\text{-}3.13)$$

Similar expressions are obtained for σ_{E2} and for the elongational viscosity η_E. It is obvious that the integrals in eq. (6-3.13) only exist if the arguments of the exponential functions are negative. This poses a limit to the possible values of the γ_i in relation to the value of the largest relaxation time λ_1. For instance, for the elongational flow defined in eq. (5-3.12) one finds

$$-\frac{1}{\lambda_1} < \gamma_E < \frac{1}{2\lambda_1} \qquad (6\text{-}3.14)$$

If γ_E does not fulfil the above inequality, the corresponding constant stretch history flow is not possible. This result does not imply that a steady elongation with rate γ_E cannot be imposed upon the material. In fact, eq. (6-3.13) gives the value that the stress ultimately attains after a constant stretching has been applied for a sufficiently long time (constant stretch history flow), i.e., it gives the steady-state stress value. Thus, if a constant stretching has been applied to the material from a certain time, the stress will eventually reach the value given by eq. (6-3.13) if inequality (6-3.14) is verified. If (6-3.14) is not fulfilled, no steady state is attained for the stress, which keeps increasing beyond any limit. The issue will be further clarified when dealing with transient elongation in chapter 7. Also, the fact that eq. (6-3.13) does not yield steady-state values for certain γ_E values is indeed a limitation of the constitutive equation (eq. (6-3.12)). In fact, polymeric materials do not have this property although it is true that for γ_E values not fulfilling condition (6-3.14) the stress reaches extraordinarily high values [7].

As for the complex viscosity which is predicted in this case, we can, in view of eq. (6-3.9), substitute the expression of $g(s)$ given by eq. (6-3.10) into eq. (5-1.28). Integration gives, directly,

$$\eta^* = G \sum_{j=1}^{z} \frac{\lambda_j}{1 + \lambda_j^2 \omega^2} - \frac{i}{\omega} G \sum_{j=1}^{z} \frac{\lambda_j^2 \omega^2}{1 + \lambda_j^2 \omega^2} \qquad (6\text{-}3.15)$$

Thus both the dynamic viscosity and the dynamic rigidity (or elastic modulus) are frequency-dependent. The dynamic viscosity decreases steadily towards zero with increasing values of frequency. The value corresponding to $\omega = 0$ must coincide with the zero-shear viscometric viscosity:

$$\eta_V(0) = G \sum_{j=1}^{z} \lambda_j \qquad (6\text{-}3.16)$$

Equation (6-3.16) can, of course, be obtained directly by substituting eq. (6-3.10) into the first of eqs (6-3.5).

Finally, the stress relaxation function, $F(\ )$, can be obtained by substituting $g(s) = -f(s)$ into eq (5-1.42):

$$F(t - t_0) = G \sum_{j=1}^{z} \exp\left(-\frac{t - t_0}{\lambda_j}\right) \qquad (6\text{-}3.17)$$

Note that if eq. (6-3.17) is assumed to hold, eqs (6-3.15) and (6-3.16) are obtained independently of the constitutive assumption of eq. (6-3.3), in view of the theorems on small deformations and slow flows valid in general for a simple fluid. Of course, this observation cannot be extended to the results of eqs (6-3.5) and (6-3.13) which deal with large deformations of arbitrary 'speed.'

The second-order approximation of the functional for simple fluids also takes the form of an integral equation. It was previously given in eq. (4-3.25), which we here rewrite for convenience in terms of tensor $\mathbf{C}^t$:

$$\tau = \int_0^\infty f(s)\mathbf{C}^t(s)\,ds + \int_0^\infty \int_0^\infty \{\alpha(s_1, s_2)\mathbf{C}^t(s_1) \cdot \mathbf{C}^t(s_2)$$
$$+ \beta(s_1, s_2)\mathbf{C}^t(s_1)\,\mathrm{tr}\,[\mathbf{C}^t(s_2)]\}\,ds_1\,ds_2 \tag{6-3.18}$$

Equation (6-3.18) can be made simpler (and, of course, less general) if one excludes coupling effects resulting from deformations occurring at different times. Taking $\alpha(s_1, s_2) = \beta(s_1, s_2) = 0$ when $s_1 \neq s_2$, eq. (6-3.18) becomes

$$\tau = \int_0^\infty \{f(s)\mathbf{C}^t + f_1(s)(\mathbf{C}^t)^2 + f_2(s)\mathbf{C}^t\,\mathrm{tr}\,(\mathbf{C}^t)\}\,ds \tag{6-3.19}$$

where

$$f_1(s) = \alpha(s, s); \qquad f_2(s) = \beta(s, s) \tag{6-3.20}$$

The third-order approximation, in which coupling effects are excluded, would be

$$\tau = \int_0^\infty \{f(s)\mathbf{C}^t + f_1(s)(\mathbf{C}^t)^2 + f_2(s)\mathrm{I}_{\mathbf{C}^t}\mathbf{C}^t + f_3(s)(\mathbf{C}^t)^3$$
$$+ f_4(s)\mathrm{I}_{\mathbf{C}^t}(\mathbf{C}^t)^2 + f_5(s)\mathrm{II}_{\mathbf{C}^t}\mathbf{C}^t\}\,ds \tag{6-3.21}$$

where $\mathrm{I}_{\mathbf{C}^t}$ is the first principal invariant or trace of $\mathbf{C}^t$ and $\mathrm{II}_{\mathbf{C}^t}$ is the second (see eq. (1-3.42)). Equation (6-3.21) can be written in a different form by using the Hamilton–Cayley theorem, eq. (1-3.49), as applied to tensor $\mathbf{C}^t$:

$$(\mathbf{C}^t)^3 - \mathrm{I}_{\mathbf{C}^t}(\mathbf{C}^t)^2 + \mathrm{II}_{\mathbf{C}^t}\mathbf{C}^t - \mathrm{III}_{\mathbf{C}^t}\mathbf{1} = \mathbf{0} \tag{6-3.22}$$

We can substitute $(\mathbf{C}^t)^3$ from eq. (6-3.22) into eq. (6-3.21) so that only $\mathbf{C}^t$ and $(\mathbf{C}^t)^2$ will appear in the latter. This procedure can be repeated for the fourth-order equation where a term containing $(\mathbf{C}^t)^4$ is present. In fact, $(\mathbf{C}^t)^4 = \mathbf{C}^t \cdot (\mathbf{C}^t)^3$ and $(\mathbf{C}^t)^3$ can be taken from eq. (6-3.22) as before.

Generally, therefore, whatever the order of the integral equation considered, by iterative application of eq. (6-3.22) one can write the constitutive equation in the form

$$\tau = \int_0^\infty \{\phi_1(s, I_{C^t}, II_{C^t})C^t + \phi_2(s, I_{C^t}, II_{C^t})(C^t)^2\}\, ds \qquad (6\text{-}3.23)$$

The third invariant, III_{C^t} does not appear as an argument of ϕ_1 and ϕ_2 because, for incompressible materials $III_{C^t} = 1$.

Equation (6-3.23) is the *most general* integral equation, provided coupling effects resulting from deformations occurring at different times are excluded. A material obeying eq. (6-3.23) is completely characterized by the material functions ϕ_1 and ϕ_2. These are functions of s and of the first and second invariants of C^t which are, in their turn, functions of s. (Note that ϕ_1 and ϕ_2 are not functionals but only composite functions.)

An equivalent way of writing eq. (6-3.23) is obtained by applying once more the Hamilton–Cayley theorem. If eq. (6-3.22) is multiplied throughout by $(C^t)^{-1}$, it becomes

$$(C^t)^2 - I_{C^t}C^t + II_{C^t}\mathbf{1} - III_{C^t}(C^t)^{-1} = 0 \qquad (6\text{-}3.24)$$

Therefore, from eq. (6-3.24) $(C^t)^2$ can be substituted into eq. (6-3.23) to yield

$$\tau = \int_0^\infty \{\psi_1(s, I_{C^t}, II_{C^t})C^t + \psi_2(s, I_{C^t}, II_{C^t})(C^t)^{-1}\}\, ds \qquad (6\text{-}3.25)$$

The symmetry of eq. (6-3.25) is better appreciated by considering that, for incompressible materials,

$$I_{C^t} = II_{(C^t)^{-1}}; \qquad II_{C^t} = I_{(C^t)^{-1}} \qquad (6\text{-}3.26)$$

which can again be obtained from eq. (6-3.22). It is also apparent that by proper choice of the functions ψ_1 and ψ_2, eq. (6-3.25) reduces to either eq. (6-3.1) or eq. (6-3.3), which have been discussed before.

A special form of eq. (6-3.25) has been derived by Bernstein, Kearsley and Zapas [8] on the basis of a physical assumption involving an elastic energy function. The theory, referred to as the BKZ theory, is previous to the general thermodynamics treatment due to Coleman and represents an attempt to extend to materials with memory some well-established concepts that are relative to perfectly elastic solids.

A perfectly elastic solid, or rather, in the terminology used by Truesdell and Noll [9], a hyperelastic material, is a material for which a strain-energy function $\sigma(\mathbf{F}_R)$ can be found such that

$$\mathbf{T} = \rho \mathbf{F}_R \cdot \left(\frac{d\sigma}{d\mathbf{F}_R}\right)^{\mathsf{T}} \qquad (6\text{-}3.27)$$

The function $\sigma(\ )$ has dimensions of energy per unit mass and depends on the choice of the reference configuration R. One may notice the formal analogy between eq. (6-3.27) and eq. (4-4.42) valid for materials with memory.

If the strain-energy function is to be frame indifferent, it is necessary that

$$\sigma(\mathbf{F}_R) = \sigma(\mathbf{U}_R) = \bar\sigma(\mathbf{C}_R) \qquad (6\text{-}3.28)$$

that is, the rotation tensor $\mathbf{R}_R$ does not affect the value of the function. If attention is restricted to isotropic solid and R is chosen as an undistorted configuration, then the function $\bar\sigma(\mathbf{C}_R)$ must be isotropic and we can write (see eq. (1-3.46))

$$\bar\sigma(\mathbf{C}_R) = U(\mathrm{I}_{\mathbf{C}_R}, \mathrm{II}_{\mathbf{C}_R}, \mathrm{III}_{\mathbf{C}_R}) \qquad (6\text{-}3.29)$$

Finally, if incompressibility is also assumed, then $\mathrm{III}_{\mathbf{C}_R} = 1$ and eq. (6-3.27) can be written as [9]

$$\mathbf{R}_R^{\mathrm{T}} \cdot \tau \cdot \mathbf{R}_R = 2\rho \frac{\partial U}{\partial \mathrm{I}_{\mathbf{C}_R}} \mathbf{C}_R - 2\rho \frac{\partial U}{\partial \mathrm{II}_{\mathbf{C}_R}} \mathbf{C}_R^{-1} \qquad (6\text{-}3.30)$$

The BKZ theory represents an extension of the above concepts to viscoelastic liquids. An energy function is postulated to exist also for these liquids which, of course, is no more a conservative property; on the contrary, it decays with passing time from the instant the deformation has been imposed. Taking the present configuration as the reference and accounting for the contribution of the deformation at all times past, this hypothesis leads to the following equation for the stress:

$$\tau = \int_0^\infty \left\{ 2\frac{\partial U}{\partial \mathrm{I}_{\mathbf{C}^t}} \mathbf{C}^t - 2\frac{\partial U}{\partial \mathrm{II}_{\mathbf{C}^t}} (\mathbf{C}^t)^{-1} \right\} ds \qquad (6\text{-}3.31)$$

where U is a function of s as well as of $\mathrm{I}_{\mathbf{C}^t}$ and $\mathrm{II}_{\mathbf{C}^t}$:

$$U = U(s, \mathrm{I}_{\mathbf{C}^t}, \mathrm{II}_{\mathbf{C}^t}) \qquad (6\text{-}3.32)$$

(While the analogy between eqs (6-3.30) and (6-3.31) is apparent, it has to be noted that the dimensions of ρU in eq. (6-3.30) are different from those of U in eq. (6-3.31). In eq. (6-3.30), ρU has dimensions of energy per unit volume while, in eq. (6-3.31), U has dimensions of energy per unit volume and time. This is because an integration with time is made in eq. (6-3.31).)

It is now clear why eq. (6-3.31) of the BKZ theory can be considered a restricted form of the integral equation (eq. (6-3.25)). The two independent functions, ψ_1 and ψ_2, of eq. (6-3.25) are related to one another in the BKZ theory. They represent (apart from numerical factors) the partial derivatives of the same energy function U with respect to the invariants of the deformation.

In order to use eq. (6-3.31), the function U remains to be specified. In the first presentation of the theory [8], a polynomial expansion of U in the invariants was proposed, carried out to the second order. Later, Zapas [10] on the basis of experimental data on elastomeric materials, suggested the following form:

$$-U = \frac{\alpha'}{2}(I_{C^t} - 3)^2 + \frac{9}{2}\beta' \ln \frac{I_{C^t} + II_{C^t} + 3}{9}$$
$$+ 24(\beta' - c') \ln \frac{I_{C^t} + 15}{II_{C^t} + 15} + c'(I_{C^t} - 3)$$

(6-3.33)

where α', β', and c' are functions of s only. Equation (6-3.33) predicts a variable viscosity and non-zero first and second normal stress differences in viscometric flows and various other features displayed by polymeric liquids. Some of the detailed calculations are shown in Illustration 6A.

Since both C^t and $(C^t)^{-1}$ appear in the integrand of eq. (6-3.25), it is clear (see the discussion following eq. (6-3.5)) that equations of this type are capable, by a proper choice of the weighing functions, to predict all possible ratios of first to second normal stress difference in viscometric flow. A previous example of this type of equation has been proposed in the literature [11].

Another integral equation which is a specialization of eq. (6-3.25) has been proposed by Tanner and Simmons [12]. With the notation used here, their constitutive equation is written as

$$\tau = \int_0^\infty \psi(s, II_{C^t})(C^t)^{-1} \, ds \qquad (6\text{-}3.34)$$

with

$$\psi(s, II_{C^t}) = f(s) \qquad \text{when} \quad II_{C^t} \leq B^2 + 3$$
$$\psi(s, II_{C^t}) = 0 \qquad \text{when} \quad II_{C^t} > B^2 + 3$$

(6-3.35)

Thus the rheological behavior of the material is determined by the function $f(s)$ and the positive number B^2. Equations (6-3.34) and (6-3.35) need to be discussed. Consider first a flow such that $II_{C^t} \leq B^2 + 3$ for all values of s (an example could be a periodic flow of sufficiently small amplitude). In such a case, eq. (6-3.34) gives the same results as eq. (6-3.3) previously considered. Conversely, consider a flow such that II_{C^t} is a monotonously increasing function of s (examples are viscometric flows or extensional flows);

then eqs (6-3.34) and (6-3.35) are equivalent to

$$\tau = \int_0^{\bar{s}} f(s)(\mathbf{C}^t)^{-1}\, ds \qquad (6\text{-}3.36)$$

where $\bar{s}$ is the value of s at which $\mathrm{II}_{\mathbf{C}^t}$ becomes equal to $B^2 + 3$. In this case, the material only 'remembers' the deformations which have occurred in the time lapse $s \le \bar{s}$, while the previous ones are entirely 'forgotten.'

Equations (6-3.34) and (6-3.35) (as well as eq. (6-3.3) previously considered) are suggested by a model of polymeric materials which describes them as 'networks.' However, in the model by Tanner and Simmons, the network 'breaks' when a scalar measure of the deformation, $\mathrm{II}_{\mathbf{C}^t}$ (or equivalently $\mathrm{I}_{(\mathbf{C}^t)^{-1}}$, see eq. (6-3.26)) reaches a limiting value, $B^2 + 3$. B is called the 'straingth' of the network. The function $f(s)$ has the ordinary meaning of a relaxation function.

Detailed predictions of eqs (6-3.34) and (6-3.35) in the rheometric flows of interest are as follows:

(i) *Viscometric flows*

$$\tau(k) = k \int_0^{B/k} s\, f(s)\, ds \qquad (6\text{-}3.37)$$

$$\sigma_1(k) = k^2 \int_0^{B/k} s^2\, f(s)\, ds \qquad (6\text{-}3.38)$$

$$\sigma_2(k) = 0 \qquad (6\text{-}3.39)$$

Equation (6-3.37) predicts a shear-dependent viscosity which decreases with increasing k. For sufficiently small k values, the viscosity becomes constant and equal to:

$$\eta_V(0) = \int_0^\infty s\, f(s)\, ds \qquad (6\text{-}3.40)$$

Equation (6-3.38) gives similar predictions for the first normal stress difference coefficient.

(ii) *Extensiometric flows*

Recalling eqs (3-5.40) and (5-3.2), eq. (6-3.36) gives, for the extensiometric material functions,

$$\sigma_{E1} = \int_0^{\bar{s}} f(s)[\exp(2\gamma_1 s) - \exp(2\gamma_2 s)]\, ds \qquad (6\text{-}3.41)$$

$$\sigma_{E2} = \int_0^{\bar{s}} f(s)[\exp(2\gamma_2 s) - \exp(2\gamma_3 s)]\, ds \qquad (6\text{-}3.42)$$

where $\bar{s}$ is the solution of the equation:

$$\exp(2\gamma_1\bar{s}) + \exp(2\gamma_2\bar{s}) + \exp(2\gamma_3\bar{s}) = B^2 + 3 \qquad (6\text{-}3.43)$$

It is worth noticing that the integrals of eqs (6-3.41) and (6-3.42) always exist because $\bar{s}$ is finite. This result can be compared with that of eq. (6-3.13), which may produce infinite values of the steady-state elongational stress.

(iii) *Periodic flows and stress relaxation*

The general results given by eqs (5-1.28) and (5-1.45) apply in this case provided the magnitude of $\mathbf{G}^t$ is such that $\mathrm{II}_{\mathbf{C}^t} \leq B^2 + 3$.

Of course, all the results above can be made more explicit by assuming a form for the function $f(s)$. In references [13] and [14] the form chosen is analogous to that of eq. (6-3.10).

Equations of the following general forms have also been proposed in the literature (e.g., [15]):

$$\tau = \int_0^\infty F(s)\dot{\mathbf{C}}^t(s)\,\mathrm{d}s \qquad (6\text{-}3.44)$$

$$\tau = \int_0^\infty F(s)\overline{(\dot{\mathbf{C}}^t)^{-1}}\,\mathrm{d}s \qquad (6\text{-}3.45)$$

Such equations reduce to either eq. (6-3.1) or eq. (6-3.3) upon integration by parts of the right-hand side. The situation is quite different when a dependence on the rate of strain is assumed for the function $F(\ )$ itself. This point is discussed in detail below.

Rate-controlled integral equations

Some integral equations which have been proposed in the literature have the following general form [16]:

$$\tau = \int_0^\infty f_1(s, \mathrm{II}_{\dot{\mathbf{C}}^t}, \mathrm{III}_{\dot{\mathbf{C}}^t})\mathbf{C}^t + f_2(s, \mathrm{II}_{\dot{\mathbf{C}}^t}, \mathrm{III}_{\dot{\mathbf{C}}^t})(\mathbf{C}^t)^{-1}]\,\mathrm{d}s \qquad (6\text{-}3.46)$$

Such equations differ from those considered before because, in the memory functions, the invariants of tensor $\dot{\mathbf{C}}^t$ instead of those of $\mathbf{C}^t$ are supposed to appear. In other words, it is assumed that the way by which the effect of deformations is forgotten (or relaxed) in the course of time depends on the *rate* by which the deformation is imposed rather than on the magnitude of the deformation. Some dispute exists on 'when' the deformation rate should be computed. Some authors [17, 18] prefer the rate of deformation at the instant of observation, i.e. $\dot{\mathbf{C}}^t(0) = 2\mathbf{D}$, others [16, 19] the rate of deformation at time s, and others [20, 21] some sort of average throughout the time interval from $s = 0$ and $s = s$. The last choice gives rise to more complex expressions for the $f_i(\ )$, but it has been proposed in order to account for coupling effects between different instants of time.

However, it is our opinion that all equations similar to eq. (6-3.46) are unlikely to represent the actual behavior of any material, in particular of the viscoelastic polymeric systems for which they have been proposed. The basis for this criticism is that they do not degenerate properly into the equation of linear viscoelasticity, eq. (4-3.24). The following discussion is divided in two parts, of which the first is more formal and is devoted to the analysis of the special topology of a functional such as that of eq. (6-3.46). In the second part, a discussion of the data by Philippoff [22] of periodic flows of polymeric materials shows convincing evidence of the inadequacy of such equations as eq. (6-3.46).

In chapter 4, we have said that the general functional of simple fluids reduces to the form given in eq. (4-3.24), i.e., to the equation of linear visco-elasticity, provided the norm of the deformation history is sufficiently small; that is, when the history of deformation is sufficiently close to the rest history. Since it is assumed that the functional $\mathscr{H}$ is Fréchet-differentiable at the rest history, the stress corresponding to a history sufficiently close to the rest history depends linearly on $\mathbf{G}^t(s)$.

Clearly, it is possible to conceive a history $\mathbf{G}^t(s)$ which is arbitrarily close to the rest history, and still has arbitrarily large strain rate values: a simple example is a periodic motion of very small amplitude but very large frequency. A constitutive equation of the form of eq. (6-3.46) would predict a non-linear dependence of τ on $\mathbf{G}^t(s)$ for such a history. In other words, eq. (6-3.46) implies that the topology of the space of histories with respect to which the functional $\mathscr{H}$ is continuous has a different nature from the one that is assumed in the formulation of simple fluid theory.

Indeed, Coleman [23] has shown that, given a space of histories with a norm of the following form:

$$\|\mathbf{G}^t\|_h = \sqrt{\left[\int_0^\infty (\mathbf{G}^t : \mathbf{G}^t) h^2(s)\, ds\right]} \qquad (6\text{-}3.47)$$

where $h(s)$ is *any* influence function going to zero fast enough as $s \to \infty$, a functional $\mathscr{H}$ which is continuous in such a space cannot depend explicitly on $\dot{\mathbf{G}}^t$, and vice versa. This point is in fact crucial in the development of the thermodynamic theory given in section 4-4 (see the discussion following eq. (4-4.29)).

Hence, although of course the topology implied by eq. (6-3.46) is not inconceivable, a material described by such an equation would not obey most of the general theorems developed in simple fluid theory, and the thermodynamic analysis in section 4-4 would not hold true. Moreover, a general theory for functionals which are continuous with respect to a topology such as implied by eq. (6-3.46) has not been developed, so that no generally valid statements for such a class of materials could be made.

It is perhaps worthwhile to clarify the issue still more. An equation such as eq. (6-3.25) can be made to be continuous in the neighborhood of the rest history by assuming that the functions ψ_1 and ψ_2 degenerate into functions of only s when the norm of G^t becomes sufficiently small, since, when $\| G^t \| \to 0$, one has

$$\mathrm{I}_{C^t} \to 3; \qquad \mathrm{II}_{C^t} \to 3 \qquad (6\text{-}3.48)$$

On the contrary, in eq. (6-3.46) it is impossible for functions $f_1(\ )$ and $f_2(\ )$ to degenerate into functions of only s when $\| G^t \| \to 0$, since the invariants of $\dot{C}^t$ do not tend to some unique value when the rest history is approached.

If experimental evidence supporting the validity of eq. (6-3.46) were available, one would of course need to develop a suitable general theory, different from the simple fluid theory; but in fact this is not the case, and indeed experimental evidence indicates that eq. (6-3.46) *does not* represent the actual behavior of polymeric materials. We shall now discuss this evidence.

The crucial point to be verified experimentally is whether the behavior predicted by linear viscoelasticity theory is approached by real materials in the limit of diminishingly small *deformations*, or in the limit of diminishingly small *deformation rates* (or perhaps when both become sufficiently small). Hence, the required evidence can only arise from periodic flow experiments carried out up to conditions where deviations from the linear viscoelastic behavior are observed.

Consider an Eulerian-periodic flow, and let ε be the amplitude of the deformation (e.g., in periodic plane shear as discussed in section 5-4, $\varepsilon = V/h\omega$). The corresponding amplitude of the rate of strain, γ_m, is related to ε by the equation:

$$\gamma_m = \varepsilon\omega \qquad (6\text{-}3.49)$$

(in the same example, $\gamma_m = V/h$). Hence, as long as experiments are carried out at some fixed frequency ω_0, the results cannot discriminate the issue, since γ_m and ε are proportional to each other. In a plot of the amplitude of stress, σ_m, versus the amplitude of the rate-of-strain, γ_m, a linear behavior will be observed up to a break-off point above which deviations are expected (see Fig. 6-1). This is indeed the behavior observed by Philippoff [22].

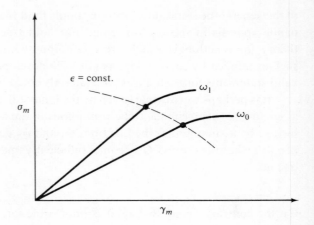

FIGURE 6-1
Deviations from linear viscoelasticity.

If now experiments are carried out at some new frequency $\omega_1 \neq \omega_0$, a linear behavior is again expected in the low γ_m region. The crucial point is that, if an equation such as eq. (6-3.46) holds (and, more generally, if the topology of the space of histories with respect to which the functional $\mathscr{H}$ is continuous is also defined in terms of strain rate), one would expect the break-off point (i.e., the point where deviations from linear behavior are observed) to correspond to some critical value of γ_m, or at the very least to depend *both* on γ_m and ε; while if the simple fluid theory's smoothness hypotheses hold, one would expect the break-off point to correspond to a critical value of the deformation ε, independently of the corresponding value of γ_m. The latter is indeed the behavior observed in the careful experiments of Philippoff [22], which clearly show that, for a variety of polymeric material, deviations from linear viscoelasticity behavior are observed at a critical value of the amplitude of deformation ε, and that this critical value is entirely independent of the value of the amplitude of the strain rate, γ_m.

This seems to us to be convincing evidence that constitutive equations of the form of eq. (6-3.46) (or analogous rate equations to be discussed in the next section) cannot be considered reliable, and more generally that all available experimental evidence supports the validity of the smoothness hypotheses underlying the theory of simple fluids with fading memory discussed in chapter 4. We may conclude by stating that constitutive equations explicitly containing the strain rate should always be considered with great care, since they may be in the same category as eq. (6-3.46), i.e., they may imply smoothness hypotheses on the constitutive functional which are in contrast with those of the theory of simple fluids. Of course, not all equations containing the strain rate present this problem; see, for example, eqs (6-3.44) and (6-3.45). The best test for a dubious equation is to calculate its predictions with regard to the break-off point for deviations from linear viscoelastic behavior, and to verify whether it is uniquely determined by the value of ε.

A final consideration is in order. Equations of the form of eq. (6-3.46) have been proposed in the literature in an effort to predict the shear-dependence of both viscosity and normal stress coefficients in viscometric flow. In doing so, an important point has been overlooked, i.e., that equations such as eq. (6-3.25) can also accommodate the observed shear-dependence by an appropriate choice of the form of the functions ψ_1 and ψ_2: the model of Tanner and Simmons, which has been discussed previously, is a typical example, see eqs (6-3.37) and (6-3.38). Hence, even if only data fitting is required, there is no need to introduce equations of the form of eq. (6-3.46), which imply conceptual difficulties of the type discussed above, and are in conflict with experimental evidence.

6-4   RATE EQUATIONS

The Maxwell equation

One of the simplest ways of modifying the constitutive equation of a Newtonian fluid (eq. (1-9.4)) in order to account for the 'elastic' properties of a given fluid, is to add a term containing a time derivative of the stress:

$$\tau + \lambda \overset{\text{\tiny 0}}{\tau} = 2\mu \mathbf{D} \qquad (6\text{-}4.1)$$

where $\overset{\text{\tiny 0}}{\tau}$ stands for some sort of time derivative of the stress tensor yet to be specified, and λ has dimensions of time.

The addition of the term containing the time derivative of τ makes the equations capable of representing the phenomenon of stress-relaxation which is characteristic of fluids with memory. In fact if, after a certain stretching has taken place and a non-isotropic stress has been built up, the stretching is suddenly set to zero, the stress will decay in time according to the differential equation:

$$\tau + \lambda \overset{\text{\tiny 0}}{\tau} = 0 \qquad (6\text{-}4.2)$$

The stress-relaxation velocity will be regulated by the value of the relaxation time λ.

Equation (6-4.1) was originally proposed by Maxwell for the case of a one-dimensional problem so that no ambiguity existed on the choice of the stress derivative which was, in fact, an ordinary derivative. For three-dimensional problems it is required that the derivative be such as to make eq. (6-4.1) frame-indifferent. Thus we must choose among the associated derivatives of the stress tensor already introduced in section 3-3.

Indeed, the most general symmetry-preserving associated derivative of a symmetric tensor $\mathbf{J}$ results from the linear operator, indicated as F_{abc}, defined as [16]:

$$F_{abc}\mathbf{J} = \overset{\circ}{\mathbf{J}} + a(\mathbf{J} \cdot \mathbf{D} + \mathbf{D} \cdot \mathbf{J}) + b(\mathbf{J}:\mathbf{D})\mathbf{1} + c\mathbf{D}\,\mathrm{tr}\,\mathbf{J} \qquad (6\text{-}4.3)$$

where a, b, and c are arbitrary numbers. For $b = c = 0$, and $a = 0, 1, -1$ one obtains the Jaumann or corotational derivative $\overset{\circ}{\mathbf{J}}$, the lower convected derivative $\overset{\triangle}{\mathbf{J}}$, and the upper convected derivative $\overset{\triangledown}{\mathbf{J}}$, respectively. In choosing $\overset{\triangle}{\tau}$, assume temporarily that $b = c = 0$. Equation (6-4.1) then becomes

$$\tau + \lambda\overset{\triangle}{\tau} + \lambda a(\tau \cdot \mathbf{D} + \mathbf{D} \cdot \tau) = 2\mu\mathbf{D} \qquad (6\text{-}4.4)$$

As usual, it is instructive to consider first what predictions are obtained from eq. (6-4.4) in viscometric flows. The appropriate calculations are carried out in Illustration 6B, yielding the following result for the viscometric functions:

$$\tau(k) = \frac{\mu k}{1 + k^2\lambda^2(1 - a^2)} \qquad (6\text{-}4.5)$$

$$\sigma_1(k) = \frac{2\mu k^2\lambda\,(1-a)}{1 + k^2\lambda^2(1 - a^2)} \qquad (6\text{-}4.6)$$

$$\sigma_2(k) = \frac{-\mu k^2\lambda(1 + a)}{1 + k^2\lambda^2(1 - a^2)} \qquad (6\text{-}4.7)$$

Let us discuss these results assuming that the parameters λ and μ have constant values (independent of k). Equation (6-4.5) shows that in general the viscosity is a function of k approaching the value μ for $k \to 0$. To ensure that the viscosity is always a positive quantity, the parameter a must be restricted to the interval $-1 \leq a \leq 1$. The viscosity is then, in general, a decreasing function of k, i.e., pseudoplastic behavior is predicted. In general the first and second normal stress differences are non-zero and show a k dependency of the corresponding coefficients.

The particular values of a (i.e., $0, 1, -1$), are now considered. For $a = 0$, i.e., when the corotational derivative is assumed in eq. (6-4.1), the first and second normal stress differences are equal in value and opposite in sign. This result is disproved by viscometric experiments on polymeric materials, thus the use of the corotational derivative in eq. (6-4.1) is not indicated. The use of the lower convected derivative—i.e., $a = 1$—which gives a zero first normal stress difference, seems to be similarly unrealistic.

For $a = -1$, i.e., using the upper convected derivative of stress in eq. (6-4.1), eqs (6-4.5)–(6-4.7) become:

$$\tau(k) = \mu k \qquad (6\text{-}4.8)$$

$$\sigma_1(k) = 2\mu\lambda k^2 \qquad (6\text{-}4.9)$$

$$\sigma_2(k) = 0 \qquad (6\text{-}4.10)$$

Thus, among the three special cases considered, the last yields the more realistic results as far as the normal stress differences are concerned. However, the viscosity becomes shear-rate independent in this case. From a phenomenological viewpoint the conclusive indication would be that constant a should better be taken within the interval $0, -1$, thus obtaining (i) a shear-rate dependent viscosity, (ii) a positive first normal stress difference with a shear-rate dependent coefficient, and (iii) a negative second normal stress difference smaller in value than the first one. All three features are commonly displayed by polymeric substances.

It can be observed that the viscometric results given in eqs (6-4.8)–(6-4.10) are equivalent to those of eqs (6-3.5) derived by the integral constitutive equation (6-3.3). This is not a coincidence because, as shown by Lodge [24], eq. (6-4.1) with the upper convected derivative for τ is in fact equal to eq. (6-3.3) with

$$g(s) = \frac{\mu}{\lambda^2} e^{-s/\lambda} \qquad (6\text{-}4.11)$$

To show this we begin by observing that eq. (6-4.1), with $\overset{\square}{\tau} = \overset{\triangledown}{\tau}$,

$$\tau + \lambda\overset{\triangledown}{\tau} = 2\mu\mathbf{D} \qquad (6\text{-}4.12)$$

is equivalent to the equation

$$\tau + \lambda\overset{\triangledown}{\tau} = \frac{\mu}{\lambda}\mathbf{1} \qquad (6\text{-}4.13)$$

In fact, substitution into eq. (6-4.12) of $\tau_1 - (\mu/\lambda)\mathbf{1}$ in place of τ yields eq. (6-4.13) for τ_1 (see eq. (3-3.34)) and, in view of incompressibility, τ and τ_1 are indistinguishable.

Next, we write eq. (6-4.13) in contravariant component form. Using eq. (1-3.31):

$$(\tau)^{ij} + \lambda(\overset{\triangledown}{\tau})^{ij} = \frac{\mu}{\lambda}g^{ij} \qquad (6\text{-}4.14)$$

Equation (6-4.14) applies for an arbitrary choice of the coordinate system. However, if a convected coordinate system is chosen (see section 3-4), it takes a particularly simple form. In fact, eq. (3-4.21) applies, and thus

$$(\overset{\triangledown}{\tau})^{ij} = \frac{d(\tau)^{ij}}{dt} \qquad (6\text{-}4.15)$$

The metric of the convected coordinate system is indicated by γ^{ij}, so that eq. (6-4.14) is rewritten as

$$(\tau)^{ij} + \lambda\frac{d(\tau)^{ij}}{dt} = \frac{\mu}{\lambda}\gamma^{ij} \qquad (6\text{-}4.16)$$

In order to integrate eq. (6-4.16) to get τ^{ij}, initial conditions are required. Assume that for $t = \bar{t}$ the stress components are known to have the values $(\bar{\tau})^{ij}$. Straightforward integration of eq. (6-4.16) then yields

$$(\tau)^{ij}(t) = \frac{\mu}{\lambda^2}\int_{\bar{t}}^t e^{-(t-t')/\lambda}\,\gamma^{ij}(t')\,dt' + (\bar{\tau})^{ij}e^{-(t-\bar{t})/\lambda} \qquad (6\text{-}4.17)$$

It is apparent from eq. (6-4.17) that, if $\bar{t}$ is sufficiently smaller than t, it is not necessary to know the actual values of $(\bar{\tau})^{ij}$, provided they are not too large. More precisely, if $(\bar{\tau})^{ij}\,e^{\bar{t}/\lambda} \to 0$ when $\bar{t} \to -\infty$, eq. (6-4.17) can be written as

$$(\tau)^{ij}(t) = \frac{\mu}{\lambda^2}\int_{-\infty}^t e^{-(t-t')/\lambda}\,\gamma^{ij}(t')\,dt' \qquad (6\text{-}4.18)$$

Putting $s = t - t'$ and recalling eq. (3-4.7), we can finally write

$$\tau = \frac{\mu}{\lambda^2}\int_0^\infty e^{-s/\lambda}(\mathbf{C}^t)^{-1}\,ds \qquad (6\text{-}4.19)$$

which, in view of incompressibility, is equal to eq. (6-3.3) with $g(s)$ given by eq. (6-4.11).

Similarly, it can be shown that the equation containing the lower convected derivative

$$\tau + \lambda\overset{\triangle}{\tau} = 2\mu\mathbf{D} \qquad (6\text{-}4.20)$$

is equivalent to the equation

$$\tau = \frac{\mu}{\lambda^2}\int_0^\infty e^{-s/\lambda}\mathbf{C}^t(s)\,ds \qquad (6\text{-}4.21)$$

However, eq. (6-4.4), which can be written in the form of a linear combination of $\overset{\triangledown}{\tau}$ and $\overset{\triangle}{\tau}$:

$$\tau + \lambda\left(\frac{1+a}{2}\overset{\triangle}{\tau} + \frac{1-a}{2}\overset{\triangledown}{\tau}\right) = 2\mu\mathbf{D} \qquad (6\text{-}4.22)$$

is not equivalent to a linear combination of eqs (6-4.19) and (6-4.21). Suffice it to say that, while eq. (6-4.4), for $-1 < a < 1$, predicts a shear-dependent viscosity (eq. (6-4.5)), a linear combination of eqs (6-4.19) and (6-4.21) gives a constant viscosity, as do the two equations separately.

Before presenting other forms of the Maxwell equation, it is useful to make the following observation. Rate type constitutive equations of the *first order*, i.e., containing no other stress derivative than the first one, if explicit in the rate, are of the general form:

$$\frac{d\tau}{dt} = f(\tau, \text{kinematic tensors}) \qquad (6\text{-}4.23)$$

where τ and the 'kinematic tensors' are all calculated at t. It follows from eq. (6-4.23) that the stress at any time t can in principle be calculated from a knowledge of the stress at a previous time $\bar{t}$ (the initial condition) plus the deformation history between $\bar{t}$ and t. In other words, knowledge of the stress at a given time $\bar{t}$ replaces the information on the deformation history previous to $\bar{t}$, with regard to calculating the stress in subsequent instants of time. This property is not displayed by the simple fluid in general, nor is it shared by most of the constitutive integral equations. For a general simple fluid with fading memory it is necessary to know the entire previous deformation history, at least up to values of s of the order of the natural time.

Should a fluid obey a constitutive rate equation of the first order, it seems that it is possible to solve, at least in principle, a number of problems which cannot be tackled for a simple fluid. For example, consider the problem of a fluid jet out of a die (generally accompanied by the well-known die-swell phenomenon). It is possible to measure the stress at the die section though measurement of force. For a rate type fluid this should be sufficient information for the subsequent evaluation of the stress–strain situation in the jet; that is, it is not necessary to know the deformation history previous to the die section.

However, there is a complication in the above scheme because generally τ in eq. (6-4.23) is a non-deviatoric extra stress, i.e., it is not traceless. For example, eq. (6-4.4) gives a τ which is traceless at equilibrium ($\mathbf{D} = 0$, $\overset{\circ}{\tau} = 0$) but is not generally so (see, for example, the results of Illustration 6B). Because the trace of an extra stress has no physical meaning, it cannot be measured. It follows that, for example, in the case of the jet considered above, determination of the total stress at the die section does not constitute sufficient information for the extra stress, and the initial condition for the subsequent integration of eq. (6-4.23) is insufficiently determined.

In this respect, it would be attractive (but by no means required on physical grounds) if rate equations were so built as to guarantee a known value of the trace (e.g., a zero value). It is, in fact, very simple to obtain rate equations that have this property. Consider, for example, eq. (6-4.4). Taking the trace of eq. (6-4.4) one obtains

$$\operatorname{tr}\tau + \lambda\frac{d}{dt}(\operatorname{tr}\tau) + 2\lambda a\operatorname{tr}(\tau\cdot\mathbf{D}) = 0 \qquad (6\text{-}4.24)$$

Equation (6-4.24) shows explicitly that, for a constitutive equation such as eq. (6-4.4), tr τ is not a constant with time but depends on $\mathbf{D}(t)$ and thus on the history of the motion. Equation (6-4.24) also suggests how to modify eq. (6-4.4) in order to obtain a constitutive equation with a constantly traceless stress. To the left-hand side of eq. (6-4.4) we must add the term

$$-\tfrac{2}{3}\lambda a\operatorname{tr}(\tau\cdot\mathbf{D})\mathbf{1} \qquad (6\text{-}4.25)$$

With this term included, the equation for the trace becomes

$$\operatorname{tr}\tau + \lambda\frac{d}{dt}(\operatorname{tr}\tau) = 0 \qquad (6\text{-}4.26)$$

and tr $\tau = 0$ is a solution of eq. (6-4.26) at all times for an arbitrary history.

It is immediately recognized that inclusion of a term such as that given by eq. (6-4.25) corresponds to choosing $b = -\tfrac{2}{3}a$ in the general derivative operator defined by eq. (6-4.3). It is also evident that with such a choice the value of c becomes irrelevant because the term containing it is identically zero. A number of rate equations have been proposed which are so built as to guarantee a traceless stress. It must be noted, however, that adding such terms as eq. (6-4.25) to a given rate equation makes the 'corrected' equation entirely different from the previous one; that is, it does not imply only a matter of making the previous stress traceless, the rheological response is indeed changed. For example, we have seen that eq. (6-4.12) predicts a constant shear viscosity (eq. (6-4.8)). The traceless modification of eq. (6-4.12) would be

$$\tau + \lambda(\overset{\triangledown}{\tau} + \tfrac{2}{3}\operatorname{tr}(\tau\cdot\mathbf{D})\mathbf{1}) = 2\mu\mathbf{D} \qquad (6\text{-}4.27)$$

It can easily be verified that eq. (6-4.27) predicts a variable shear viscosity:

$$\tau(k) = \frac{\mu k}{1 + \tfrac{2}{3}\lambda^2 k^2} \qquad (6\text{-}4.28)$$

Generalized Maxwell equations and their topology

Let us now consider a class of possible generalizations of the Maxwell equation. We have seen that the Maxwell equation with the upper convected derivative is equivalent to eq. (6-3.3), which is given below:

$$\tau = \int_0^\infty g(s)\mathbf{H}^t(s)\,ds \qquad (6\text{-}4.29)$$

for the particular choice of $g(s)$ given by eq. (6-4.11). Let us now assume that $g(s)$ is given by

$$g(s) = \frac{\mu_1}{\lambda_1^2}e^{-s/\lambda_1} + \frac{\mu_2}{\lambda_2^2}e^{-s/\lambda_2} \qquad (6\text{-}4.30)$$

It is evident that eq. (6-4.29), with the particular choice for $g(s)$ given by eq. (6-4.30), is equivalent to

$$\tau = \tau^{(1)} + \tau^{(2)} \qquad (6\text{-}4.31)$$

$$\tau^{(1)} = \frac{\mu_1}{\lambda_1^2}\int_0^\infty e^{-s/\lambda_1}\mathbf{H}^t(s)\,ds \qquad (6\text{-}4.32)$$

$$\tau^{(2)} = \frac{\mu_2}{\lambda_2^2}\int_0^\infty e^{-s/\lambda_2}\mathbf{H}^t(s)\,ds \qquad (6\text{-}4.33)$$

Equations (6-4.32) and (6-4.33) are equivalent to

$$\tau^{(1)} + \lambda_1\overset{\triangledown}{\tau}{}^{(1)} = 2\mu_1\mathbf{D} \qquad (6\text{-}4.34)$$

$$\tau^{(2)} + \lambda_2\overset{\triangledown}{\tau}{}^{(2)} = 2\mu_2\mathbf{D} \qquad (6\text{-}4.35)$$

Equations (6-4.31), (6-4.34), and (6-4.35) can now be combined to yield†

$$\tau + (\lambda_1 + \lambda_2)\overset{\triangledown}{\tau} + \lambda_1\lambda_2\overset{\triangledown}{\tau}_2 = 2(\mu_1 + \mu_2)\mathbf{D} + 2(\lambda_1\mu_1 + \lambda_2\mu_2)\overset{\triangledown}{\mathbf{D}} \qquad (6\text{-}4.36)$$

Hence, we have shown that eq. (6-4.36) is equivalent to eq. (6-4.29) with the choice for $g(s)$ given by eq. (6-4.30). Analogously, if one chooses for $g(s)$ a sum of n exponentials, the equivalent rate equation has the form

$$\tau + \sum_1^n \alpha_N\overset{\triangledown}{\tau}_N = 2\mu\left[\mathbf{D} + \sum_1^{n-1}\beta_N\overset{\triangledown}{\mathbf{D}}_N\right] \qquad (6\text{-}4.37)$$

---

† We recall that $\overset{\triangledown}{\tau}_2$ is the *second* upper convected derivative of τ; also, $2\overset{\triangledown}{\mathbf{D}} = -\mathbf{B}_2$, the second White–Metzner tensor.

where the α_N and β_N are simply related to the μ_N and λ_N. Similarly, eq. (6-3.1), with $f(s)$ given by a sum of n exponentials, is equivalent to†

$$\tau + \sum_{1}^{n} \alpha_N \overset{\triangle}{\tau}_N = 2\mu \left[\mathbf{D} + \sum_{1}^{n-1} \beta_N \overset{\triangle}{\mathbf{D}}_N \right] \qquad (6\text{-}4.38)$$

It is important to point out, for future discussion, that the sum on the right-hand side of eqs (6-4.37) and (6-4.38) runs up to n, while the sum on the left-hand side runs up to $n - 1$.

Equations of the same form as eqs (6-4.37) and (6-4.38), but with associated derivatives of stress and strain rate different from the upper or lower convected ones, are not equivalent to simple integral equations. Nonetheless, it remains true that an equation of the general form:

$$\tau + \sum_{1}^{n} \alpha_N F_{abc}^N \tau = 2\mu \left[\mathbf{D} + \sum_{1}^{n-1} \beta_N F_{abc}^N \mathbf{D} \right] \qquad (6\text{-}4.39)$$

(where by F_{abc}^N we mean the operator F_{abc} applied N times in succession) is equivalent to the following formulation:

$$\tau = \sum_{1}^{n} \tau^{(N)} \qquad (6\text{-}4.40)$$

$$\tau^{(N)} + \lambda_N F_{abc} \tau^{(N)} = 2\mu_N \mathbf{D} \qquad (6\text{-}4.41)$$

Equations of this general type have been proposed in the literature [25, 26, 27]. Before discussing these equations in some detail, a general discussion on 'linear' equations of this type is in order.

We have derived eqs (6-4.37) and (6-4.38) from the linear viscoelasticity equations, regarded as describing some real materials that are also outside the small deformations limit. In view of this, eqs (6-4.37) and (6-4.38) predict different rheological behaviors, although they are equivalent in the limit of small deformations (see the discussion following eq (6-3.1)). On the other hand, equations of the same type can be obtained by considering simple one-dimensional models of 'springs' and 'dashpots,' and generalizing to a properly frame-invariant three-dimensional form the relative mechanical equations. It may be worthwhile illustrating this approach, which is useful in understanding the topological properties of the resulting functionals.

We begin by the simple model in Fig. 6-2, which consists of a dashpot of viscosity μ and a spring of stiffness μ/λ, connected in series. The mechani-

---

† We recall that $2\overset{\triangle}{\mathbf{D}}_N = A_N$, the Nth Rivlin–Ericksen tensor.

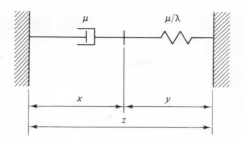

FIGURE 6-2
The simplest viscoelastic model.

cal equations for the force F acting along the axis of the model are

$$F = \frac{\mu}{\lambda}(y - y_0) = \mu \dot{x} \qquad (6\text{-}4.42)$$

$$z = x + y \qquad (6\text{-}4.43)$$

which can easily be combined to yield the following equation relating F to z:

$$F + \lambda \dot{F} = \mu \dot{z} \qquad (6\text{-}4.44)$$

It is evident that eq. (6-4.1) is the three-dimensional equivalent of eq. (6-4.44); in this sense, Fig. 6-2 gives the equivalent model of the Maxwell fluid. Of course, at the one-dimensional model level, the distinction between eqs (6-4.12) and (6-4.20) (or (6-4.37), (6-4.38), and (6-4.39)) disappears.

More complicated models than the one in Fig. 6-2 give rise to generalized forms of the Maxwell equation. By way of example, the model in Fig. 6-3 obviously corresponds to eqs (6-4.40) and (6-4.41), and thus to eq. (6-4.39).

An obvious observation is that, if a one-dimensional model permits movement from one extreme to the other without passing over a dashpot, it is the model of a *solid* rather than a fluid; its mechanical equation would contain the length itself rather than only its time derivatives.

A more important point concerns models which permit movement from one extreme to the other without passing over a spring; for example, see the model in Fig. 6-4. The equation for the force is, in this case,

$$F + \lambda_1 \dot{F} = (\mu_1 + \mu_2)\dot{z} + \lambda_1 \mu_2 \ddot{z} \qquad (6\text{-}4.45)$$

and an equivalent three-dimensional model could be

$$\tau + \lambda_1 \overset{\triangledown}{\tau} = 2\mu[\mathbf{D} + \lambda_2 \overset{\triangledown}{\mathbf{D}}] \qquad (6\text{-}4.46)$$

An equation of the form of eq. (6-4.46), with additional terms added in order to make τ traceless, has been proposed by Williams and Bird [28]. The parameter λ_2 is usually referred to as the retardation time. Equation (6-4.46) looks superficially similar to the general type of eq. (6-4.39), but one may notice

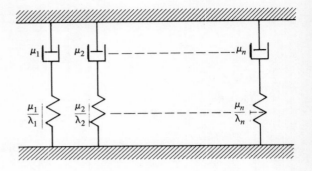

FIGURE 6-3
Model for eq. (6-4.39).

that the highest derivative on the right-hand side is of the same order as the highest one on the left-hand side. Equation (6-4.46) generalizes to the following form:

$$\tau + \sum_{1}^{n} \alpha_N F_{abc}^N \tau = 2\mu \left[\mathbf{D} + \sum_{1}^{n} \beta_N F_{abc}^N \mathbf{D} \right] \qquad (6\text{-}4.47)$$

The difference between such equations as (6-4.39) and (6-4.47) is by no means minor. In fact, a sudden jump in strain would, for a material described by eq. (6-4.39), result in a sudden jump in stress, while it would result in an infinite stress in a material described by eq. (6-4.47). This can be easily understood by considering that the model in Fig. 6-4 does not allow an instantaneous change of z, while the model in Fig. 6-3 does. On a more formal level, one may observe that eq. (6-4.29) allows an instantaneous jump in strain, which would result in a jump in stress; the same property is, of course, shared by eq. (6-4.37). Addition of an Nth time derivative of the strain rate on the right-hand side of eq. (6-4.37) changes the topology of the constitutive functional. Thus, equations like eq. (6-4.47) do not allow jumps in strain, hence making the thermodynamic theory discussed in section 4-4 inapplicable.

Indeed, equations of this type have possibly been suggested by a general line of reasoning which has been formulated explicitly by Oldroyd [29]. Excerpts from Oldroyd's statement are quoted below:

'A set of equations of state, to be physically acceptable, must determine the deformation history of a material element, including its configuration at the current instant, when its whole stress and temperature histories are given.... It is in principle possible to apply stresses to a small piece of matter experimentally in an arbitrary manner over a period of time,

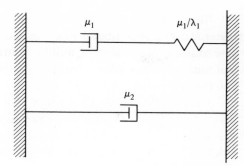

FIGURE 6-4
Model for a fluid incapable of absorbing a
jump in strain.

simultaneously controlling its temperature, and to observe the resulting deformation over the same period. In particular, the complete set of equations of state for a material must specify what happens at a discontinuity in applied stress, whether (for example) the deformation or the first, or a higher rate of strain is discontinuous. . . .

The work of Coleman and Noll on simple fluids . . . is based on an inverse argument, which is less easy to accept physically because of the nature of the experimental procedures available to us. It is postulated . . . that the stress in an element . . . should be determined completely by the deformation history. . . . It is not self-evident that this postulate is true of all real fluids. For example, it may well be possible . . . to impose a discontinuous jump in stress that produces no instantaneous change in deformation, but only a change in the first, or a higher, rate of strain . . . indeed an ideally incompressible Newtonian viscous liquid would (so behave).'

The viewpoint expressed by Oldroyd's statement deserves a detailed discussion. First, it may be a matter of possibly unlimited debate whether it is closer to 'the nature of experimental procedures available to us' to assume that the deformation is determined by the history of stress, or the other way around. This would be a futile debate, since the two viewpoints are equivalent to each other as long as no smoothness hypotheses are laid down.

In fact, consider the classical equation for the mechanical theory of simple fluids, eq. (4-3.12). As long as no smoothness hypotheses are formulated concerning the functional $\mathscr{H}$, one cannot determine whether a jump in strain (and hence an infinite value of the instantaneous strain rate) would correspond to a finite or an infinite value of the instantaneous stress. When smoothness hypotheses, such as discussed in section 4-4, are formulated, then it is implicitly assumed that jumps in strain and stress correspond to each other, i.e., that infinite values of the instantaneous strain rate are possible.

Oldroyd's statement points out the possibility that, for some real materials, a discontinuity in stress may correspond to a discontinuity in the rate of strain rather than to a discontinuity in the strain itself; this, indeed, would contradict the smoothness hypotheses of the theory of simple fluids (see the discussion following 4-4.41).

Oldroyd's quoting the example of a Newtonian fluid is indeed appropriate, since Newtonian fluids (as well as any material where the free energy depends explicitly on the instantaneous value of the rate of change of the independent variables) do not fulfil the smoothness hypotheses of the theory of simple fluids (see section 4-4). It is a matter of conjecture whether *real* materials exist which behave as ideal Newtonian fluids under an ideally discontinuous stress history; one may also think (as we are inclined to do) that any real material behaving as a Newtonian fluid is simply a material with an extremely short natural time, which would show deviations from Newtonian behavior under ideally discontinuous stress histories, i.e., under a jump in stress taking place in a time interval which is truly zero, and therefore shorter than the natural time of the material, no matter how small it may be.

Now, in considering non-Newtonian fluids, appropriate smoothness hypotheses must be made. The simple fluid theory yields certain results insofar as it *does* make assumptions concerning the smoothness properties of the constitutive functionals. Of course, one can imagine the existence of materials which do not obey such smoothness hypotheses; but an alternative theory is not available, since an alternate set of smoothness hypotheses has not been formulated, let alone the derivation of the implications of such an alternative set. A few results (such as which material functions arise in certain constant stretch history flows) can be obtained without actually formulating *any* smoothness hypotheses, and that is as far as one can go without either accepting the simple fluid theory or formulating an alternative one.

We have seen that equations of the type of eq. (6-4.47), which do not allow strain impulses, are not included in the general theory of simple fluids with fading memory, and in view of the discussion above, they are not included in *any* general theory. Of course, they are legitimate as written, and define their own topology. The question that arises is whether there are any real non-Newtonian fluids which are described by this type of equation.

The most directly observable property which distinguishes fluids described by eq. (6-4.47) and simple fluids with fading memory is the behavior under a sudden change in applied stress. In a recoil experiment, the motion of a fluid is observed after a sudden cessation of applied stress. Neglecting inertia, a purely viscous fluid would stop deforming after cessation of stress;

a simple fluid with the smoothness properties in section 4-4 would show some instantaneous recoil (i.e., a jump in strain would correspond to the jump in stress); a fluid described by eq. (6-4.47) would exhibit no instantaneous recoil but only some delayed recoil (i.e., the jump in stress would cause a jump in the strain rate). Unfortunately, inertia *cannot* be neglected if instantaneous recoil tends to take place, and therefore a clearcut experimental evidence is not available. Still, materials exhibiting recoil do show some very rapid recoil—a result which gives credit to the smoothness hypotheses underlying the simple fluid theory.

It must, however, be observed that there are molecular arguments which suggest that very dilute solutions of polymers might exhibit a stress which depends both on the deformation history and on the instantaneous value of the strain rate with the viscous part of the behavior contributed by the solvent, the contribution being non-negligible in view of the very low concentration of polymer. Thus, eq. (6-4.47) is probably useful mainly when applied to dilute polymer solutions.

Specific rate type equations

Let us now turn back to a more detailed analysis of the properties of rate equations which have been proposed in the literature. Oldroyd [25] has studied the predictions of equations of the type (6-4.39) or (6-4.47) for the particular case where $a = b = c = 0$, i.e., when the corotational derivative is considered on both sides:

$$\tau + \sum_1^{n_1} \alpha_N \overset{\circ}{\tau}_N = 2\mu \left[\mathbf{D} + \sum_1^{n_2} \beta_N \overset{\circ}{\mathbf{D}}_N \right] \qquad (6\text{-}4.48)$$

He has shown that, for arbitrary values of n_1 and n_2, values of the first and second normal stress difference are predicted which are equal in value and opposite in sign—a prediction which is in contrast with experimental evidence.

Spriggs [27] has considered a model of the type of eqs (6-4.40) and (6-4.41), with the following choice for λ_N, μ_N, and F_{abc}:

$$\lambda_N = \frac{\lambda_1}{N^\alpha} \qquad (6\text{-}4.49)$$

$$\mu_N = \mu \frac{N^{-\alpha}}{\sum_1^\infty N^{-\alpha}} \qquad (6\text{-}4.50)$$

$$F_{abc} = F_{-(1+\varepsilon), \frac{2}{3}(1+\varepsilon), 0} \qquad (6\text{-}4.51)$$

$$n \to \infty \qquad (6\text{-}4.52)$$

The model has four constant parameters $(\lambda_1, \alpha, \mu, \varepsilon)$. It is suggested by microrheological theories [30, 31], according to which one should have $\alpha = 2$. If α is not too different from 2, the contribution of the $\tau^{(N)}$ become rapidly negligible as N increases. The constant ε is introduced in order to obtain a non-zero second normal stress difference in viscometric flow. Note that if $\alpha = 2$ and $\varepsilon = 0$, the Spriggs equation becomes equivalent to eq. (6-3.12) with a set of relaxation times given by eq. (6-3.11).

A subsequent modification of the model is the Spriggs 6-constant equation [32], which introduces on the right-hand side of eq. (6-4.41) a term containing a retardation time and a derivative of **D**. To this equation apply the same topology considerations already discussed in connection with eq. (6-4.47).

Finally, we shall consider a number of rate type equations which have the form of the Maxwell equation or of a generalized Maxwell equation, i.e., containing a set of relaxation times, but in which the 'constants' (typically λ and μ) are replaced by 'functions.' The argument of these functions is taken to be an invariant of the deformation rate, typically the second invariant. Examples of equations in this category can be found in references [33] and [34].

These equations correspond to those integral equations in which the memory functions are taken to depend on the deformation rate. To these equations the same criticism can be advanced as illustrated in the last part of the previous section, i.e., although they may be useful in correlating a number of data of different experiments, they fail to degenerate properly into the linear viscoelasticity response because of the peculiar topology that characterizes them (see the discussion at the end of section 6-3).

We conclude this section with a remark concerning rate equations in general. In quite general terms, a rate equation does not define a single material, i.e., a single functional which determines the present stress once the previous deformation history is assigned. Let us consider the analogous case of a function. When a function is assigned by means of a differential equation, the initial conditions must also be given. If the initial conditions are not assigned, the differential equation defines a set of functions, not just one. In the same way, generally, if no other specifications are made, a constitutive rate equation, defines simultaneously a set of functionals, i.e., a set of different materials. It is even possible that in a set of materials defined in this way, fluids and solids are simultaneously represented.

However, for the equations explicitly considered in this section the situation is much less ambiguous than in the general case considered by Truesdell and Noll [35]. In fact, all equations considered here are *linear* in the stress tensor so that the following arguments apply. Consider, for example, either eq. (6-4.39) or eq. (6-4.47), which are three-dimensional equivalents of composite systems of springs and dashpots. Assume that any such system is assigned and that it represents a 'fluid,' i.e., it allows infinitely large deformations. To such a system add a spring in parallel so as to transform it into a solid. The differential equation of the modified system would again be of the same form, with an extra term on the right-hand side proportional to the deformation (and not to the deformation rate or higher derivatives). Thus, at first sight it might seem that 'solids' equations are distinguishable because of the presence of a deformation term. Yet, by differentiating such an equation, the deformation would be made to disappear, except that, simultaneously (and because of linearity) the term containing the stress on the left-hand side of the equation would also be transformed into a term containing the stress derivative. Thus, in conclusion, eqs (6-4.39) and (6-4.47), are indeed constitutive equations of fluids, since stress appears explicitly.

There still remains the problem of whether eq. (6-4.39), with assigned values of the parameters, defines a single fluid or a set of fluids. At first sight it might seem that, depending on the initial conditions which are arbitrarily assigned, different functionals, i.e., different fluids, are obtained from the same equation. However, the structure of the equation is such that it already contains the property of fading memory; that is, if the instant of time at which the initial conditions are assigned is displaced further and further in the past, the resulting functional becomes progressively independent of the initial conditions. An example of this property was shown in obtaining eq. (6-4.19) from eq. (6-4.12). Thus, the conclusion is reached that, provided the initial conditions are supposed to be stated far in the past, they are of no consequence and the equations considered in this section define unequivocally a single fluid material.

6-5   ILLUSTRATIONS

ILLUSTRATION 6A : *Predictions of the BKZ theory in viscometric flows.*

The function U of the BKZ theory has been given by Zapas [10] the form of eq. (6-3.33). For viscometric flows eq. (6-3.33) takes a simpler form because I_{C^t} and II_{C^t} have the following expressions:

$$I_{C^t} = 3 + k^2 s^2 \qquad (6\text{-}5.1)$$

$$II_{C^t} = I_{(C^t)^{-1}} = 3 + k^2 s^2 \qquad (6\text{-}5.2)$$

Equations (6-5.1) and (6-5.2) are immediately obtained from the kinematic description of viscometric flows (see Illustration 3A, p. 110).

In order to obtain the stress, we need to calculate the derivatives of the function U appearing in eq. (6-3.31). We have, from eq. (6-3.33),

$$-\frac{\partial U}{\partial I_{C^t}} = \alpha'(I_{C^t} - 3) + \frac{9}{2}\beta' \frac{1}{I_{C^t} + II_{C^t} + 3} + \frac{24(\beta' - c')}{I_{C^t} + 15} + c' \qquad (6\text{-}5.3)$$

$$-\frac{\partial U}{\partial II_{C^t}} = \frac{9}{2}\beta' \frac{1}{I_{C^t} + II_{C^t} + 3} - \frac{24(\beta' - c')}{II_{C^t} + 15} \qquad (6\text{-}5.4)$$

Thus, using eqs (6-5.1) and (6-5.2):

$$2\frac{\partial U}{\partial I_{C^t}} = -2\alpha' k^2 s^2 - \frac{\beta'}{1 + \frac{2}{9}k^2 s^2} - \frac{48(\beta' + c')}{18 + k^2 s^2} - 2c' \qquad (6\text{-}5.5)$$

$$-2\frac{\partial U}{\partial II_{C^t}} = \frac{\beta'}{1 + \frac{2}{9}k^2 s^2} - \frac{48(\beta' - c')}{18 + k^2 s^2} \qquad (6\text{-}5.6)$$

Equations (6-5.5) and (6-5.6), together with the matrices of C^t and $(C^t)^{-1}$ given in Illustration 3A, allow explicit calculations of the stress to be made through (6-3.31).

For the tangential component we obtain:

$$\tau(k) = k \int_0^\infty \left\{ 2\alpha' k^2 s^3 + \frac{2\beta' s}{1 + \frac{2}{9}k^2 s^2} + 2c' s \right\} ds \qquad (6\text{-}5.7)$$

The integration cannot be performed explicitly because α', β', and c' are unknown functions of s. However, we reckon that the second term in the integral is a decreasing function of the parameter k, while the first one is an increasing function which will predominate for sufficiently large values of k. If we confine ourselves to consider only polymer melts or polymer solutions which are generally pseudoplastic in behavior, we infer that for these substances

$$\alpha' = 0 \qquad (6\text{-}5.8)$$

Thus, the viscometric viscosity is given by

$$\eta_V(k) = 2 \int_0^\infty \left\{ \frac{\beta' s}{1 + \frac{2}{9}k^2 s^2} + c' s \right\} ds \qquad (6\text{-}5.9)$$

The zero-shear viscosity is then given by

$$\eta_V(0) = 2 \int_0^\infty (\beta' + c') s \, ds \qquad (6\text{-}5.10)$$

while, for very large k values,

$$\eta_V(\infty) = 2 \int_0^\infty c's \, ds \qquad (6\text{-}5.11)$$

The following expressions are obtained for the first and second normal stress differences:

$$\sigma_1(k) = 2k^2 \int_0^\infty \left\{ \frac{\beta's^2}{1 + \frac{2}{9}k^2s^2} + c's^2 \right\} ds \qquad (6\text{-}5.12)$$

$$\sigma_2(k) = -k^2 \int_0^\infty \left\{ \frac{\beta's^2}{1 + \frac{2}{9}k^2s^2} + \frac{48(\beta' - c')}{18 + k^2s^2} + 2c's^2 \right\} ds \qquad (6\text{-}5.13)$$

The first normal stress difference is predicted to be positive and the second one negative, but the latter is perhaps too large. Probably, the third term in the expression for the energy function, eq. (6-3.33), needs some modification.

ILLUSTRATION 6B: *Calculation of the viscometric functions for a Maxwell-type fluid.*

We refer, in particular, to the constitutive equation (eq. (6-4.4)). Only the instantaneous kinematic description is needed in order to make the calculations, and in particular the tensors **D** and **W**. The latter appears in the expression of the corotational derivative $\overset{\circ}{\tau}$. With reference to a lineal Couette flow described in an appropriate Cartesian coordinate system, the matrices of **D** and **W** are given by (see Illustration 2A, p. 70):

$$[\mathbf{D}] = \begin{Vmatrix} 0 & k/2 & 0 \\ k/2 & 0 & 0 \\ 0 & 0 & 0 \end{Vmatrix} \qquad (6\text{-}5.14)$$

$$[\mathbf{W}] = \begin{Vmatrix} 0 & k/2 & 0 \\ -k/2 & 0 & 0 \\ 0 & 0 & 0 \end{Vmatrix} \qquad (6\text{-}5.15)$$

Because the flow has a constant stretch history, $[\overset{\cdot}{\tau}] = 0$ and thus (see eq. (3-3.38)):

$$\overset{\circ}{\tau} = \tau \cdot \mathbf{W} + (\tau \cdot \mathbf{W})^T \qquad (6\text{-}5.16)$$

$$\tau \cdot \mathbf{W} = \begin{Vmatrix} -(k/2)\tau^{12} & (k/2)\tau^{11} & 0 \\ -(k/2)\tau^{22} & (k/2)\tau^{12} & 0 \\ -(k/2)\tau^{23} & (k/2)\tau^{13} & 0 \end{Vmatrix} \qquad (6\text{-}5.17)$$

$$[\overset{\circ}{\tau}] = \begin{Vmatrix} -k\tau^{12} & \frac{k}{2}(\tau^{11} - \tau^{22}) & -\frac{k}{2}\tau^{23} \\ \frac{k}{2}(\tau^{11} - \tau^{22}) & k\tau^{12} & \frac{k}{2}\tau^{13} \\ -\frac{k}{2}\tau^{23} & \frac{k}{2}\tau^{13} & 0 \end{Vmatrix} \qquad (6\text{-}5.18)$$

We also have

$$[\boldsymbol{\tau} \cdot \mathbf{D} + \mathbf{D} \cdot \boldsymbol{\tau}] = \begin{Vmatrix} k\tau^{12} & \frac{k}{2}(\tau^{11} + \tau^{22}) & \frac{k}{2}\tau^{23} \\ \frac{k}{2}(\tau^{11} + \tau^{22}) & k\tau^{12} & \frac{k}{2}\tau^{13} \\ \frac{k}{2}\tau^{23} & \frac{k}{2}\tau^{13} & 0 \end{Vmatrix} \qquad (6\text{-}5.19)$$

Thus eq. (6-4.4) gives the following set of simultaneous algebraic equations:

11: $$\tau^{11} - k\lambda\tau^{12} + k\lambda a\tau^{12} = 0$$

12: $$\tau^{12} + \frac{k\lambda}{2}(\tau^{11} - \tau^{22}) + \frac{k\lambda}{2}a(\tau^{11} + \tau^{22}) = \mu k$$

13: $$\tau^{13} - \frac{k\lambda}{2}\tau^{23} + \frac{k\lambda}{2}a\tau^{23} = 0$$

22: $$\tau^{22} + k\lambda\tau^{12} + k\lambda a\tau^{12} = 0 \qquad (6\text{-}5.20)$$

23: $$\tau^{23} + \frac{k\lambda}{2}\tau^{13} + \frac{k\lambda}{2}\tau^{13} = 0$$

33: $$\tau^{33} = 0$$

The 1–3 and 2–3 of these equations give immediately $\tau^{13} = \tau^{23} = 0$ as expected. The 1–1 and 2–2 are used to express τ^{11} and τ^{22} in terms of τ^{12}:

$$\tau^{11} = k\lambda(1 - a)\tau^{12} \qquad (6\text{-}5.21)$$

$$\tau^{22} = -k\lambda(1 + a)\tau^{12} \qquad (6\text{-}5.22)$$

Equations (6-5.21) and (6-5.22) are substituted into the 1–2 equation of the set, which gives

$$\tau^{12} = \frac{\mu k}{1 + k^2\lambda^2(1 - a^2)} \qquad (6\text{-}5.23)$$

Thus, eq. (6-4.5) is obtained. Equations (6-4.6) and (6-4.7) are then immediately obtained.

Note that tr τ is not zero. In fact,

$$\text{tr } \tau = -\frac{2\mu k^2 \lambda a}{1 + k^2 \lambda^2 (1 - a^2)} \qquad (6\text{-}5.24)$$

PROBLEMS

6-1 Derive eqs (6-2.9) and (6-2.10).

6-2 Show that eq. (6-3.2) is phenomenologically indistinguishable from eq. (6-2.4) on the basis of only viscometric rheometry.

6-3 Calculate the expression for η_E predicted by eqs (6-3.31) and (6-3.33).

6-4 Calculate the expression for η_E predicted by eq. (6-4.4).

6-5 Check whether eq. (6-4.27) implies the smoothness hypotheses of the simple fluid theory.

6-6 Derive eq. (6-4.28).

6-7 Determine the one-dimensional model corresponding to eq. (6-4.48) with $n_1 = 3$, $n_2 = 2$. Discuss why any one-dimensional model corresponds either to $n_1 = n_2$ or to $n_1 = n_2 + 1$.

6-8 Discuss the topology implied by eq. (6-4.48) when $n_1 > n_2 + 1$.

BIBLIOGRAPHY

1. TRUESDELL, C., and NOLL, W.: *The Non-linear Field Theories*, p. 92. Springer-Verlag, Berlin (1965).

2. RIVLIN, R. S., and ERICKSEN, J. L.: *J. Ratl Mech. Anal.*, **4**, 323 (1955).

3. COLEMAN, B. D., and MARKOVITZ, H.: *J. appl. Phys.*, **1**, 1 (1964).

4. MARKOWITZ, H., and COLEMAN, B. D.: *Adv. Appl. Mech.*, **8**, 69 (1964).

5. LODGE, A. S.: *Rheol. Acta*, **7**, 379 (1968); *Trans. Farad. Soc.*, **52**, 20 (1956).

6. FERRY, J. D.: *Viscoelastic Properties of Polymers*. Wiley, New York (1961).

7. ASTARITA, G., and NICODEMO, L.: *Chem. Engng Jl*, **1**, 58 (1970).

8. BERNSTEIN, B., KEARSLEY, E. A., and ZAPAS, L. J.: *Trans. Soc. Rheol.*, **7**, 391 (1963).

9. TRUESDELL, C., and NOLL, W.: *op. cit.*, p. 175.

10. ZAPAS, L. J.: *Res. Natl Bur. Std*, **70A**, 525 (1966).

11. WARD, A. F. H., and JENKINS, G. M.: *Rheol. Acta*, **1**, 110 (1958).

12. TANNER, R. I., and SIMMONS, J. M.: *Chem. Engng Sci.*, **22**, 1803 (1957).

13. TANNER, R. I., and BALLMAN, R. L.: *Ind. Engng Chem. Fund.*, **8**, 588 (1969).

14. MARRUCCI, G.: *Ind. Engng Chem. Fund.*, **9**, 514 (1970).

15. FREDICKSON, A. G.: *Chem. Engng Sci.*, **17**, 155 (1962).

16. SPRIGGS, T. W., HUPPLER, J. D., and BIRD, R. B.: *Trans. Soc. Rheol.*, **10**, 191 (1966).

17. LODGE, A. S.: *Elastic Liquids*, p. 121. Academic Press, London (1964).

18. LODGE, A. S.: (1965) as quoted in reference [16].

19. BIRD, R. B., and CARREAU, P. J.: *Chem. Engng Sci.*, **23**, 427 (1968).

20. BOGUE, D. C.: *Ind Engng Chem. Fund.*, **5**, 253 (1966).
21. CARREAU, P. J.: *Trans. Soc. Rheol.* (in press).
22. PHILIPPOFF, W.: *Trans. Soc. Rheol.*, **10**, 317 (1966).
23. COLEMAN, B. D.: *Arch. Ratl Mech. Anal.*, **17**, 1 (1964).
24. LODGE, A. S.: Proc. 5th Int. Cong. on Rheology, vol. 4. University Park Press (1970).
25. OLDROYD, J. D.: *Second-order Effects in Elasticity, Plasticity and Fluid Dynamics*, p. 520. Pergamon Press, Oxford (1964).
26. ROSCOE, R.: *Br. J. Appl. Phys.*, **15**, 1095 (1965).
27. SPRIGGS, T. W.: *Chem. Engng Sci.*, **20**, 931 (1965).
28. WILLIAMS, M. C., and BIRD, R. B.: *Phys. Fluids*, **5**, 1126 (1962); *A.I.Ch.E. Jl*, **8**, 378 (1962); *Ind. Engng Chem. Fund.*, **3**, 42 (1964).
29. OLDROYD, J. G.: *Trans. Roy. Soc. London*, *A*, **200**, 523 (1950); **283**, 115 (1965).
30. ROUSE, P. E., Jr.: *J. Chem. Phys.*, **21**, 1272 (1953).
31. ZIMM, B. H.: *J. Chem. Phys.*, **24**, 269 (1956).
32. SPRIGGS, T. W.: Ph.D. Thesis, University of Wisconsin (1966).
33. TANNER, R. I.: *A.S.L.E. Trans.*, **8**, 179 (1965).
34. WHITE, J. L., and METZNER, A. B.: *J. Polymer Sci.*, **7**, 1867 (1963).
35. TRUESDELL, C., and NOLL, W.: *op. cit.*, p. 98.

7
FLUID MECHANICS

7-1 FLOW CLASSIFICATION

Classical (i.e., Newtonian) isothermal fluid mechanics of incompressible fluids consists essentially in obtaining solutions for physically significant sets of boundary conditions imposed on the Navier–Stokes equations:

$$\rho \frac{D\mathbf{v}}{Dt} = -\nabla p + \mu \nabla^2 \mathbf{v} + \rho \mathbf{g} \qquad (7\text{-}1.1)$$

Equation (7-1.1) is coupled with the continuity equation:

$$\nabla \cdot \mathbf{v} = 0 \qquad (7\text{-}1.2)$$

Although the program of classical fluid mechanics is easy to state, carrying it out is a quite difficult task, due to the analytical complexity of the system of non-linear, second-order partial differential equations ((7-1.1) and (7-1.2)). In practice, rigorous or approximate solutions can be obtained only if either the boundary conditions are of an extremely simple form, or if some simplifications are made at the beginning. Indeed, several categories of fluid mechanical problems can be classified, according to the type of simplification carried out. The classification of a specific problem to one of these categories is based essentially on dimensional analysis.

The comparable problem in the case of non-Newtonian fluid mechanics is several orders of magnitude more difficult. In fact, the following three new points need to be considered:

(i) Not only is the basic equation of motion much more complex than eq. (7-1.1), since the internal stress term does not take the simple form $\mu \nabla^2 \mathbf{v}$, but in fact its form is not known, because a constitutive equation which is known to describe accurately the rheological behavior of a specific material under *all* flow conditions is not available.

(ii) Although one would expect that at least some of the physical justification for the approximations used in classical fluid mechanics also applies to non-Newtonian fluids, the classification in flow categories undoubtedly needs further detail, and its nature is not generally understood.

(iii) Dimensional analysis is based on the appropriate grouping of physical parameters, and can be used in the classical sense only when the fundamental constitutive assumptions are linear, so that the behavior of materials is unequivocally determined by the value of some material parameters. Its extension to non-Newtonian fluid mechanics poses a series of problems which are not yet entirely solved.

In the following we illustrate the classification scheme for problems of classical fluid mechanics, attempting to bring into focus the physical implications rather than the mathematical formalism. This allows us to place in proper perspective the largely unsolved analogous problems of non-Newtonian fluid mechanics, some of which are discussed in more general terms in the next section.

Gravity forces

Let us first examine the relevance of the body force term $\rho \mathbf{g}$ in eq. (7-1.1). If V is some characteristic velocity of the flow field considered, L a characteristic dimension, and g the modulus of $\mathbf{g}$, the Froude number is defined as

$$Fr = \frac{V^2}{Lg} = \frac{\text{inertia}}{\text{gravity}} \qquad (7\text{-}1.3)$$

If the Froude number is very large, gravity forces are negligible as compared to inertia forces (say $|\rho \mathbf{g}| \ll |\rho\, D\mathbf{v}/Dt|$), and the term $\rho \mathbf{g}$ can be dropped from the equations of motion. Since this procedure is based on comparing two terms of eq. (7-1.1) which also have the same form for non-Newtonian fluids, its extension to the latter is obviously legitimate.

Very few problems of fluid mechanics give rise to very large Froude numbers. Nonetheless, gravity forces often do not enter explicitly into the solution of a specific problem for an entirely different reason. In fact, eq. (7-1.1) can be written in the form

$$\rho \frac{D\mathbf{v}}{Dt} = -\nabla \mathscr{P} + \mu \nabla^2 \mathbf{v} \qquad (7\text{-}1.4)$$

where $\mathscr{P}$ is the 'extra pressure' introduced in section 5-1:

$$\mathscr{P} = p + \rho\psi$$

If the gravity forces do not appear explicitly into the boundary conditions (when these are expressed in terms of $\mathscr{P}$ rather than p), the solution of the boundary-value problem is obtained as a distribution of velocity and extra pressure, and gravity forces do not appear explicitly at any stage of the procedure. This is true when the flow field considered is bounded by rigid surfaces (say solids). When free boundaries (i.e., interfaces with a gaseous phase) are present, the boundary conditions on these surfaces are expressed in terms of p, and gravity forces appear explicitly in the solution of the boundary-value problem. Again, this reasoning also applies to non-Newtonian fluid mechanics.

The Euler equations

The fundamental dimensionless group of Newtonian fluid mechanics is the Reynolds number:

$$Re = \frac{LV\rho}{\mu} = \frac{\text{inertia forces}}{\text{viscous forces}} \qquad (7\text{-}1.5)$$

When the Reynolds number is very large (say much larger than unity), the term $\mu \nabla^2 \mathbf{v}$ can be dropped from eq. (7-1.3) which reduces to[†]

$$\rho \frac{D\mathbf{v}}{Dt} = -\nabla \mathscr{P} \qquad (7\text{-}1.6)$$

Equation (7-1.6) is the so-called Euler equation, or the equation of motion for a perfect fluid (i.e., a fluid where $\mu = 0$, so that the stress is always isotropic, $\mathbf{T} = -p\mathbf{1}$). The literature on the solution of boundary-value problems for eq. (7-1.6) is very large, and constitutes the body of classical hydrodynamics. One of the best treatises on the subject is the work of Lamb [1].

---

† An important exception to this statement is the case of laminar flows, to be discussed below.

Of course, for non-Newtonian fluids also, one can easily conceive flow situations where the internal stresses effects are negligible as compared to inertia forces, say,

$$|\nabla \cdot \tau| \ll |\rho \, D\mathbf{v}/Dt| \qquad (7\text{-}1.7)$$

In such flow conditions, the term $\nabla \cdot \tau$ could be dropped from the equation of motion, which would again degenerate into Euler's equation ((7-1.6)). Indeed, this argument has been used in the discussion of a specific problem of non-Newtonian fluid mechanics [2]. The problem is that, while for Newtonian fluids the condition for the validity of eq. (7-1.6) is well known, i.e.,

$$Re \gg 1 \qquad (7\text{-}1.8)$$

no comparable condition can be written *in general* for a non-Newtonian fluid unless a specific constitutive equation containing a finite number of material parameters is assumed.

An important point needs to be discussed here, requiring the introduction of the rotation vector **w**, which is defined as (the symbol ✗ indicates the vector product):

$$\mathbf{v} \times \mathbf{w} = -2\mathbf{W} \cdot \mathbf{v} \qquad (7\text{-}1.9)$$

It can be shown in general that *any* solution of eq. (7-1.6) (i.e., any velocity distribution satisfying Euler's equation) fulfils the following condition:

$$\frac{D\mathbf{w}}{Dt} = \nabla\mathbf{v} \cdot \mathbf{w} \qquad (7\text{-}1.10)$$

In particular, eq. (7-1.10) implies that (i) in any motion starting from a state of rest, the rotation is always zero, and (ii) if a steady flow field is such that all trajectories come from infinity, and the rotation is zero at infinity, it is zero throughout the flow field.

These two conclusions imply that irrotational motions, i.e., flow fields where $\mathbf{w} = 0$, constitute a very important class of solutions of Euler's equations. Note that in a flow field where $\mathbf{w} = 0$, then $\mathbf{W} = 0$.

Since the following identity can be proven in general:

$$\nabla \cdot \mathbf{W} = \nabla \cdot \mathbf{D} - \nabla(\nabla \cdot \mathbf{v}) \qquad (7\text{-}1.11)$$

it follows that *in an irrotational flow field of an incompressible Newtonian fluid the divergence of the stress is zero.*† *Hence, solutions of eq. (7-1.6) with* $\mathbf{W} = 0$ *are also exact solutions of the complete equations of motion (eq. (7-1.4) in the case of an incompressible Newtonian fluid).*

The point made above gives Euler's equation a status in Newtonian incompressible fluid mechanics which exceeds the limitations imposed by condition (7-1.8). In fact, but for the problems arising in the vicinity of rigid boundaries (which will be discussed below), eq. (7-1.6) would yield a large class of solutions of the general equation of motion, which are valid also at moderately low Reynolds numbers.

This is no more true in the case of non-Newtonian fluids. In fact, with *any* constitutive equation different from the Newtonian one, eq. (7-1.11) *would not* imply that the divergence of stress is zero for incompressible fluids, and hence irrotational flow fields satisfying eq. (7-1.6) *would not* be solutions of the complete equations of motion. Hence, classical hydrodynamics is applicable to non-Newtonian fluids only when the limitations imposed by eq. (7-1.7) are fulfilled.

Boundary layers

We need here to discuss in general a problem arising in connection with Euler's equation. In classical Newtonian fluid mechanics, the equation of motion (eq. (7-1.4)) is a second-order differential equation. The second-order terms arise only in the viscous term $\mu \nabla^2 \mathbf{v}$; hence, when this term is neglected, the resulting equation (i.e., Euler's equation) is a first-order differential equation. This implies that not all boundary conditions which can be imposed on the complete equation of motion can also be imposed on Euler's equation. Since the boundary conditions represent actual physical conditions which are known to be correct, solutions of Euler's equation, even when they are also solutions of the complete equations of motions, do not satisfy all the boundary conditions which are physically imposed on the flow field. In particular, the tangential velocity on rigid bounding surfaces cannot be imposed as zero on

† The physical significance of the fact that $\nabla \cdot \boldsymbol{\tau} = 0$ should be well understood. In perfect fluid theory, one assumes $\mu = 0$, and hence $\boldsymbol{\tau} = 0$, and $\nabla \cdot \boldsymbol{\tau} = 0$ trivially. For a Newtonian incompressible fluid in irrotational flow, $\nabla \cdot \boldsymbol{\tau} = 0$ (i.e., the resultant force of the stresses on any closed surface is zero), but the stresses themselves are not zero. That the divergence of the stress may be zero although the stress is not should not be surprising; indeed, in chapter 5 this was shown to be the case in, for example, elongational flow. Note that the energy dissipation, $\boldsymbol{\tau} : \nabla \mathbf{v}$, is always zero in a perfect fluid, but is non-zero for a Newtonian fluid even if it undergoes an isochoric irrotational flow where $\nabla \cdot \boldsymbol{\tau} = 0$; indeed, an interesting problem of Newtonian fluid mechanics was originally solved [2, 3] by calculating the total rate of energy dissipation in an irrotational flow field satisfying eq. (7-1.6).

Euler's equation, solutions of which therefore predict finite tangential velocities on such surfaces. This leads to a major difficulty of the classical hydrodynamic theory, whose most conspicuous manifestation is the so-called d'Alembert paradox.†

The difficulty is related to the fact that neglecting viscous forces, even at very large Reynolds numbers, is not a valid assumption in the vicinity of a rigid boundary. In fact, since on a rigid boundary the velocity is zero while the velocity gradient is finite, viscous forces are always predominant in that region. Therefore, one needs to analyze the flow near rigid boundaries on the basis of eq. (7-1.4) even if the Reynolds number is large. The region close to the boundary where the validity of eq. (7-1.6) breaks down is called the boundary layer.

The classical theory of boundary layer flows of Newtonian fluids is well developed, and the best treatise on the subject is the work of Schlichting [4]. We here wish to discuss very briefly only some fundamental concepts relating to two-dimensional boundary layers, in order to analyze the possible extensions of the theory to non-Newtonian fluids.

In two-dimensional boundary layers one can show, by an order-of-magnitude analysis, that the thickness of the boundary layer, δ, is related to the distance from the stagnation point, x, by the equation

$$\frac{\delta}{x} = 0\left(\frac{1}{\sqrt{(Re)_x}}\right) \quad (7\text{-}1.12)$$

where $(Re)_x$ is a Reynolds number in which the characteristic dimension is taken as x. Equation (7-1.12) implies that, with the exception of a small region near the stagnation point, $\delta \ll x$. This conclusion is of major importance since it allows simplification of the equations of motion by a simple order-of-magnitude analysis, from which the following results are obtained.

In the direction orthogonal to the bounding surface (which will be indicated by y), inertia forces are negligible, and internal stresses are balanced by the pressure gradient. The equation of motion in the y direction reduces to:‡

$$\frac{\partial \mathscr{P}}{\partial y} = \mu \frac{\partial^2 v_y}{\partial y^2} \quad (7\text{-}1.13)$$

---

† This paradox is as follows: a fluid flowing around a submerged object with a shape symmetrical with respect to the flow direction, exerts a zero force on it. This, of course, is known to be incorrect, however small the viscosity of the fluid may be.

‡ Since $\delta \ll x$, one can regard a coordinate system with x in the tangential direction and y in the direction orthogonal to the bounding surface as quasi-Cartesian.

On the basis of eq. (7-1.13), it is seen that pressure variations across the boundary layer are negligible compared to forces acting in the direction of flow, x. Hence, in the equation of motion in the x direction, the pressure distribution can be taken as being equal to the one obtained from the solution of Euler's equation:

$$\frac{\partial \mathscr{P}}{\partial x} = -\rho U \frac{dU}{dx} \qquad (7\text{-}1.14)$$

where U is the tangential velocity satisfying Euler's equation. U is regarded as depending only on x, since the variations of tangential velocity across the boundary layer predicted from Euler's equation are negligible. In view of eq. (7-1.14), the equation of motion in the x direction reduces to

$$\rho \left[v_x \frac{\partial v_x}{\partial x} + v_y \frac{\partial v_x}{\partial y} - U \frac{dU}{dx} \right] = \mu \frac{\partial^2 v_x}{\partial y^2} \qquad (7\text{-}1.15)$$

Equation (7-1.15), together with the continuity equation

$$\frac{\partial v_x}{\partial x} + \frac{\partial v_y}{\partial y} = 0 \qquad (7\text{-}1.16)$$

are the fundamental two-dimensional boundary layer equations, for which exact or approximate solutions are available for a variety of geometries [4].

This same general approach has been extended to non-Newtonian fluids by White and Metzner [5]. In this case, no equation analogous to eq. (7-1.12) can be written in general, and the whole argument about relative orders of magnitude is much more vague. Nonetheless, the conclusions stated above (but not the equations) are also approximately valid for non-Newtonian fluids, for which again physical intuition suggests that one can envisage flow situations where Euler's equation breaks down only in a thin layer adjacent to the rigid boundaries. The equation of motion in the x direction takes the form

$$\rho \left[v_{xx} \frac{\partial v_x}{\partial x} + v_y \frac{\partial v_x}{\partial y} - U \frac{dU}{dx} \right] = \frac{\partial \tau_{xy}}{\partial y} + \frac{\partial}{\partial x} (\tau_{xx} - \tau_{yy}) \qquad (7\text{-}1.17)$$

Again, eqs (7-1.16) and (7-1.17) can be regarded as the fundamental two-dimensional boundary layer equations for non-Newtonian flow. Of course, their solution requires assumption of a specific constitutive equation.

It may be worthwhile mentioning that Astarita [6] has brought up a qualitative argument according to which the region near the stagnation point where the boundary layer approximations break down is, in the case of non-Newtonian fluids, possibly much larger than in the case of Newtonian fluids.

Laminar flows

In this book we use the term laminar flow to indicate a flow field which (i) satisfies the complete equation of motion and all the relevant boundary conditions, and (ii) is such that $D\mathbf{v}/Dt = 0$.

Apart from the trivial cases of hydrostatics and of rigid-body steady translation, laminar flows are in practice possible only for steady flows in long conduits of constant cross-section. In laminar flows, of course, one *cannot* neglect viscous forces as compared to inertia forces even if the Reynolds number is large, since inertia forces are identically zero.

It is well known that laminar flows are unstable at very large Reynolds numbers, where the flow degenerates into turbulent flow. This implies that, although a laminar flow field is a solution of the complete equations of motion satisfying all the boundary conditions, it is not a unique solution, since, of course, the turbulent flow field also satisfies both the differential equation of motion and the boundary conditions.

All laminar flow fields are viscometric flows (although the converse is not true: in chapter 5, several viscometric flow fields have been discussed for which inertia forces were not identically zero). Although laminar flows are possible in non-Newtonian fluids, it has been shown [7] that steady rectilinear flow down a constant-section conduit is not possible in general for non-Newtonian fluids, with the exception of very few geometries of the cross-section (such as circular tubes or infinite slits). Secondary flows, i.e., circulations in the cross-section plane, take place as soon as deviations from Newtonian behavior are taken into account.

Creeping flow

For Newtonian fluids, the inertia forces can be dropped from the equation of motion, which reduces to

$$\nabla \mathscr{P} = \mu \nabla^2 \mathbf{v} \qquad (7\text{-}1.18)$$

whenever the Reynolds number is very small, say, sufficiently smaller than unity. Flow fields satisfying eq. (7-1.18) are called creeping flows. It should be borne in mind that $D\mathbf{v}/Dt$ *is not* zero in creeping flows.

Of course, creeping flows can also be envisaged in the case of non-Newtonian fluids, and indeed most of the flows discussed in chapter 5 are creeping flows (with *corrections* for taking into account the minor contribution of inertia forces). Indeed, since non-Newtonian fluids are in general quite more 'viscous' than ordinary Newtonian fluids (in the sense that, at comparable kinematic conditions, they exhibit larger internal stresses), creeping flows are possibly the most interesting flow category for non-Newtonian fluids.

While dropping inertia forces from the equation of motion yields, for Newtonian fluids, eq. (7-1.18), which is *linear* (since the only non-linear term in eq. (7-1.4) is the inertia force term), the same is not true for any non-Newtonian fluid, for which the creeping flow equation is non-linear. This is true whatever the form of the constitutive equation assumed; not even the form of the internal stress term is known for non-Newtonian fluids in general.

Turbulent flows

It is well known that turbulent flows of Newtonian fluids represent the most difficult category of flow types to analyze. Since a complete description of the velocity distribution in a turbulent flow field would not only be impossible to achieve, but would be of limited pragmatic use, turbulent flows are generally described in terms of the average values of both velocity and pressure:

$$\bar{\mathbf{v}} = \frac{1}{T} \int_0^T \mathbf{v} \, dt \qquad (7\text{-}1.19)$$

$$\bar{\mathscr{P}} = \frac{1}{T} \int_0^T \mathscr{P} \, dt \qquad (7\text{-}1.20)$$

where T is a time interval which is long enough to accommodate a statistically significant sample of the fluctuations.

The fluctuating velocity $\mathbf{v}'$ and pressure $\mathscr{P}'$ are defined as

$$\bar{\mathbf{v}} = \mathbf{v} - \mathbf{v}' \qquad (7\text{-}1.21)$$

$$\bar{\mathscr{P}} = \mathscr{P} - \mathscr{P}' \qquad (7\text{-}1.22)$$

When eqs (7-1.21) and (7-1.22) are substituted into eq. (7-1.4), and all terms are averaged over an interval T, one obtains

$$\rho \frac{D\bar{\mathbf{v}}}{Dt} = -\nabla \bar{\mathscr{P}} + \mu \nabla^2 \bar{\mathbf{v}} - \nabla \cdot (\rho \overline{\mathbf{v}'\mathbf{v}'}) \qquad (7\text{-}1.23)$$

where the tensor $-\rho \overline{\mathbf{v}'\mathbf{v}'}$ is called the Reynolds stress tensor. It can be seen that the equation of motion for the average velocity $\bar{\mathbf{v}}$ differs from that for the instantaneous velocity $\mathbf{v}$ only for the Reynolds stress term, which arises from the only non-linear term in eq. (7-1.4), i.e., from the term $\nabla \mathbf{v} \cdot \mathbf{v}$. In fact, since averaging (in the sense of eqs (7-1.19) and (7-1.20)) is a linear operation, it commutes with all linear operators appearing in eq. (7-1.4).

When the same procedure is applied to a general non-Newtonian fluid, one obtains

$$\rho \frac{D\bar{\mathbf{v}}}{Dt} = -\nabla \bar{\mathscr{P}} + \nabla \cdot (\bar{\tau} - \overline{\rho \mathbf{v}'\mathbf{v}'}) \qquad (7\text{-}1.24)$$

but, since τ *is not* a linear function of velocity, one would not expect the equation of motion for $\bar{\mathbf{v}}$ to differ from the one for $\mathbf{v}$ only for the presence of the Reynolds stress term, since additional terms would arise from the non-linearity of the constitutive equation.

Oseen-type expansions

The analysis of flow of Newtonian fluids past submerged objects is carried out, in the limit of diminishingly small Reynolds numbers, with the creeping flow approximation which entirely neglects the inertia term $\rho\, D\mathbf{v}/Dt$ in the equations of motion. When the Reynolds number is not very small, a correction to the creeping flow solution which accounts for inertia, albeit in an approximate way, can be obtained by an Oseen-type expansion. This is based on the following considerations.

The ratio of inertia forces to viscous forces increases with increasing distance from the submerged object; indeed, a *local* Reynolds number should be based on the distance from the center of the object rather than on its linear dimension. At great distances from the object, the velocity is approximately equal to the unperturbed velocity vector $\mathbf{v}_\infty$, and one can therefore write, for a steady flow,

$$\rho \frac{D\mathbf{v}}{Dt} \cong \nabla \mathbf{v} \cdot \mathbf{v}_\infty \qquad (7\text{-}1.25)$$

Hence, if one seeks a solution to the equations of motion for situations where inertia is negligible in the neighborhood of the submerged object, one can write the equations of motion for a Newtonian fluid in the form

$$\rho \nabla \mathbf{v} \cdot \mathbf{v}_\infty = -\nabla \mathscr{P} + \mu \nabla^2 \mathbf{v} \qquad (7\text{-}1.26)$$

The great advantage of eq. (7-1.26) over eq. (7-1.4) lies in the fact that it is linear.

An analogous procedure can also be applied to non-Newtonian fluids, since the reasoning leading to eq. (7-1.25) is independent of the rheological behavior of the fluid considered (provided that internal stresses do not depend on the kinematics of motion in such a way as to invalidate the concept that the ratio of inertia to internal stresses increases with increasing distance from the object). Yet the equation of motion which is obtained, i.e.,

$$\rho \nabla \mathbf{v} \cdot \mathbf{v}_\infty = -\nabla \mathscr{P} + \nabla \cdot \boldsymbol{\tau} \qquad (7\text{-}1.27)$$

is not linear, in contrast with eq. (7-1.26), since non-linearities arise from the term $\nabla \cdot \boldsymbol{\tau}$. An Oseen-type expansion is therefore possible only if one is willing to make the additional (and largely unwarranted) approximation of linearizing the constitutive equation also. Of course, some problems may be dealt with which also allow non-linearities in the stress term, i.e., by obtaining a first-order correction for inertia to a creeping flow solution with a linearized form for the inertia term.

7-2 DIMENSIONAL ANALYSIS

In classical Newtonian fluid mechanics, there are essentially six dimensional parameters to be considered. Of these, three are characteristic of the specific flow problem which is being considered, namely a velocity V, a linear dimension L, and (for unsteady flows) a characteristic flow time T_f; one is gravity, g, and two are characteristic of the fluid, namely density, ρ, and viscosity, μ. For incompressible fluids, the rheological behavior (i.e., the constitutive equation) is entirely determined by the value of the viscosity. These six quantities give rise to the classical dimensionless groups of Newtonian fluid mechanics, say.:

(i) the Strouhal number,

$$Sr = \frac{V T_f}{L} \qquad (7\text{-}2.1)$$

(ii) the Froude number,

$$Fr = \frac{V^2}{Lg} \qquad (7\text{-}2.2)$$

(iii) the Reynolds number,

$$Re = \frac{V L \rho}{\mu} \qquad (7\text{-}2.3)$$

For all steady flows the Strouhal number becomes irrelevant. Since the Froude number is also irrelevant in many cases for the reasons discussed in section 7-1, most of classical Newtonian fluid mechanics is based on one dimensionless group, i.e., the Reynolds number.

It is indeed easy to show that, for Newtonian fluids, the ratio of inertial to viscous forces is of the same order of magnitude as the Reynolds number, say,

$$\frac{\rho \, |\mathbf{Dv}/\mathbf{D}t|}{|\mathbf{V} \cdot \tau|} = O(Re) \qquad (7\text{-}2.4)$$

a conclusion which was made use of in many of the points discussed in the preceding section. (Of course, laminar flows, as discussed in the previous section, are an exception to this general statement.)

Let us now turn our attention to non-Newtonian incompressible fluids. Equation (7-2.4) of course breaks down, since the viscosity μ is not defined. The stress τ is given by

$$\tau = \underset{s=0}{\overset{s=\infty}{\mathscr{H}}} \, [\mathbf{G}^t(s)] \qquad (7\text{-}2.5)$$

where both $\mathbf{G}^t$ and $\mathscr{H}$ are dimensional operators.

It is perhaps worthwhile to discuss in general terms the significance of the dimensions of an operator. If either the argument or the value, or both, of an operator are dimensional quantities, the operator is dimensional in the sense that the units chosen for the argument (and/or the value) determine the analytical form of the operator. When an operator is linear (tensors are a good example), they can be assigned precise dimensions, for example those of the value divided by those of the argument; hence, if the value and the argument have the *same* dimensions, a linear operator is dimensionless. Non-linear operators are dimensionless only when *both* the argument and the value are dimensionless, since only in this case is their analytical form independent of any choice of units.

One can easily construct two dimensionless operators, $\hat{\mathbf{G}}^t$ and $\hat{\mathscr{H}}$, by introducing two appropriate dimensional constants, which are characteristic of the material considered. These are:

(i) A 'natural time,' Λ, which in some sense is the length of the memory of the fluid. This has been already discussed in chapter 4; its evaluation for a specific material may be a rather ambiguous task, which will be discussed later. A dimensionless time lapse $\hat{s}$ is immediately defined as:

$$\hat{s} = s/\Lambda \qquad (7\text{-}2.6)$$

and, according to the definition given above for Λ, one would expect the history at $\hat{s} \gg 1$ to be largely irrelevant as far as the present stress is concerned.

(ii) A 'natural viscosity,' μ. Note that μ/Λ has the dimensions of a stress, and therefore a dimensionless stress $\hat{\tau}$ can be defined as:

$$\hat{\tau} = \frac{\Lambda}{\mu}\tau \qquad (7\text{-}2.7)$$

The natural viscosity should be chosen so that, when the magnitude of the value of $\hat{G}^t$ is of order one at $\hat{s} \leq 1$, the magnitude of $\hat{\tau}$ is also of order one, so that μ/Λ measures the magnitude of the stress corresponding to a deformation history of unit magnitude. Again, the task of evaluating μ for a specific material is an ambiguous one, and will be discussed later.

The dimensionless operators $\hat{G}^t$ and $\hat{\mathscr{H}}$ are now easily defined as follows:

$$\hat{G}^t(\hat{s}) = G^t(s) \qquad (7\text{-}2.8)$$

$$\hat{\tau} = \hat{\mathscr{H}}[\hat{G}^t(s)] = \frac{\Lambda}{\mu}\mathscr{H}(G^t(s)) \qquad (7\text{-}2.9)$$

Dimensional analysis of non-Newtonian fluid mechanics problems differs from its Newtonian counterpart in two very important points. First, rather than only one, there are two dimensional parameters arising from the constitutive equation. Furthermore, two fluids characterized by the *same* values of both μ and Λ are not equal in their rheological response, i.e., they do not have the same constitutive equation, since the form of the dimensionless functional $\hat{\mathscr{H}}$ may differ from one fluid to the other. Thus, the values of μ and Λ do not *entirely* determine the behavior of the fluid, and dimensional analysis based on these two parameters can at best give only qualitative indications.

The second difficulty is one which cannot be circumvented, unless one wishes to assume some specific constitutive equation.† Hence, as long as one wishes to maintain a certain degree of generality, one is bound to assume that one is dealing with a class of materials which are characterized by the *same* dimensionless functional $\hat{\mathscr{H}}$; in the following, such materials will be called 'homologous.' The remainder of this section is restricted to an analysis applying separately to each class of homologous materials (of course, all Newtonian fluids are homologous).

---

† If this is done, a number of parameters exceeding two may appear. These can always be transformed to μ, Λ, and a set of dimensionless parameters, which will then appear as such in any listing of the relevant dimensionless groups.

The natural viscosity μ

The definition given above for the natural viscosity μ is somewhat intuitive, and a specific definition which leads itself to a concrete possibility of measurement of μ is a matter of choice. Since it is known (see section 4-3) that *all* simple fluids with fading memory behave as Newtonian fluids in the limit of slow flows, it seems appropriate to identify the natural viscosity with the limiting Newtonian viscosity of the fluid, say

$$\mu = \lim_{k \to 0} \eta_V(k) \qquad (7\text{-}2.10)$$

where η_V is the viscometric viscosity.

Indeed, such a definition fulfils the requirements of the intuitive definition given before. In fact, developing $\hat{\mathbf{G}}^t$ in a Taylor series about $\hat{s} = 0$, one has

$$|\hat{\mathbf{G}}^t(\hat{s} = 1)| \cong |\hat{\mathbf{G}}^t(s = 0)| = \Lambda\,|\dot{\hat{\mathbf{G}}}^t(0)| = 2\Lambda\,|\mathbf{D}| \qquad (7\text{-}2.11)$$

and since in the limit of Newtonian behavior one has

$$\tau = 2\mu\mathbf{D} \qquad (7\text{-}2.12)$$

the ratio μ/Λ is seen to be indeed of the order of magnitude of the stress corresponding to a deformation history for which $|\hat{\mathbf{G}}^t(1)| = 1$. Note that the value of μ as defined in eq. (7-2.10) is also related to well-defined values of other material functions; for example,

$$\mu = \lim_{\omega \to 0} \eta' = -\int_0^\infty f(s)s\,\mathrm{d}s = \tfrac{1}{3} \lim_{\gamma_E \to 0} \eta_E \qquad (7\text{-}2.13)$$

and therefore the natural viscosity μ can be obtained both from viscometric rheometry (eq. (7-2.10)) and from dynamic or elongational rheometry (eq. (7-2.13)).

The natural time Λ

There is no exact definition of Λ which is so intuitively appropriate as the one given for μ. One would prefer to obtain Λ from the linear viscoelasticity function $f(s)$, since the natural viscosity is defined in terms of the same function, see eq. (7-2.13). We choose (rather arbitrarily) to base our definition on the dynamic modulus G'. From eq. (5-1.28), one has

$$\lim_{\omega \to 0} G' = -\frac{\omega^2}{2} \int_0^\infty f(s)s^2\,\mathrm{d}s \qquad (7\text{-}2.14)$$

which shows that, when $\omega \to 0$, G' approaches zero proportionally to ω^2.

By defining a dimensionless frequency $\hat{\omega}$ as

$$\hat{\omega} = \omega\Lambda \qquad (7\text{-}2.15)$$

one sees that the ratio $G'/\hat{\omega}^2$ has a finite limit as $\hat{\omega} \to 0$; the same ratio has the dimensions of a stress, and hence the following definition of Λ is permissible:

$$\frac{\mu}{\Lambda} = \lim_{\hat{\omega}\to 0} \frac{G'}{\hat{\omega}^2} = -\frac{1}{2\Lambda^2}\int_0^\infty f(s)s^2\,ds \qquad (7\text{-}2.16)$$

By combining eqs (7-2.13) and (7-2.16) one obtains

$$\Lambda = \frac{\int_0^\infty f(s)s^2\,ds}{2\int_0^\infty f(s)s\,ds} = \lim_{\omega\to 0}\frac{G'}{\eta'\omega^2} \qquad (7\text{-}2.17)$$

Of course, different definitions of Λ, also based on the function $f(s)$, could be chosen, such as, for example,

$$\frac{\int_0^\infty f(s)s\,ds}{\int_0^\infty f(s)\,ds}; \quad \frac{\int_0^\infty f(s)\,ds}{f(0)}; \quad \frac{f(0)}{f'(0)}; \text{ etc.} \qquad (7\text{-}2.18)$$

Equation (7-2.17) seems preferable on the grounds of molecular theories of polymeric materials: in fact, from eq. (6-3.10), it is seen that while definition (7-2.17) yields a value of Λ which does not depend on Z, and is of the order of the *largest* of the λ_j, any of definitions (7-2.18) would not have such desirable properties.

Note that the values of μ and Λ defined above *do not* completely identify the behavior of the material even in the limit of small deformations, say when eq. (4-3.24) holds.

It is easy to show that eq. (7-2.16) implies that

$$\frac{\mu}{\Lambda} = \lim_{k\to 0}\frac{\sigma_1(k)}{\hat{k}^2} \qquad (7\text{-}2.19)$$

where $\sigma_1(\ )$ is the first normal stress viscometric function, and $\hat{k}$ is a dimensionless viscometric shear rate, $\hat{k} = k\Lambda$. Thus, one has

$$\Lambda = \lim_{k\to 0}\frac{\sigma_1(k)}{k\tau(k)} = \lim_{k\to 0}\frac{\psi_1(k)}{\eta_V(k)} \qquad (7\text{-}2.20)$$

which shows that Λ can also be obtained both from viscometric rheometry (eq. (7-2.20)) and from dynamic rheometry (eq. (7-2.17)).

Dimensionless groups of non-Newtonian fluid mechanics

When the mechanics of a class of homologous non-Newtonian fluids are considered, the dimensional parameters to be taken into account are the same as for the class of Newtonian fluids $(V, L, T_f, g, \rho, \mu)$ plus the natural time Λ. From a strictly mathematical viewpoint, only one new dimensionless group should be considered, since there is only one new dimensional parameter. Nonetheless, several quite different dimensionless groups have been considered in the literature, each one with a specific physical interpretation. Of course, these groups are not independent of each other. We here attempt to list the most important groups which have appeared in the literature, to illustrate their physical significance, and to discuss the relationships among different groups.

The basic dimensionless group for non-Newtonian fluid mechanics is the Weissenberg number, We. Since the behavior of any fluid approaches that of a Newtonian fluid for slow flows, one wishes to define a dimensionless number which is a measure of the 'non-slowness' of flow, say of the importance of non-Newtonian effects.

Let us first restrict attention to viscometric flows, for which the following definition of the Weissenberg number appears satisfactory:

$$We = \Lambda k \qquad (7\text{-}2.21)$$

where k is the shear rate. One may notice that, from the definition of Λ (eq. (7-2.20)), one has

$$\lim_{k \to 0} We = \frac{\sigma_1(k)}{\tau(k)} \qquad (7\text{-}2.22)$$

i.e., the Weissenberg number approaches the ratio of the first normal stress difference to the tangential stress; more generally, the Weissenberg number can be regarded as a measure of the relative importance of normal and tangential stresses.

Although a Weissenberg number could be defined for all constant stretch history flows (e.g., in elongational flow one could consider the product $\Lambda\gamma_E$), its usefulness appears basically only when the flow considered is—at least approximately—viscometric. For a general quasi-viscometric flow, the Weissenberg number needs to be defined in terms of some equivalent shear rate, V/D, where V is some characteristic velocity of the flow and D is a characteristic linear dimension *in the direction along which the velocity varies.* Thus, one has

$$We = \frac{\Lambda V}{D} \qquad (7\text{-}2.23)$$

as a general definition for the Weissenberg number.

Since, in quasi-viscometric flows, the Reynolds number measures the ratio of inertia forces to tangential stresses, the ratio of normal stresses to inertia forces is obtained as the ratio of the Weissenberg and Reynolds numbers:

$$\frac{We}{Re} = \frac{\Lambda\mu}{\rho D^2} = El_1 \qquad (7\text{-}2.24)$$

This dimensionless group has not been used systematically in the literature, and we here propose the name of 'first elasticity number' and the symbol El_1. It may be worthwhile mentioning that, in the analysis of specific problems of non-Newtonian fluid mechanics based on special constitutive assumptions, both normal stresses and inertia forces often turn out to be proportional to the square of the velocity; neglect of one or the other is based on the magnitude of El_1.

In chapters 4 and 5 it has been shown that constant stretch history flows are almost the only ones for which a rigorous analysis is possible. One would therefore wish to define a dimensionless group which, in some sense, measures the 'closeness' of a general flow field to a constant stretch history flow. This leads to the introduction of the Deborah number, De, which is defined as [8]:

$$De = \Lambda/T_f \qquad (7\text{-}2.25)$$

where T_f is a time scale of the flow, i.e., the time (measured following a material point) required for a non-negligible change of kinematic conditions. Although a rather detailed analysis of the best possible evaluation of T_f for any given flow pattern has been published [9], for most order-of-magnitude analyses it is sufficient to take the ratio of *a characteristic linear dimension in the flow direction*, L, to the characteristic velocity. Thus, one has

$$De = \frac{\Lambda V}{L} \qquad (7\text{-}2.26)$$

as a generally acceptable definition of the Deborah number for steady flow fields. For flow fields which have a characteristic time scale (such as unsteady flows or periodic flows), eq. (7-2.25) can be used directly.

Note that for steady flows, the ratio of the Deborah and other dimensionless groups such as the Weissenberg number equals a shape factor of the flow field, and is thus a constant within any class of geometrically similar flow fields. For unsteady flows, the ratio of the Weissenberg to the Deborah number equals the Strouhal number.

The three dimensional parameters characterizing a class of homologous non-Newtonian fluids (μ, Λ, and ρ) can be combined to yield a *natural velocity* *of the fluid, V_w*:

$$V_w = \sqrt{(\mu/\Lambda\rho)} \qquad (7\text{-}2.27)$$

This velocity is related [10, 11] to the speed of propagation of discontinuities in the fluid. One can thus define a dimensionless group (which will be referred to as the second elasticity number El_2) as the ratio of the characteristic flow velocity to the fluid's natural velocity V_w:

$$El_2 = \frac{V}{V_w} = \sqrt{\frac{\Lambda\rho V^2}{\mu}} \qquad (7\text{-}2.28)$$

In some as yet unspecified sense, supercritical conditions are to be expected at El_2 values exceeding unity. The following relationship exists among the Weissenberg, Reynolds, and El_2 groups:

$$El_2 = \sqrt{(We\ Re)} \qquad (7\text{-}2.29)$$

which shows that El_2 can be regarded as the ratio of the geometric average of inertia and elastic forces to the viscous forces.

Finally, there are problems of fluid mechanics where some characteristic stress, τ_0, can be identified: a good example is the steady motion of suspended particles under the influence of gravity, where the net gravity force divided by the particle's surface is the characteristic stress. Since μ/Λ is the 'natural' stress of the fluid, one can define the third elastic number, El_3, as [2]:

$$El_3 = \frac{\tau_0\Lambda}{\mu} \qquad (7\text{-}2.30)$$

In classical Newtonian fluid mechanics, the flow field's characteristic stress τ_0 appears in the definition of the Euler number:

$$Eu = \frac{\tau_0}{\rho V^2} \qquad (7\text{-}2.31)$$

The groups El_2 and El_3 are related to each other by the following equation:

$$El_2 = \sqrt{(El_3/Eu)} = \sqrt{(We\ Re)} \qquad (7\text{-}2.32)$$

7-3 SECONDARY AND SUPERIMPOSED FLOWS

In the analysis of several flow fields which have been discussed in chapter 5, the kinematics of motion have been assumed at the start, as suggested by the known boundary conditions and, in more general terms, by physical intuition.

The next step is that of calculating the stress field from an appropriate constitutive equation; in chapter 5 the general equation for simple fluids with fading memory was considered, but the steps in the procedure are essentially the same when a more specific constitutive equation is assumed. Indeed, the type of constitutive equation which may be used is often suggested by the assumed kinematic pattern, which is recognized as one that is well described by a certain type of constitutive equation. The third step is the substitution of the velocity and stress fields into the equations of motion, from which both the pressure field and some parameters of the kinematic description that had been left unspecified in the first step are obtained.

When this procedure is applied, it may turn out that the flow pattern considered is not controllable; nonetheless, often either neglect of inertia (i.e., a creeping flow approximation), or some geometrical approximation allow the analysis that has been carried out to be regarded as 'almost' correct. This implies that the actual flow pattern differs from the assumed one by a superimposed motion which one assumes to be, in some sense, 'small.' A specific example may clarify the issue.

Suppose one considers steady rectilinear flow down a long pipe with a cross-section of non-circular shape; say, for example, an elliptical pipe. If the analysis given in chapter 5 for Poiseuille flow is duplicated for this case, no rectilinear flow pattern is controllable: the distribution of $\mathscr{P}$ in a cross-section is not constant along the θ coordinate of an elliptical coordinate system, indicating that a driving force for a non-zero velocity distribution in the cross-sectional plane would exist. Nonetheless, one is willing to assume (for certain problems) that this 'secondary flow' is not very important: for example, no great influence would be expected on the value of f, the pressure drop per unit length of pipe.

Still, there are some problems for which the secondary flow cannot be neglected. Suppose, for instance, that one is interested, for the flow field discussed above, in the rate of heat transfer from the walls of the pipe to the flowing material. The secondary flow would induce a convective mechanism of heat transfer in the cross-section plane, which may greatly enhance the purely conductive mechanism which would prevail in the absence of secondary flow.†

The analysis of secondary flows superimposed on a basic constant stretch history flow can be carried out with a certain degree of mathematical

---

† Indeed, this is the situation encountered in screw extrusion of polymers: the flow in the helical region between screw and barrel is basically a laminar flow down a conduit with an approximately rectangular cross-section (if curvature is neglected); but heat transfer to the barrel would be greatly influenced by any secondary flow pattern.

rigor. In fact, consider a constant stretch history flow characterized by the tensor $\mathbf{N}$ appearing in eq. (3-5.21); let $\mathbf{G}_0^t$ be the corresponding history of deformation, which is obtained from eq. (3-5.24):

$$\mathbf{G}_0^t = \mathbf{P}(t) \cdot [\exp(-k\mathbf{N}^T s) \cdot \exp(-k\mathbf{N}s)] \cdot \mathbf{P}^T(t) - \mathbf{1} \qquad (7\text{-}3.1)$$

Now consider a flow pattern for which the history $\mathbf{G}^t$ is given by:

$$\mathbf{G}^t = \mathbf{G}_0^t + \varepsilon\mathbf{G}_1^t \qquad (7\text{-}3.2)$$

where the norm of $\mathbf{G}_1^t$ is finite. The concept that the secondary flow (as described by $\varepsilon\mathbf{G}_1^t$) is 'small' can be given a precise meaning by assuming that $\varepsilon \to 0$, and that one seeks a solution for the secondary flow pattern only to within terms of order one in ε. If one assumes that the functional $\mathscr{H}$ is Fréchet-differentiable around $\mathbf{G}_0^t$, the stress τ for the flow field considered can be written as:

$$\tau = \mathscr{H}[\mathbf{G}_0^t] + \varepsilon\delta\mathscr{H}[\mathbf{G}_0^t|\mathbf{G}_1^t] + O(\varepsilon^2) \qquad (7\text{-}3.3)$$

where the last term on the right-hand side can now be neglected, since one wishes to carry out the analysis only to $O(\varepsilon)$. The first term on the right-hand side of eq. (7-3.3) is the stress corresponding to the basic constant stretch history flow, which will be indicated by τ_0, while the second term is linear in its second argument.

Thus we can write

$$\tau = \tau_0 + \varepsilon\int_0^\infty \mathscr{L} : \mathbf{G}_1^t \, ds \qquad (7\text{-}3.4)$$

where $\mathscr{L}:$ is a linear transformation of tensors into tensors. Note that $\mathscr{L}$ depends both on s and on the nature of the basic flow, i.e., on tensor $\mathbf{N}$:

$$\mathscr{L} = \mathscr{L}[\mathbf{N}, s] \qquad (7\text{-}3.5)$$

The requirement of objectivity implies that

$$\mathbf{Q} \cdot [\mathscr{L}(\mathbf{N}, s) : \mathbf{G}_1^t(s)] \cdot \mathbf{Q}^T = \mathscr{L}(\mathbf{Q} \cdot \mathbf{N} \cdot \mathbf{Q}^T, s) : [\mathbf{Q} \cdot \mathbf{G}_1^t(s) \cdot \mathbf{Q}^T] \qquad (7\text{-}3.6)$$

for *any* orthogonal tensor $\mathbf{Q}$. Pipkin and Owen [12] have investigated in detail the implications of eq. (7-3.6).

Note that, in the degenerate case where the basic flow is a state of rest or a rigid body rotation, $\mathbf{N} = 0$, and eq. (7-3.6) implies that $\mathscr{L}$ is an isotropic linear transformation. In this case, eq. (7-3.4) degenerates into eq. (4-3.24). When a small deformation is superimposed on a non-zero basic flow, the linear transformation $\mathscr{L}$ is not isotropic, as shown by eq. (7-3.6). The physical interpretation of this point is that an isotropic material undergoing a flow field does not respond isotropically to the superposition of a small perturbation, although it does respond linearly to it.

The only flow field of the type discussed above which has been analyzed in any detail for a general simple fluid is the superposition of small periodic deformations on a viscometric flow [13]. For such a case, the contribution of the second Fréchet differential of $\mathscr{H}$ has also been taken into account; this contribution also appears in the *mean* value of the stress, while the contribution of the linear term is, of course, only observable in the instantaneous value of the stress [14].

When analyzing flow fields of the type described by eq. (7-3.2) (with ε a small number, and calculations performed only to within terms of the first order in ε), relations between some integrals (over $0 < s < \infty$) of the components of $\mathscr{L}$ and the derivatives of the material functions of the basic flow can be derived. Such relations are called the 'consistency relations,' and are obtainable by considering that *any* constant stretch history flow can be regarded as the superposition of an appropriate small perturbation over a constant stretch history flow of the same type. Let k and $\mathbf{N}$ identify the basic constant stretch history flow, and let $k + \varepsilon k$, $\mathbf{N}$ identify the 'perturbed' constant stretch history flow; an easy calculation shows that, for the perturbed flow, eq. (7-3.2) holds, with $\mathbf{G}_1^t$ given by

$$\mathbf{G}_1^t = \mathbf{P}(t) \cdot \exp(-ks\mathbf{N}^T) \cdot [-ks(\mathbf{N} + \mathbf{N}^T)] \cdot \exp(-ks\mathbf{N}) \cdot \mathbf{P}^T(t) \qquad (7\text{-}3.7)$$

with $\mathbf{P}(t)$ the *same* orthogonal tensor function appearing in eq. (3-5.24). Since, for the perturbed flow, one can write:

$$\tau = \mathbf{H}(k, \mathbf{N}) + \varepsilon \frac{\partial \mathbf{H}(k, \mathbf{N})}{\partial k} + O(\varepsilon^2) \qquad (7\text{-}3.8)$$

where $\mathbf{H}(\ )$ is the tensor-valued function introduced in section 5-1 whose components are the material functions characteristic of the flow considered, one has

$$\frac{\partial \mathbf{H}(k, \mathbf{N})}{\partial k} = \int_0^\infty \mathscr{L}[\mathbf{N}, s] : \mathbf{G}_1^t(s) \, ds \qquad (7\text{-}3.9)$$

When eq. (7-3.7) is substituted into eq. (7-3.9), the consistency relations for every type of constant stretch history flow are obtained.

Apart from the special problems discussed above, the basic result of the theory of secondary flows is embodied in eq. (7-3.3), i.e., that the analysis of secondary flows can be carried out on the basis of a constitutive equation such as eq. (7-3.4), which assigns a linear (though not isotropic) relationship between the additional stress and the secondary flow's history of deformation.

One should remember that the form of this relationship depends on the basic flow pattern.

7-4 FLOWS AROUND SUBMERGED OBJECTS AND BOUNDARY LAYERS

In this section we discuss the analysis of flows of non-Newtonian fluids around submerged objects. The discussion is divided into two parts: we first consider low Reynolds number flows, i.e., flows for which inertia forces are not predominant over internal stresses; next, we consider a boundary layer analysis, which is of interest in high Reynolds number flows for which the kinematics outside of the boundary layer and wake are determined by the Euler equations (eq. (7-1.6)).

Low Reynolds number flows

The analysis to be given below is due to Ultman and Denn [15]. We start by considering an Oseen-type expansion as discussed in section 7-1; this is known to be reasonably accurate up to Reynolds number values of the order of unity in the case of Newtonian fluids. We choose a Cartesian coordinate system with the x axis oriented along the unperturbed velocity, which is thus given by $V\mathbf{e}_x$, V being the modulus of the unperturbed velocity. Equation (7-1.27) is thus written as

$$-\nabla\mathscr{P} + \nabla\cdot\mathbf{\tau} = \rho V\nabla\mathbf{v}\cdot\mathbf{e}_x = \rho V\frac{\partial\mathbf{v}}{\partial x} \qquad (7\text{-}4.1)$$

A constitutive equation now needs to be assumed. If either a first-order rate equation such as the Maxwell equation, or a simple integral equation, is assumed, and if it is linearized with respect to the velocity perturbation $V\mathbf{e}_x - \mathbf{v}$, one obtains

$$\mathbf{\tau} + \Lambda V\frac{\partial\mathbf{\tau}}{\partial x} = 2\mu\mathbf{D} \qquad (7\text{-}4.2)$$

where Λ and μ have the meaning discussed in section 7-2. One hopes that the essential qualitative features of the problem are describable in terms of eq. (7-4.2). Combination of eqs (7-4.1) and (7-4.2) yields the following equation:

$$\left(1 + \Lambda V\frac{\partial}{\partial x}\right)\nabla\mathscr{P} = \mu\nabla^2\mathbf{v} - \rho V\frac{\partial\mathbf{v}}{\partial x} - \rho V^2\Lambda\frac{\partial^2\mathbf{v}}{\partial x^2} \qquad (7\text{-}4.3)$$

where the last term on the right-hand side and the second term on the left-hand side account for the non-Newtonian character of the fluid.

Equation (7-4.3) can be made dimensionless by introducing the following dimensionless variables:

$$\mathbf{v}' = \mathbf{v}/V \qquad (7\text{-}4.4)$$

$$\mathscr{P}' = R\mathscr{P}/\mu V \qquad (7\text{-}4.5)$$

$$x', y', z' = \frac{x}{R}, \frac{y}{R}, \frac{z}{R} \qquad (7\text{-}4.6)$$

$$\nabla' = R\nabla; \qquad \nabla^{2'} = R^2\nabla^2 \qquad (7\text{-}4.7)$$

where R is some characteristic linear dimension of the submerged object, such as its radius in the case of a sphere. With these substitutions, eq. (7-4.3) becomes

$$\left(1 + We\frac{\partial}{\partial x'}\right)\nabla'\mathscr{P}' = \nabla^{2'}\mathbf{v}' - Re\frac{\partial \mathbf{v}'}{\partial x'} - El_2^2\frac{\partial^2 \mathbf{v}'}{\partial x'^2} \qquad (7\text{-}4.8)$$

where the Weissenberg number We, Reynolds number Re, and second elasticity number El_2, are defined as in section 7-2 on the basis of the linear dimension R.

The solution of eq. (7-4.8), together with the equation of continuity, is of the following form:

$$\mathscr{P}' = Re\frac{\partial \phi}{\partial x'} \qquad (7\text{-}4.9)$$

$$\mathbf{v}' = \mathbf{e}_x - \nabla'\phi + \frac{1}{Re}\left\{\nabla' - El_2^2\mathbf{e}_x\frac{\partial}{\partial x'} - Re\,\mathbf{e}_x\right\}\psi \qquad (7\text{-}4.10)$$

where the functions ϕ and ψ satisfy the following differential equations:

$$\nabla^{2'}\phi = 0 \qquad (7\text{-}4.11)$$

$$\left(1 - El_2^2\right)\frac{\partial^2 \psi}{\partial x'^2} + \frac{\partial^2 \psi}{\partial y'^2} + \frac{\partial^2 \psi}{\partial z'^2} - Re\frac{\partial \psi}{\partial x'} = 0 \qquad (7\text{-}4.12)$$

Equation (7-4.12) should be analyzed in some detail. First, it is apparent that the condition

$$El_2 = 1 \qquad (7\text{-}4.13)$$

is the borderline between two quite distinct situations. In the subcritical case ($El_2 < 1$), eq. (7-4.12) is elliptic, and admits continuous solutions. In the supercritical case ($El_2 > 1$), eq. (7-4.12) is hyperbolic, and has solutions with discontinuities along wave fronts. One would therefore expect that, when plotted against V, macroscopically observable quantities such as the drag coefficient exhibit discontinuities at some critical velocity V_{cr} corresponding to the fulfilment of eq. (7-4.13). This has indeed been observed, and will be discussed below.

Acosta and James [16] have measured heat transfer coefficients from very thin wires (such as are used in hot-wire velocity probes) to non-Newtonian fluids. The heat transfer coefficient increases with increasing velocity up to a critical velocity V_{cr} beyond which it becomes independent of V. These results have been discussed in detail by Ultman and Denn [17] in their analysis of the supercritical case. It is interesting to note that the observed value of V_{cr} is independent of the wire diameter—a result which supports an interpretation based on eq. (7-4.13), since the value of El_2 is independent of the linear dimension R.

It should be pointed out that the low Reynolds number approximation embodied in eq. (7-4.1) is not in contrast with an analysis extending to the supercritical case, i.e., to large values of El_2: in fact, if the Weissenberg number is large enough, El_2 may be larger than unity for a low Reynolds number flow, see eq. (7-2.29).

It is important to realize that the analysis given above is based on eq. (7-4.3), i.e., an equation which is linear, although it does account for memory-type effects through the terms containing Λ. Indeed, for simple geometries of the submerged object (such as spheres and cylinders), the solution of eq. (7-4.3) can be obtained up to an explicit calculation of the drag coefficient [15, 17]. A conceptually much simpler problem, that is, calculation of the drag coefficient for flow of generalized Newtonian fluids (i.e., fluids for which the stress is given by eq. (2-4.1)), is in practice more difficult to solve because of the non-linear form of the viscous stress term: even for the simplest geometry (sphere), only non-coincident upper and lower bounds to the solution have been obtained [18].

High Reynolds number flows

We shall now consider flows for which inertia forces are predominant over internal stresses. On the basis of the discussion in section 7-2, this implies that both El_1 and $1/Re$ are very small as compared to unity:

$$El_1 = \frac{We}{Re} \ll 1; \qquad \frac{1}{Re} \ll 1 \qquad (7\text{-}4.14)$$

or, equivalently,

$$Re \gg \max(1, We) \qquad (7\text{-}4.15)$$

where We is taken as $\Lambda V/D$, D being the characteristic linear dimension of the object. Under such conditions the flow in the 'external' field can be assumed to be described by the appropriate solution of the Euler equation (eq. (7-1.6)), and a boundary layer analysis needs to be carried out for the flow near the submerged object. The boundary layer equation of motion, which has been introduced in section 7-1, is reported below:

$$\rho \left[v_x \frac{\partial v_x}{\partial x} + v_y \frac{\partial v_x}{\partial y} - U \frac{dU}{dx} \right] = \frac{\partial \tau_{xy}}{\partial y} + \frac{\partial}{\partial x}(\tau_{xx} - \tau_{yy}) \qquad (7\text{-}4.16)$$

The usual order of magnitude analysis for boundary layer flows shows that the left-hand side of eq. (7-4.16) is $O(\rho U^2/x)$, while

$$\frac{\partial \tau_{xy}}{\partial y} = O\left(\mu \frac{U}{\delta^2} \right) \qquad (7\text{-}4.17)$$

$$\frac{\partial}{\partial x}(\tau_{xx} - \tau_{yy}) = O\left(\frac{\Lambda \mu U^2}{x \delta^2} \right) = O\left(We \, \mu \frac{U}{\delta^2} \right) \qquad (7\text{-}4.18)$$

where eq. (7-2.22) has been assumed to hold, at least as a very crude first approximation. One thus sees that there are two subcases to be considered:

(i) *'Viscous' boundary layer.* If We is at most of order 1, the term $(\partial/\partial x)(\tau_{xx} - \tau_{yy})$ is not predominant, and eq. (7-1.12) is still true. In the boundary layer, inertia forces and pressure forces (as measured by $\rho U(dU/dx)$) are balanced essentially by tangential stresses, and the influence of normal stresses can be analyzed in terms of a perturbation around an essentially Newtonian solution of the problem.

(ii) *'Elastic' boundary layer.* If $We \gg 1$, the term $\partial \tau_{xy}/\partial y$ can be neglected in eq. (7-4.16). This appears to give rise to a situation where inertia and pressure forces in the boundary layer are balanced essentially by normal stresses, say:

$$\frac{\rho U^2}{x} \simeq \frac{\Lambda \mu U^2}{x \delta^2} \qquad (7\text{-}4.19)$$

Thus, one would expect the boundary layer thickness δ to be independent of both U (and hence V) and x:

$$\delta \simeq \sqrt{(\Lambda \mu / \rho)} \qquad (7\text{-}4.20)$$

or, equivalently,

$$\frac{\delta}{R} \simeq \sqrt{\frac{\Lambda\mu}{\rho R^2}} = \sqrt{El_1} \qquad (7\text{-}4.21)$$

Note that the thickness of an elastic boundary layer is in any case small, since eq. (7-4.14) is assumed to hold; hence the approximation $\partial \mathscr{P}/\partial x \simeq \rho U(dU/dx)$ is justified.

Analyses of boundary layer flows of non-Newtonian fluids are quite abundant in the literature; all of them seem to refer, implicitly or explicitly, to the viscous case. Srivastava and Maiti [19] have analyzed the boundary layer flow of a second-order fluid; the choice of such a constitutive equation is obviously suggestive of a low-Weissenberg number approximation, say of a viscous boundary layer. Their main result is proof that the separation point moves closer to the forward stagnation point as the value of We is increased.

A more detailed and rather general analysis has been carried out by Denn [20], who also discusses a number of results of previous investigations. Denn starts by assuming a constitutive equation of the second-order fluid type, but the coefficients η_0, β_0, and γ_0 are taken as functions of the magnitude of $\mathbf{D}$. Apart from the conceptual difficulties that such an equation raises (these have been discussed in chapter 6), the results of the analysis are not very encouraging. A differential equation for $v_x(x, y)$ is obtained, with non-Newtonian terms being multiplied by an 'elastic parameter' ε, defined as follows:

$$\varepsilon = We\, Re\, \beta \qquad (7\text{-}4.22)$$

with β a correction factor which arises from the non-constancy of the constitutive equation's coefficients:

$$\beta = \left(\frac{\tau_{xx} - \tau_{yy}}{\tau_{xy}^2}\right)_{k = V/R} \bigg/ \left(\frac{\tau_{xx} - \tau_{yy}}{\tau_{xy}^2}\right)_{k = 0} \qquad (7\text{-}4.23)$$

Similarity solutions are then sought for wedge flows, i.e., flows for which the external field is governed, on the solid surface, by the following equation:

$$U = U_0 x^a \qquad (7\text{-}4.24)$$

It turns out that similarity solutions are possible only under very special assumptions concerning the dependence of the constitutive equation's coefficients on the magnitude of $\mathbf{D}$; non-similar solutions can only be obtained in the limit of small ε (i.e., a viscous boundary layer). For stagnation flow ($a = 1$), a similarity solution is obtained for *constant* coefficients of the constitutive equation, i.e., for a second-order fluid. Macroscopic quantities of interest, such as the drag coefficient on a flat plate at zero incidence, depend on the non-Newtonian character of the fluid in a way which is strongly dependent on the assumptions made concerning the dependence of the coefficients on the magnitude of **D**.

It is worthwhile mentioning that, although flow of non Newtonian fluids around submerged *solid* objects, as discussed above, is not very well understood, even less is known concerning flow around *fluid* objects, such as gas bubbles or liquid drops, and only some very qualitative analyses have been published [2, 21].

7-5   TURBULENT FLOWS

Even in the case of Newtonian fluids, turbulent flows are very difficult to analyze, since no entirely satisfactory phenomenological theory is available for the calculation of the Reynolds stress term $\mathbf{V} \cdot (\rho \overline{\mathbf{v}'\mathbf{v}'})$ appearing in eq. (7-1.23). In the case of non-Newtonian liquids, the non-linearity of the constitutive equation introduces additional major difficulties, and the analysis is necessarily very qualitative in character.

For purely viscous fluids, satisfactory correlations for the pressure drop in turbulent flow through circular tubes are available [22]. A generalized Reynolds number is defined in such a way as to force laminar flow data to lie on the Newtonian line in a friction factor–Reynolds number diagram, see eq. (2-5.25); in turbulent flow, the friction factor turns out to depend both on the Reynolds number and on the parameter n' defined in eq. (2-5.13) and evaluated at the shear stress level prevailing at the wall.

The relevance of any elasticity that a fluid may have in a given flow field depends on how large Λ is, and on how small is some characteristic time scale of flow. In turbulent flows, the time scale of the flow is very small indeed [23], and major anomalies of behavior are observed even in only slightly elastic liquids, such as very dilute solutions of polymers [24]. In fact, a characteristic time scale of turbulent flow may be taken as the inverse of the frequency of the largest dissipative eddies [25]:

$$T_f = (1/v_\lambda)_{Re_\lambda = 1} \qquad (7\text{-}5.1)$$

$$Re_\lambda = \frac{V_\lambda \lambda \rho}{\mu} \qquad (7\text{-}5.2)$$

where λ is the dimension of the eddy, V_λ its characteristic velocity, and $v_\lambda = \lambda/V_\lambda$ its frequency. Values of T_f calculated from eq. (7-5.1) are typically of

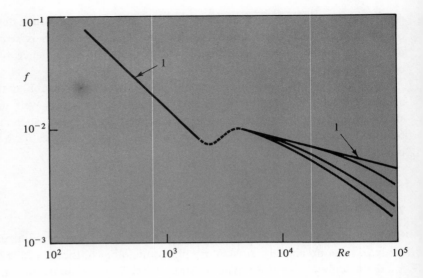

FIGURE 7-1
Drag reduction in dilute solutions.

the order of 10^{-5} s, so that even slightly elastic liquids (with a natural time of 10^{-5} s) may exhibit Deborah numbers of the order of unity. No other pragmatically interesting flow pattern exhibits such low values of T_f.

Before entering into a discussion of the analysis of turbulent flows of elastic liquids, it is worthwhile giving a phenomenological description of the behavior that has been observed. The pressure drop observed for turbulent flow of dilute polymer solutions through circular tubes is often dramatically lower than would be observed at the same flow rate of the pure solvent, although the viscosity of the solution is larger than that of the solvent. This phenomenon is referred to as drag reduction. Similar phenomena are observed in the flow around submerged objects, when a polymer is injected into the boundary layer.

There are several possible ways of presenting drag reduction data, and often what seems conflicting evidence is simply due to a different choice of plotting. Let us consider a plot of friction factor versus Reynolds number, such as in Figs 7-1 and 7-2. The lines labelled 1 are relative to Newtonian liquids; the left parts are the Poiseuille law valid for laminar flow, while the right parts are the usually accepted smooth tube correlations.

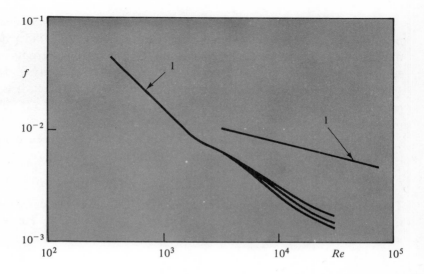

FIGURE 7-2
Drag reduction in concentrated solutions.

When data for polymer solutions are plotted in this form, a question arises as to the proper definition of the Reynolds number, because the viscometric viscosity of these solutions is usually dependent on the shear rate. The definition of Reynolds number that is usually accepted is such that the laminar flow correlation holds for the polymer solution [26], so that, by definition, no drag reduction occurs at Reynolds numbers below 2100 (transition to turbulence is never observed at values less than 2100). In effect, the pressure drop in laminar flow of the solution is *larger* than at equal flow rate of the pure solvent, because the solution's viscosity, albeit variable, is larger than the (constant) solvent viscosity.

With the Reynolds number definition discussed above, the typical behavior observed in dilute solutions is shown in Fig. 7-1, although different behaviors have been reported [27, 28]. At equal Reynolds numbers the friction factor depends on the tube diameter, approaching the Newtonian value at very large tube diameters. For more concentrated solutions, the behavior reported in Fig. 7-2 is often observed; the diameter effect is still present, and the transition from laminar to turbulent flow is not easily identified, though generally a small kink can be discerned in the neighborhood of $Re = 2100$.

The existence of a diameter effect implies that the friction factor depends not only on the Reynolds number, but also on some other independent dimensionless group. Such a group can only be obtained by introducing an additional parameter to tube diameter, velocity, density, viscosity, and pressure drop; the natural time seems an obvious choice. In effect, it is now generally accepted that drag reduction is in some way related to the elastic properties of the fluid.

A satisfactory correlation of drag reduction data has been obtained [29] by considering a Deborah number defined as

$$De = \Lambda(v_\lambda)_{Re_\lambda = 1} \qquad (7\text{-}5.3)$$

The lowest frequency of dissipative eddies, $(v_\lambda)_{Re_\lambda = 1}$, can be evaluated following a line of thought originally suggested by Levich [3] and by Hinze [23], and the following result is obtained:

$$(v_\lambda)_{Re_\lambda = 1} \cong \frac{V}{D}(Re)^{0.75} \qquad (7\text{-}5.4)$$

so that the Deborah number is given by

$$De = \frac{V\Lambda}{D}(Re)^{0.75} = We(Re)^{0.75} \qquad (7\text{-}5.5)$$

Equation (7-5.5) fully justifies the qualitative character of the behavior observed in drag reduction. In fact, consider an increase in diameter at constant Reynolds number: the value of De calculated from eq. (7-5.5) would correspondingly decrease, and therefore one would expect that Newtonian behavior is approached, which indeed is the case. A correlation based on eq. (7-5.5) is also satisfactory quantitatively. Figure 7-3 is a plot of the ratio f/f^0 (where f^0 is the 'Newtonian' friction factor) versus the quantity v/v_{cr}, where

$$v = \frac{V}{D}(Re)^{0.75} \qquad (7\text{-}5.6)$$

and v_{cr} is that value of v for which $f/f^0 = 0.5$. The data in Fig. 7-3 cover a range of tube diameters from 0.69 to 2.05 cm, a range of Reynolds numbers from 10^3 to 10^5, and a range of polymer concentrations from 0.0125 to 0.2 per cent by weight. Figure 7-4 is a plot of $1/v_{cr}$, which can be taken as an indirect measure of Λ, versus the polymer concentration [29].

A somewhat different viewpoint has often been presented in the literature, which is based on consideration of boundary layer thickening in drag-reducing liquids; attention is in this case focused on the structure of wall

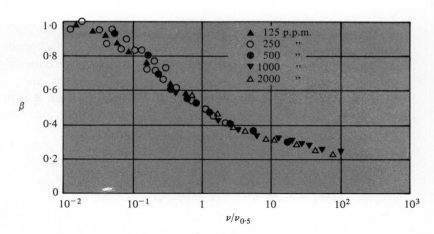

FIGURE 7-3
Correlation of drag reduction.

turbulence rather than on the rate of energy dissipation in the entire flow field. This approach has suggested the relevance of experiments on drag reduction in rough pipes, but results so far published are somewhat contradictory. Correlations based on this approach are frequent in the literature, and are often expressed in terms of a critical wall shear stress, τ_{cr}, below which no drag reduction is observed. If a Blasius type of equation is used for the friction factor in the absence of drag reduction, the condition for the latter can be written as:

$$\tau = 0.04 \, \rho V^2 (Re)^{-0.25} < \tau_{cr} \qquad (7\text{-}5.7)$$

Simple algebraic manipulation yields

$$\frac{\mu}{\tau_{cr}} \frac{V}{D} (Re)^{0.75} < \frac{1}{0.04} = 25 \qquad (7\text{-}5.8)$$

where the left-hand side is seen to coincide with the expression for the Deborah number obtained from the former approach, provided that τ_{cr} is interpreted as μ/Λ, i.e., as a sort of elastic modulus of the fluid. It seems, therefore, that overall drag reduction results cannot easily discriminate between the two approaches; proponents of the boundary layer thickening approach generally use velocity distribution data to support their analysis [30]. It is unfortunate that many published velocity distributions are open to considerable doubt, due to the anomalous response of velocity probes in drag-reducing liquids which has been amply demonstrated; only recently

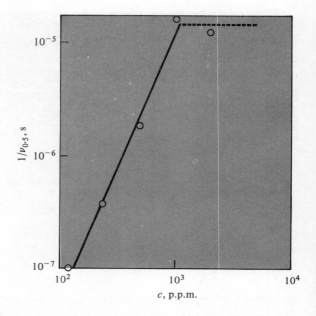

FIGURE 7-4
Equivalent natural time versus concentration.

have velocity distributions been obtained under conditions of sufficient reliability of the measuring device [31].

Another concept has been introduced in the analysis of drag reduction, namely, the fact that liquid filaments are continuously stretched in a turbulent flow field. Since it is known that elastic liquids have a high resistance to stretching, this has been taken as a possible cause for a reduced level of turbulent intensity in such fluids. Again, when a quantitative formulation of this approach is tried out, a grouping of variables such as shown on the right-hand side of eq. (7-5.5) is obtained. It is interesting to observe that an approach based on shear wave considerations would produce the El_2 group and therefore, according to eq. (7-2.29), a slightly different dependence on the Reynolds number would be predicted. This could, in principle, be subjected to experimental test. In this regard, it is interesting to observe that values of the critical stress at the wall, τ_{cr}, reported in the literature are typically of the order of 50 dyn/cm². This would correspond, if τ_{cr} is interpreted as μ/Λ, to a wave velocity of about 7 cm/s (see eq. (7-2.27)). Indirect evidence supporting this value for the wave velocity (see section 7-4) gives some quantitative support to a shear wave interpretation of drag reduction.

Because of its implications in other problems of non-Newtonian fluid mechanics, it is perhaps worthwhile to discuss the dynamics of the flow in the dissipative vortices. Consider a plane elliptical vortex, and let ε be the ratio of the minor to the major axis of the vortex [32], x^1 the direction of the major, and x^2 the direction of the minor axis. The flow field in the vortex can be described as

$$v^1 = \gamma x^2 \qquad (7\text{-}5.9)$$

$$v^2 = -\varepsilon^2 \gamma x^1 \qquad (7\text{-}5.10)$$

$$v^3 = 0 \qquad (7\text{-}5.11)$$

A standard calculation yields the matrix of tensor $\mathbf{F}^t$ with respect to the natural basis of the x^i system:

$$[\mathbf{F}^t] = \begin{Vmatrix} \cos(\varepsilon\gamma s) & -\sin(\varepsilon\gamma s)/\varepsilon & 0 \\ \varepsilon \sin(\varepsilon\gamma s) & \cos(\varepsilon\gamma s) & 0 \\ 0 & 0 & 0 \end{Vmatrix} \qquad (7\text{-}5.12)$$

The flow field is therefore identified as a constant stretch history flow, with $k\mathbf{N} = \nabla\mathbf{v}$ and $k = \gamma\sqrt{(1 + \varepsilon^4)}$. Also, the flow is a periodic flow of the type discussed in section 5-4, with

$$\mathbf{G}^t = \text{Re}\,\{\mathbf{\psi}^*[1 - e^{-ivs}]\} \qquad (7\text{-}5.13)$$

where $v = 2\varepsilon\gamma$ is the frequency of the vortex, and $\mathbf{\psi}^*$ is a constant complex tensor:

$$\mathbf{\psi}^* = \mathbf{\psi}_\mathbf{R} + \frac{2i}{\omega}\mathbf{D}; \qquad [\mathbf{\psi_R}] = \begin{Vmatrix} (\varepsilon^2 - 1)/2 & 0 & 0 \\ 0 & (1 - \varepsilon^2)/2\varepsilon^2 & 0 \\ 0 & 0 & 0 \end{Vmatrix} \qquad (7\text{-}5.14)$$

The flow considered is controllable, since $\nabla \cdot \mathbf{\tau} = 0$ and $D\mathbf{v}/Dt$ is expressible as the gradient of a scalar field:

$$\frac{D\mathbf{v}}{Dt} = \nabla\left(-\frac{\varepsilon^2\gamma^2}{2}[(x^1)^2 + (x^2)^2]\right) \qquad (7\text{-}5.15)$$

Since the flow is a constant stretch history flow, the rate of dissipation is

$$D_\mathbf{M} = \frac{1}{\rho}\mathbf{\tau}:\mathbf{D} = \frac{1}{\rho}\tau_{12}(1 - \varepsilon^2)\gamma \qquad (7\text{-}5.16)$$

Evaluation of D_M can only be based on some specific constitutive equation, and by way of example we consider the Maxwell equation:

$$\tau = 2\mu\mathbf{D} - \Lambda\overset{\triangledown}{\tau} \qquad (7\text{-}5.17)$$

which predicts a constant viscometric viscosity, $\eta_V = \mu$. The value of D_M is calculated as

$$D_M = \frac{\mu\gamma^2(1 - \varepsilon^2)^2}{(1 + v^2\Lambda^2)\rho} \qquad (7\text{-}5.18)$$

Equation (7-5.18) shows that, unless $v\Lambda \ll 1$, the dissipation is smaller than would be calculated by assuming that the viscometric viscosity can be used to calculate τ_{12}, say

$$D_M^0 = \frac{\mu\gamma^2(1 - \varepsilon^2)^2}{\rho} \qquad (7\text{-}5.19)$$

The decrease factor D_M/D_M^0, which can perhaps be regarded as an approximation for f/f^0, is given by

$$\frac{D_M}{D_M^0} = \frac{1}{1 + v^2\Lambda^2} \qquad (7\text{-}5.20)$$

The form of eq. (7-5.20) justifies a correlation of drag reduction data such as in Fig. 7-3, and the evaluation of the De number given in eq. (7-5.3).

Note that, for the flow considered, one has

$$\operatorname{tr} \mathbf{N}^2 = -\frac{2\varepsilon^4}{1 + \varepsilon^4} \qquad (7\text{-}5.21)$$

i.e., $\operatorname{tr} \mathbf{N}^2$ is negative. This should be compared with eqs (5-3.5) and (5-3.6), which show that $\operatorname{tr} \mathbf{N}^2$ is zero for viscometric flows, and is positive for extensional flows. In some sense, elliptical flows, viscometric flows, and extensional flows are in order of increasing 'extensionality,' and a model such as the Maxwell model (eq. (7-5.17)) predicts a viscosity which is decreasing with increasing rate of strain in elliptical flows (eq. (7-5.18)), constant in viscometric flows, and increasing in extensional flows.

7-6 EXTENSIONAL FLOWS

Many flows which are encountered in practical applications are of a type which can be called extensional. In broad language, these are flows where the differences in velocity develop mostly in the direction of the velocity itself rather than in the direction orthogonal to it, as in the shearing flows. Examples of these flows are encountered in the processes of fiber spinning and film forming where the liquid material, a melt or a solution, is drawn out of a die. Also within the extruder head, in the converging field towards the die section, the flow field may be essentially extensional in character, although some shearing must also be present.

After Coleman [33], a precise definition of an extensional flow is as follows: A motion has been an extension up to time t if there exists an orthogonal basis $\mathbf{e}_i$, independent of s, such that the matrix of the components of $\mathbf{U}^t$ (or of $\mathbf{C}^t$, $(\mathbf{C}^t)^{-1}$, $\mathbf{H}^t$) with respect to such a basis is diagonal, viz.:

$$[\mathbf{U}^t(s)] = \left\| \begin{array}{ccc} \alpha_1^t(s) & 0 & 0 \\ 0 & \alpha_2^t(s) & 0 \\ 0 & 0 & \alpha_3^t(s) \end{array} \right\| \qquad (7\text{-}6.1)$$

The functions α_1^t, α_2^t, and α_3^t are the principal stretch ratios with respect to the reference configuration (which has been chosen at the present time t). For an incompressible material, one has

$$\alpha_1^t \alpha_2^t \alpha_3^t = 1 \qquad (7\text{-}6.2)$$

For a simple fluid in general, it can be shown that the matrix of the stress tensor at time t, with respect to the same orthonormal basis, is also diagonal:

$$[\mathbf{T}(t)] = \left\| \begin{array}{ccc} T_{11} & 0 & 0 \\ 0 & T_{22} & 0 \\ 0 & 0 & T_{33} \end{array} \right\| \qquad (7\text{-}6.3)$$

Thus the basis vectors lie on the so-called principal directions of stress.

It is obvious that, because in general $\mathbf{R}^t(s) \neq \mathbf{0}$, the principal directions of stress are different at different instants of time, t (i.e., the basis $\mathbf{e}_i$ is independent of s but depends on t). The principal directions of stress are fixed only if a corotating frame is considered.

Incompressible simple fluids undergoing extensional flows are completely characterized by two scalar-valued material functionals which give the first and second normal stress differences once any two of the three α^t

are assigned. For example,

$$\tau_{11} - \tau_{22} = \mathscr{H}_1[\alpha_1^t, \alpha_2^t] \qquad (7\text{-}6.4)$$

$$\tau_{22} - \tau_{33} = \mathscr{H}_2[\alpha_1^t, \alpha_2^t] \qquad (7\text{-}6.5)$$

It is easily shown that

$$\mathscr{H}_1[\alpha, \alpha] = 0 \qquad (7\text{-}6.6)$$

$$\mathscr{H}_2[\alpha, 1/\sqrt{\alpha}] = 0 \qquad (7\text{-}6.7)$$

Equations (7-6.6) and (7-6.7) express the property of symmetry by which an uniaxial extension ($\alpha_i^t = \alpha_j^t$) in a simple fluid does not generate normal stress differences among directions orthogonal to that of stretch.

The constant stretch history extensional flows already considered in sections 3-5 and 5-3 correspond to the following form of the stretch histories:

$$\alpha_i^t = \exp(\gamma_i s) \qquad (7\text{-}6.8)$$

where the γ_i are constant in time and obey the incompressibility condition:

$$\sum_1^3 \gamma_i = 0 \qquad (7\text{-}6.9)$$

The material functionals of eqs (7-6.4) and (7-6.5) then reduce to the material functions given in eqs (5-3.9) and (5-3.10).

A class of extensional flows which is likely to be approximated by real flows upstream of duct or die entrances is that of sink flows [34]. They may be steady with respect to the laboratory frame, but also in this case they are not constant stretch history flows. The stretching keeps increasing along the stream up to the sink. The analysis of sink flows for the case of an incompressible simple fluid has been carried out [34] both for spherical and cylindrical symmetry. An approximately spherical symmetric sink flow is likely to occur in the flow of viscoelastic liquids upstream of a pipe or circular die entrance [35, 36]. The cylindrical symmetry is expected for the analogous case upstream of a slot or rectangular channel.

By particularizing general results due to Coleman [33] on the inflation of a spherical or cylindrical shell, the analysis carried out by Marrucci and Murch [34] shows that the stresses arising in a steady sink flow of assigned symmetry only depend on the instantaneous value of the stretching, Γ. This occurs because the deformation history, although not a constant stretch history, is entirely determined by Γ.

In the spherically symmetric sink flow, one material function characterizes the stress response of the material. This gives the difference between the normal stress in the direction of flow and that in any direction orthogonal to it as a function of Γ:

$$T_{11} - T_{22} = \bar{\sigma}(\Gamma) \qquad (7\text{-}6.10)$$

where Γ, the local stretching, is the ratio between the sink strength and the third power of the distance from the (real or virtual) point sink.

The second-order approximation of the function $\sigma(\Gamma)$ is given by

$$\bar{\sigma}(\Gamma) = 3\mu\Gamma + 3(\beta + \tfrac{5}{2}\gamma)\Gamma^2 \qquad (7\text{-}6.11)$$

The sink flow with cylindrical symmetry is characterized by two material functions. Taking the sink line as the z axis of a cylindrical coordinate system, the material functions are the relationships

$$T\langle rr \rangle - T\langle \theta\theta \rangle = \bar{\sigma}_1(\Gamma) \qquad (7\text{-}6.12)$$

$$T\langle \theta\theta \rangle - T\langle zz \rangle = \bar{\sigma}_2(\Gamma) \qquad (7\text{-}6.13)$$

where Γ is the ratio between the sink strength per unit length of the sink line and r^2.

The second-order approximations of $\bar{\sigma}_1$ and $\bar{\sigma}_2$ are given by

$$\bar{\sigma}_1(\Gamma) = 4\mu\Gamma + 8\gamma\Gamma^2 \qquad (7\text{-}6.14)$$

$$\bar{\sigma}_2(\Gamma) = -2\mu\Gamma + 4\beta\Gamma^2 \qquad (7\text{-}6.15)$$

As always, apart from the slow flow approximations, the material functions cannot be determined within the framework of the general simple fluid theory. However, they are easily determined once a particular constitutive equation has been chosen.

By way of example, we can calculate $\sigma(\Gamma)$ for the spherically symmetric sink flow by assuming the simple Maxwell equation (eq. (6-4.12)). As already shown, the Maxwell equation coincides with the integral constitutive equation (eq. (6-4.19)). The matrix of the tensor $\mathbf{C}^t(s)$ for the sink flow has been calculated in Illustration 3B (p. 113). Straightforward integration gives the expression

$$\sigma(\Gamma) = 2\frac{\mu}{\Lambda}\left[\Lambda\Gamma + \tfrac{1}{2}\Lambda^2\Gamma^2 + \frac{1 - \Lambda^2\Gamma^2}{3(\Lambda\Gamma)^{2/3}} \right.$$
$$\left. \cdot \exp\left(\frac{2}{3\Lambda\Gamma}\right) \int_0^{\Lambda\Gamma} \xi^{-1/3} \exp\left(-\frac{2}{3\xi}\right) d\xi \right] \qquad (7\text{-}6.16)$$

It may be observed that eq. (7-6.16) predicts a finite value of the stress for all values of the stretching Γ. This result may be compared with that of the extensiometric flow discussed in section 6-4 where, for the same constitutive assumption, infinite stresses were predicted for values of the stretching larger than a limiting value. This difference in behavior is related to the unsteady nature, in a Lagrangian sense, of the sink flow and may be better understood by considering in some detail the behavior of other types of unsteady elongational flows.

Let us consider the kinematics known as simple extension. In a suitable Cartesian coordinate system this is given by

$$v_x = \gamma_E x$$
$$v_y = -\tfrac{1}{2}\gamma_E y \qquad (7\text{-}6.17)$$
$$v_z = -\tfrac{1}{2}\gamma_E z$$

Assume that γ_E as a function of time is a step function, i.e., it is zero up to $t = 0$ and has a constant value, also called γ_E, afterwards.

It is evident that, for a simple fluid with fading memory, the stress response to such kinematics becomes eventually (i.e., for $t \to \infty$) that of the constant stretch history flow already considered in section 5-3, which is fully determined by the material function $\eta_E(\ )$ defined by eq. (5-3.16). However, we are here interested in the transient stress response, which is obtained before the ultimate value, if it exists, has been attained.

Once more, in order to obtain definite predictions, we use specific constitutive assumptions. The use of a Maxwell equation (6-4.12) or equivalently of the integral equation (6-4.19) gives the results which are reported by Denn and Marrucci [37]. These results can be summarized as follows.

The transient elongational viscosity can be defined in the same way as for the constant stretch history case, eq. (5-3.16):

$$\eta_{Et}(\gamma_E, t) = \frac{\tau_{xx} - \tau_{yy}}{\gamma_E} \qquad (7\text{-}6.18)$$

The following expression for η_{Et} is obtained [37]:

$$\eta_{Et} = \frac{3\mu}{(1 - 2\gamma_E \Lambda)(1 + \Lambda\gamma_E)} - \frac{2\mu}{1 - 2\Lambda\gamma_E} \exp\left(-\frac{1 - 2\Lambda\gamma_E}{\Lambda}t\right)$$
$$- \frac{\mu}{1 + \Lambda\gamma_E} \exp\left(-\frac{1 + \Lambda\gamma_E}{\Lambda}t\right) \qquad (7\text{-}6.19)$$

Equation (7-6.19) shows that two entirely different behaviors are expected for values of $\Lambda\gamma_E$ smaller or larger than 0.5. In the former case the viscosity, starting with a zero value at $t = 0$, increases towards an asymptotic value which is the constant stretch history result. Conversely, in the latter case the viscosity, although always finite, increases beyond any limit with increasing time and no ultimate value is ever reached.

This result explains the meaning of the infinite value for the elongational viscosity obtained in section 6-4. The infinite value for the constant stretch history elongational viscosity simply means that the steady state for the stress is unattainable under those conditions. The stress keeps increasing in any experiment of finite duration if $\Lambda\gamma_E > 0.5$. Therefore, in any actual elongational experiment, no infinite stresses are ever expected.

The duration of an elongational flow is always limited in practice because of the large extension ratios which are soon reached. Moreover, the larger is γ_E and thus $\Lambda\gamma_E$, the shorter will be the time allowed for the experiment. Thus, in many cases the values of elongational viscosity which are encountered, being relative to the initial part of the transient, are even smaller than the shear viscosity.

It is true, however, that in other instances exceptionally large values have been reported [38, 39].

7-7 WAVES AND INSTABILITY

In this section we discuss in very general terms two categories of fluid mechanical problems which have lately been the subject of numerous investigations. Since both types of problem require special mathematical techniques for their solution, a detailed discussion is outside of the scope of this book. Therefore, we limit the discussion to very few points, and we try to focus attention on those special points which arise in the case of non-Newtonian fluids.

Waves

The theory of propagation of discontinuities in elastic solids is well developed; the same is true for ideal fluids (i.e., fluids which can exhibit only isotropic internal stresses). Both theories do not allow for *damping* of disturbances, since the assumed constitutive equations are those of non-dissipative materials (i.e., the work of internal stresses equals the accumulation of elastic energy).

In contrast with this, the classical theory of incompressible Newtonian fluids (which would, of course, allow for damping) predicts that no discontinuities can exist. This difficulty has been partly circumvented by the introduction of a 'quasi-wave,' i.e., a thin region where, although no discontinuity exists, some variable has a very large rate of change.

The considerations above do not imply that any damping mechanism results in the impossibility of discontinuities; indeed, the damped-wave differential equation (eq. (7-7.10) below) does admit discontinuous solutions of any order. The theory of simple fluids with fading memory, with the smoothness hypotheses for the constitutive functionals which have been discussed in section 4-4, has in fact been applied to the study of wave propagation, and very interesting results have been obtained [40]. Not only do such fluids admit propagating disturbances, but the amplitude of such disturbances may be both damped or increased as it travels. The reader is referred to the monograph by Coleman *et al.* [40] for a detailed study of this problem.

We here wish to discuss a very simple limiting case. Let us consider a semi-infinite body of a fluid at rest, which is bounded by a plane solid surface at $x^1 = 0$ (a Cartesian coordinate system is considered).

At time $t = 0$, the solid surface is suddenly set to move at a constant velocity V in the direction x^2. We analyze the propagation of the disturbance (in this case, a non-zero velocity component in the x^2 direction) along the x^1 direction.

Let us first recall the well-known solution to this problem for the case of a Newtonian fluid. The equation of motion in the x^2 direction is

$$\frac{\partial v^2}{\partial t} = \frac{\mu}{\rho} \frac{\partial^2 v^2}{\partial (x^1)^2} \qquad (7\text{-}7.1)$$

with the boundary conditions:

$$t = 0, \qquad x^1 > 0, \qquad v^2 = 0 \qquad (7\text{-}7.2)$$

$$t > 0, \qquad x^1 = 0, \qquad v^2 = V \qquad (7\text{-}7.3)$$

$$t \geq 0, \qquad x^1 \to \infty, \qquad v^2 \text{ is bounded} \qquad (7\text{-}7.4)$$

One should note that, since there are only two dimensional parameters to the problem (the kinematic viscosity μ/ρ and the velocity V), there is no way of defining three dimensionless variables (for v^2, x^1, and t) independently of each other; hence, the problem admits a similarity solution. The similarity variable is $x^1/\sqrt{(4\mu t/\rho)}$, and the solution is:

$$\frac{v^2}{V} = \text{erfc}\left(\frac{x^1}{\sqrt{(4\mu t/\rho)}}\right) \qquad (7\text{-}7.5)$$

Equation (7-7.5) implies that, at any $t > 0$, the velocity v^2 is non-zero at all values of x^1; i.e., the discontinuity that has been applied at $t = 0$ and $x^1 = 0$ has propagated with infinite speed along the x^1 axis. Indeed, the tangential stress τ_{12} at $x^1 = 0$ is infinity at $t = 0$, which implies that in fact it is impossible to set the solid surface in motion suddenly, i.e., that no discontinuity can exist.

Let us now consider the case of a non-Newtonian fluid. Since we now have a third dimensional parameter to be considered, i.e., a natural time Λ, no similarity solution is expected. We may introduce the following three dimensionless variables:

$$v = v^2/V \qquad (7\text{-}7.6)$$

$$\theta = t/\Lambda \qquad (7\text{-}7.7)$$

$$\eta = x^1/\sqrt{(\mu\Lambda/\rho)} \qquad (7\text{-}7.8)$$

and try to determine the form of the velocity distribution $v(\eta, \theta)$. In order to do so, we need to assume a constitutive equation, and we choose a Maxwell-type equation (which is known to imply the smoothness properties underlying the theory of simple fluids), say:

$$\tau + \Lambda\overset{\triangledown}{\tau} = 2\mu\mathbf{D} \qquad (7\text{-}7.9)$$

The equation of motion in the x^2 direction becomes

$$\frac{\partial v}{\partial \theta} + \frac{\partial^2 v}{\partial \theta^2} = \frac{\partial^2 v}{\partial \eta^2} \qquad (7\text{-}7.10)$$

Equation (7-7.10) is a damped-wave equation [41, 42], and is known to admit discontinuous solutions. For the problem at hand, we need to add to the boundary conditions (7-7.2)–(7-7.4) the following one (since the equation now contains a second-order time derivative):

$$t = 0, \qquad x^1 > 0, \qquad \partial v^2/\partial t = 0 \qquad (7\text{-}7.11)$$

The solution of eq. (7-7.10), subject to the boundary conditions (7-7.2)–(7-7.4) and (7-7.11), is

$$v = \psi(\theta - \eta)\,e^{-\theta/2}\left\{I_0\!\left(\frac{\sqrt{(\theta^2 - \eta^2)}}{2}\right)\right.$$

$$\left. + \frac{1}{2}\int_\eta^\theta I_0\!\left(\frac{\sqrt{(\xi^2 - \eta^2)}}{2}\right)\left[I_0\!\left(\frac{\theta - \xi}{2}\right) + I_1\!\left(\frac{\theta - \xi}{2}\right)\right]d\xi\right\} \qquad (7\text{-}7.12)$$

where $\psi(\xi)$ is the unit step function ($\psi = 1$ when $\xi > 0$, and $\psi = 0$ when $\xi < 0$), and $I_0(\ )$ and $I_1(\ )$ are modified Bessel functions of the first kind.

Equation (7-7.12) should be discussed in some detail. First, one should note that, at $\eta > \theta$ (i.e., at $x^1 > \sqrt{(\mu/\rho\Lambda)}t$), the velocity is zero; i.e., the discontinuity travels with a finite speed given by $\sqrt{(\mu/\rho\Lambda)}$:

at $\eta > \theta$, $\qquad\qquad\qquad\qquad\qquad\qquad\qquad v = 0 \qquad (7\text{-}7.13)$

As the disturbance propagates, its amplitude, i.e., the velocity at the point of discontinuity, is damped out exponentially:

at $\qquad\qquad\qquad\qquad\qquad \eta = \theta, \qquad v = e^{-\theta/2} \qquad$ (7-7.14)

At the wave front, both the velocity and the strain are discontinuous in space and in time. This is a general property of wave fronts (i.e., it can be shown in general that a discontinuity in velocity corresponds to a discontinuity of strain), so that the interesting conclusion is drawn that those constitutive equations which do not allow strain discontinuities, some of which have been discussed in section 3-4, would not allow a velocity discontinuity of the type obtained here. Indeed, Tanner [43] has shown, for the problem at hand, that addition of a term containing even a small retardation time in the constitutive equation smoothes out the discontinuity.

It may at first sight seem disturbing that the Newtonian constitutive equation, which arises as an asymptotic solution for the general theory of simple fluids (and is obtained from eq. (7-7.9) as $\Lambda \to 0$), should predict a behavior which, in regard to the propagation of discontinuities, is qualitatively different from that predicted by simple fluid theory. This is in fact only an apparent paradox, since the procedure by means of which the Newtonian equation is obtained from the simple fluid theory requires the restriction to deformation histories which are continuous at the instant of observation (see the discussion following eq. (6-2.3))—a condition which is grossly violated in the problem at hand. Indeed, similar difficulties are encountered with any nth-order type of constitutive equation, as discussed in detail by Coleman et al. [44] for the second-order fluid. The equation of motion for the second-order fluid in the flow considered here is

$$\frac{\partial v}{\partial \theta} = \frac{\partial^2 v}{\partial \eta^2} - \frac{\partial^3 v}{\partial \eta^2 \, \partial \theta} \qquad (7\text{-}7.15)$$

solutions of which lead to a number of paradoxes whenever discontinuous boundary conditions are considered.

When the problem of wave propagation in simple fluids with fading memory is studied in general, the rate of propagation is seen to be equal to the square root of the ratio of an elastic modulus and density; the elastic modulus is to be evaluated locally, and is given by μ/Λ only when the wave is propagating in a medium at rest. Acceleration waves (i.e., discontinuities of the acceleration, which correspond to discontinuities of the strain rate) may be damped as they travel, but may also increase in amplitude and transform into shock waves (velocity discontinuities) in a finite time. The latter situation is encountered provided the initial amplitude of the wave is sufficiently large, and provided the constitutive equation is sufficiently non-linear. It is interesting that a wave propagating in a material which is undergoing a shear flow has a critical amplitude which decreases with increasing rate of shear [45]; this basic instability of shear flows may be responsible for some effects observed in extrusion of polymer melts, such as the so-called melt fracture.

Stability

The problem of the stability of fluid flow is well known in classical fluid mechanics. The problem may be stated in general terms as follows: given a well-defined boundary-value problem, an exact solution of the equations of motion satisfying all the boundary conditions may exist (and may actually have been obtained explicitly), which is steady in an Eulerian sense $(\partial v/\partial t = 0)$. Yet, such a solution may be unstable, in the sense that, if a small perturbation is superimposed on it at a certain time, the perturbation itself tends to grow in time rather than being damped out. This implies that another (possibly unsteady) solution to the equations of motion exists, and that the flow pattern which will be observed in practice is not the steady one, since of course it is impossible to avoid any perturbation in a real case. The typical example is the one of turbulent flow down a constant-section conduit, where the steady but unstable flow pattern is laminar.

The analysis of stability is typically carried out as follows. Let $\mathbf{v}^0(\mathbf{X})$ represent the basic (and possibly unstable) steady flow. It is assumed that the actual velocity distribution is of the form

$$\mathbf{v}(\mathbf{X}, t) = \mathbf{v}^0(\mathbf{X}) + \mathbf{v}'(\mathbf{X}, t) \qquad (7\text{-}7.16)$$

where $\mathbf{v}'(\mathbf{X}, t)$ is the perturbation. If the assumption is made that the perturbation is small, the equations of motion can be linearized with respect to the perturbation variables; solutions are then searched for which show an exponential time dependence of the form $\exp(-i\sigma t)$. When the imaginary part of σ is zero, the perturbation's amplitude does not change in time, and the flow is called 'neutrally stable.' When $\text{Im}(\sigma) > 0$, the disturbance grows in time and hence the flow is unstable; when $\text{Im}(\sigma) < 0$, the disturbance decays in time and the flow is stable.

Typically, a neutral stability curve is obtained in a wave number–Reynolds number plane, separating the unstable from the stable regions. (The wave number is a dimensionless measure of the wavelength of the disturbance.) Reynolds numbers which correspond to stable flows for *all* wave numbers correspond to actually stable conditions.

When this procedure is extended to non-Newtonian flows, a number of problems arise. First, a constitutive equation needs to be assumed; the choice is by itself a dubious one. Analyses have been carried out for second-order [46–48] and for Maxwell-type fluids [41, 49, 50].

The linearization step introduces additional problems. In fact, since the constitutive equation is non-linear, this step implies not only neglect of the $\rho \, \mathbf{V v'} \cdot \mathbf{v'}$ term as in the Newtonian case, but also a linearization of the stress term. As discussed by Porteous and Denn [50], such a linearization corresponds to a constitutive assumption: indeed, in the limit of small values of the dimensionless group $El_1 = \Lambda \mu / \rho D^2$, the second-order fluid and the Maxwell fluid yield exactly the same results.

This point has been the cause of some paradoxical results which have been obtained [47, 48], as discussed by Craik [51]. In fact, a second-order fluid constitutive assumption would not be expected to yield significant results for large wave numbers, which correspond to very small time scales of the disturbance. Hence, a Maxwell-type constitutive equation, *when linearized*, would also be expected to give rise to problems when the Deborah number of the disturbance is not small. On the other hand, if the stress term is not linearized, no extension of the classical techniques of stability analysis can be used, since the governing equations become non-linear in the perturbation variables.

It should also be pointed out that the classical linear analysis of stability can only yield results concerning the stability to infinitesimal disturbances.

Since a flow pattern which is stable in regard to infinitesimal disturbances may be unstable in regard to finite disturbances, the linear analysis provides, at best, an upper bound for the stability criterion. This is, of course, true for both Newtonian and non-Newtonian fluids.

Finally, stability of a flow pattern may well depend on the geometry of the disturbance. While, for Newtonian fluids, certain general conclusions can be drawn concerning the geometry of disturbances that is most likely to result in instabilities, the same is not true for non-Newtonian fluids.

The above considerations imply that the entire field of stability analysis for non-Newtonian fluids is still in a very primitive stage. Since flow instabilities of several types have indeed been reported for non-Newtonian fluids—and also in flow fields which are known to be stable in the case of Newtonian fluids—this is an outstanding problem in the fluid mechanics of non-Newtonian fluids.

BIBLIOGRAPHY

1. LAMB, H.: *Hydrodynamics* (6th edn). Cambridge University Press, Cambridge (1932).
2. ASTARITA, G.: *Ind. Engng Chem. Fund.*, **5**, 548 (1966).
3. LEVICH, V. G.: *Physico-Chemical Hydrodynamics*. Prentice-Hall, Englewood Cliffs, N.J. (1962).
4. SCHLICHTING, H.: *Boundary Layer Theory*. McGraw-Hill, New York (1960).
5. WHITE, J. L., and METZNER, A. B.: *A.I.Ch.E. Jl*, **11**, 322 (1965).
6. ASTARITA, G.: *Rend. Acc. Lincei*, **8-41**, 355 (1966).
7. LANGLOIS, W. E., and RIVLIN, R. S.: *Rend. Mat.*, **22**, 169 (1963).
8. REINER, M.: *Physics Today*, **17**, 62 (Jan. 1964).
9. MARRUCCI, G., and ASTARITA, G.: *Meccanica*, **2**, 141 (1967).
10. COLEMAN, B. D., GURTIN, M. E., and HERRERA, I.: *Arch. Ratl Mech. Anal.*, **19**, 1 (1965).
11. ULTMANN, J. S., and DENN, M. M.: 40th Meeting of the Society of Rheology, St Paul (1969).
12. PIPKIN, A. C., and OWEN, D. R.: *Phys. Fluids*, **10**, 836 (1967).
13. MARKOWITZ, H.: Proc. 5th Int. Congr. on Rheology, vol. 1, p. 499. University Park Press (1968).
14. JONES, T. E. R.: *J. Phys. A*, **4**, 85 (1971).
15. ULTMAN, J. S., and DENN, M. M.: *Chem. Engng Jl*, **2**, 81 (1970).
16. JAMES, D. F.: Ph.D. Thesis, Calif. Inst. Techn., Pasadena (1967).
17. ULTMAN, J. S., and DENN, M. M.: *Trans. Soc. Rheol.*, **14**, 307 (1970).
18. WASSERMAN, M. L., and SLATTERY, J. C.: *A.I.Ch.E. Jl*, **10**, 383 (1964).
19. SRIVASTAVA, A. C., and MAITI, M. K.: *Phys. Fluids,* **9**, 462 (1966).
20. DENN, M. M.: *Boundary Layer Flows for a Class of Elastic Liquids*, in press.
21. ASTARITA, G., and APUZZO, G.: *A.I.Ch.E. Jl*, **11**, 815 (1965).
22. SKELLAND, A. H. P.: *Non-Newtonian Flow and Heat Transfer*. Wiley, New York (1967).
23. HINZE, J. O.: *Turbulence*. McGraw-Hill, New York (1958).
24. WELLS, C. S. (ed.): *Viscous Drag Reduction*. Plenum Press, New York (1969).
25. ASTARITA, G.: *Ind. Engng Chem. Fund.*, **4**, 354 (1965).
26. METZNER, A. B., and REED, J. C.: *A.I.Ch.E. Jl*, **1**, 434 (1955).
27. HERSHEY, H. C., and ZAKIN, J. L.: *Chem. Engng Sci.*, **22**, 1847 (1967).
28. SAVINS, J. G.: *Trans. A.S.M.E.*, **231**, 203 (1964).
29. ASTARITA, G., GRECO, G., and NICODEMO, L.: *A.I.Ch.E. Jl*, **15**, 564 (1969).
30. MEYER, W. A.: *A.I.Ch.E. Jl*, **12**, 522 (1966).
31. NICODEMO, L., ACIERNO, D., and ASTARITA, G.: *Chem. Engng Sci.*, **24** (1969).
32. ASTARITA, G., and MARRUCCI, G.: 1st National Congress on the Theory of Applied Mechanics, Udine (June 1971).
33. COLEMAN, B. D.: *Proc. Roy. Soc. London A*, **306**, 449 (1968).
34. MARRUCCI, G., and MURCH, R. E.: *Ind. Engng Chem. Fund.*, **9**, 498 (1970).
35. GIESEKUS, H.: *Rheol. Acta*, **7**, 127 (1968).

36. UEBLER, E. A.: Ph.D. Thesis, University of Delaware, Newark, Delaware (1966).
37. DENN, M. M., and MARRUCCI, G.: *A.I.Ch.E. Jl*, **17**, 101 (1971).
38. ASTARITA, G., and NICODEMO, L.: *Chem. Engng Jl*, **1**, 57 (1970).
39. ACIERNO, D., GRECO, R., and NICODEMO, L.: Meeting of Italian Society of Rheology, Siena (1971).
40. COLEMAN, B. D., GURTIN, M. E., HERRERA, I., and TRUESDELL, C.: *Wave Propagation in Dissipative Materials*. Springer-Verlag, New York (1965).
41. FREDRICKSON, A. G.: *Principles and Applications of Rheology*. Prentice-Hall, Englewood Cliffs, N.J. (1964).
42. DENN, M. M., and PORTEOUS, K. C.: *Chem. Engng Jl*, **2**, 280 (1971).
43. TANNER, R. I.: *Zeit. angewis. Math. Phys.*, **13**, 573, (1962).
44. COLEMAN, B. D., DUFFIN, R. J., and MIZEL, U. J.: *Arch. Ratl Mech. Anal.*, **19**, 100 (1965).
45. COLEMAN, B. D., and GURTIN, M. E.: *J. Fluid Mech.*, **33**, 165 (1968).
46. CHAN MAN FONG, C. F.: Ph.D. Dissertation, University of Wales (1965).
47. GUPTA, A. S.: *J. Fluid Mech.*, **28**, 17 (1967).
48. PLATTEN, J., and SCHECHTER, R. S.: *Phys. Fluids*, **13**, 832 (1970).
49. TLAPA, G., and BERNSTEIN, B.: *Phys. Fluids*, **13**, 565 (1970).
50. PORTEOUS, K. C., and DENN, M. M.: University of Delaware Water Research Center, Contribution No. 16 (1971).
51. CRAIK, A. D. D.: *J. Fluid Mech.*, **33**, 33 (1968).

INDEX

Printed by J. W. Arrowsmith Ltd. Bristol 3.